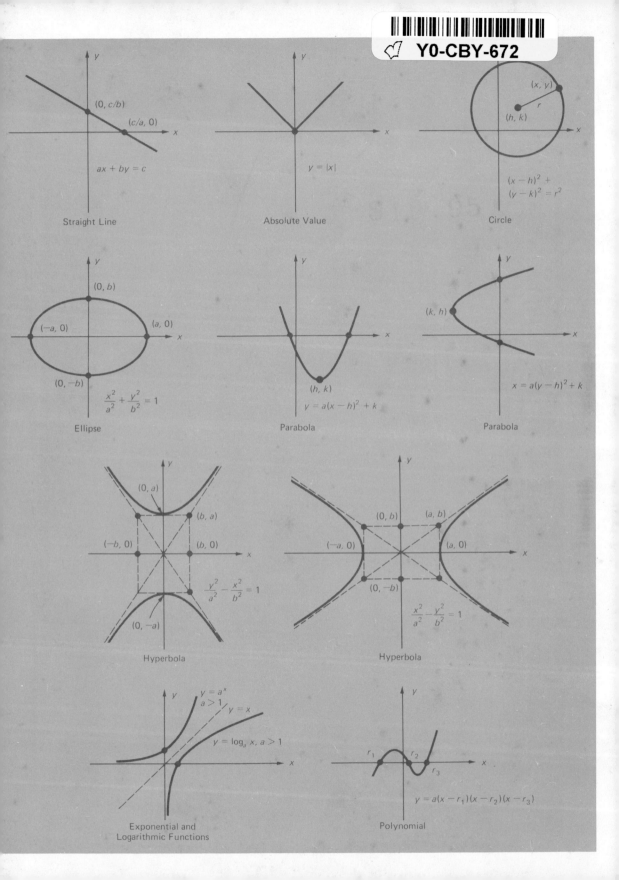

Straight Line

$ax + by = c$

Absolute Value

$y = |x|$

Circle

$(x - h)^2 + (y - k)^2 = r^2$

Ellipse

$\dfrac{x^2}{a^2} + \dfrac{y^2}{b^2} = 1$

Parabola

$y = a(x - h)^2 + k$

Parabola

$x = a(y - h)^2 + k$

Hyperbola

$\dfrac{y^2}{a^2} - \dfrac{x^2}{b^2} = 1$

Hyperbola

$\dfrac{x^2}{a^2} - \dfrac{y^2}{b^2} = 1$

Exponential and Logarithmic Functions

$y = a^x$, $a > 1$

$y = x$

$y = \log_a x$, $a > 1$

Polynomial

$y = a(x - r_1)(x - r_2)(x - r_3)$

College Algebra

College

JIMMIE GILBERT

JAMES SPENCER

LINDA GILBERT

Louisiana Tech University

Algebra

Prentice-Hall, Inc., Englewood Cliffs, New Jersey 07632

Library of Congress Cataloging in Publication Data

GILBERT, JIMMIE D (date)
  College algebra.

  Includes index.
  1. Algebra. I. Spencer, James D., joint author.
II. Gilbert, Linda P., joint author. III. Title.
QA1542.G52   512.9   80-20213
ISBN   0-13-141804-1

COLLEGE ALGEBRA
Jimmie Gilbert
James Spencer
Linda Gilbert

© *1981 by Prentice-Hall, Inc., Englewood Cliffs, N.J. 07632*

Printed in the United States of America

10 9 8 7 6 5 4 3 2 1

Editorial/production supervision by Kathleen M. Lafferty
Interior and cover design by Mark A. Binn
Manufacturing buyer: John B. Hall

Prentice-Hall International, Inc., London
Prentice-Hall of Australia Pty. Limited, Sydney
Prentice-Hall of Canada, Ltd., Toronto
Prentice-Hall of India Private Limited, New Delhi
Prentice-Hall of Japan, Inc., Tokyo
Prentice-Hall of Southeast Asia Pte. Ltd., Singapore
Whitehall Books Limited, Wellington, New Zealand

To our children—
Donna, Lisa, Martin, Dan, and Beckie
—Jimmie and Linda Gilbert

Leslie and Matthew
—James Spencer

Contents

Preface   xiii

Fundamentals

1-1   *Sets   1*
1-2   *The Field Properties of the Real Numbers   7*
1-3   *Properties of Real Numbers   12*
1-4   *Ordering of the Real Numbers   17*
1-5   *Absolute Value   24*
          Practice Test for Chapter 1   29

Polynomials and Rational Integral Expressions

2-1   *Integral Exponents   30*
2-2   *Polynomials   35*
2-3   *Factoring Polynomials   39*
2-4   *Division of Polynomials   42*
2-5   *Rational Integral Expressions   47*
          Practice Test for Chapter 2   53

Radicals and Rational Exponents

3-1 Radicals 54

3-2 Rational Exponents 57

3-3 Elementary Operations on Radicals 59

3-4 Simplest Radical Form 63

Practice Test for Chapter 3 68

Linear and Quadratic Equations in One Variable

4-1 Linear Equations 69

4-2 Applications (Optional) 74

4-3 Linear and Absolute Value Inequalities 79

4-4 Ratio, Proportion, and Variation (Optional) 84

4-5 Quadratic Equations 90

4-6 The Quadratic Formula 95

4-7 Equations Involving Radicals 101

Practice Test for Chapter 4 105

Linear and Quadratic Functions and Relations

5-1 Functions and Relations 106

5-2 Linear Functions 115

5-3 The Straight Line 120

5-4 Quadratic Relations (Parabolas) 124

5-5 Nonlinear Inequalities in One Variable 131

Practice Test for Chapter 5 141

ix

Contents

Conic Selections and Systems of Equations and Inequalities

6-1 Introduction 142
6-2 The Circle 143
6-3 The Ellipse 149
6-4 The Hyperbola 155
6-5 Systems of Equations 162
6-6 Systems of Inequalities 170
6-7 Inverse Functions 176
  Practice Test for Chapter 6 183

Matrices

7-1 Definitions and Basic Properties 184
7-2 Matrix Multiplication 191
7-3 Solution of Linear Systems of n Equations in n Unknowns by Gaussian Elimination 199
7-4 Calculation of Inverses 207
7-5 Solution of Linear Systems by Inverses 213
  Practice Test for Chapter 7 219

Determinants

8-1 The Definition of a Determinant 220
8-2 Evaluation of Determinants 227
8-3 Cramer's Rule 234
  Practice Test for Chapter 8 240

Theory of Polynomials

9-1 Complex Numbers *241*

9-2 The Factor and Remainder Theorems *249*

9-3 The Fundamental Theorem of Algebra and Descartes' Rule of Signs *256*

9-4 Rational Zeros *263*

9-5 Approximation of Real Zeros *270*

9-6 Graphs of Polynomial Functions *273*

Practice Test for Chapter 9 *280*

Exponential and Logarithmic Functions

10-1 Exponential Functions *281*

10-2 Logarithmic Functions *288*

10-3 Common Logarithms (Optional) *295*

10-4 Computations with Logarithms (Optional) *302*

10-5 Exponential and Logarithmic Equations *306*

Practice Test for Chapter 10 *310*

Sequences and Series

11-1 Definitions *311*

11-2 Arithmetic Progressions *316*

11-3 Geometric Progressions *322*

11-4 Infinite Geometric Progressions *327*

11-5 Mathematical Induction *332*

Practice Test for Chapter 11 *338*

Further Topics

12-1 *The Counting Principle 339*

12-2 *Permutations 342*

12-3 *Combinations 348*

12-4 *The Binomial Theorem 352*

12-5 *Partial Fractions 357*

12-6 *Graphs of Rational Functions 363*

Practice Test for Chapter 12 372

Solutions for Practice Tests A1

Answers to Selected Exercises A23

Index A53

Preface

This book is intended as a text for a one-semester or one-quarter college algebra course at the freshman level. It includes more material than would usually be taught in such a course, with the idea that selections and omissions of topics may be made to fit the needs of different classes. A primary purpose is to provide the student with a knowledge of background material that is necessary for further study in mathematics. In particular, the basic skills and manipulative techniques which are essential to the study of calculus are developed.

A complete treatment of the traditional material in a college algebra text is provided, together with a few specialized topics such as partial fractions.

The first three chapters furnish a review of such fundamentals as the real number system, absolute value, factoring, and operations with radicals. This material is presented in such a manner that it can be covered thoroughly, briefly reviewed, or skipped completely, depending on the needs of the students. Considering the varied backgrounds of the students currently entering college algebra, this flexibility seems to be desirable, if not necessary.

With the exception of Chapter 1, a formal theorem–proof format is avoided. In those cases where the material in Chapter 1 is to be taught with thoroughness, the theorem–proof format was adopted to help separate "what we know" from "what we can use" and to emphasize the difference between intuitive thinking and deductive reasoning.

After Chapter 1, the approach to the material is by discussion and illustration, with emphasis on understanding of concepts and techniques. Most topics are introduced with an intuitive motivation and discussion, followed by precise statements of theorems, rules, or procedures. Formal definitions are carefully stated, and detailed examples are provided, frequently even for special cases. The formal statements of definitions and theorems are made to indicate their importance, and they are numbered for ease of reference.

The content and treatment of Chapters 4 through 8 are fairly standard. Matrices and determinants are presented separately in two chapters. Inequalities, which always seem to be troublesome, are treated with exposures on four different occasions in Chapters 1, 4, 5, and 6. This repeated exposure is intended to produce better mastery of the topic.

Each of Chapters 9 through 12 is independent of the others, and any one or more of these chapters may be omitted. Logarithms are presented so that the computational work can easily be omitted, if this is desired. Since partial fractions and graphs of rational functions are frequently not included in a college algebra course, they were placed at the end of Chapter 12. They can be taught at an earlier point if the instructor wishes.

Many examples and exercises involving applications of the material have been included at points where it was practical to do so without unduly increasing the size of the text.

Each set of exercises contains enough problems to furnish two to four assignments. In almost all cases, there is an odd–even pairing of problems, with answers provided to the odd-numbered problems. As a general rule, the problems are arranged in order of increasing difficulty. Asterisks are used to indicate problems which are either specialized or more difficult. They can be omitted if the instructor wishes.

The most important features of the book include:

1. a clear and readable presentation of the material to the student;

2. a format which presents material in an easily referenced form and allows flexibility in the selection of topics;

3. over 400 examples, many involving applications;

4. about 2750 exercises, with odd–even pairing and answers for odd-numbered problems;

5. starred exercises which can be used as challenge problems;

6. repeated exposure to concepts, whenever possible;

7. practice tests with worked-out solutions at the end of each chapter.

An instructor's manual containing sample tests with answer keys and answers to even-numbered problems is provided. A study guide for the student is also available with solutions to many odd-numbered problems and practice tests with detailed solutions.

ACKNOWLEDGMENTS

For many helpful comments and suggestions made in reviews, we wish to express our sincere appreciation to Edward B. Anders, Northwestern State University of Louisiana; Richard W. Ball, Auburn University; James Calhoun, Western Illinois University; Alan M. Johnson, University of Arkansas at Little Rock; Donald Marsian, Hillsborough Community

College; William L. Perry, Texas A & M University; and Monty Strauss, Texas Tech University.

Special thanks are due to Margaret Dunn for her expert typing of the entire manuscript and to Wanda Provance for her assistance throughout the writing of this book.

We also wish to express our gratitude to Susan Formilan for her help in processing the manuscript, to Kathleen Lafferty for her cooperation and supervision of the final production of the text, to Mark Binn for his excellent design work, to Robin Bartlett and Ken Quinn for their help in promotion, and to Robert Sickles for his efforts in the final stages of production. Finally, we wish to thank Harry Gaines, Executive Editor, Prentice-Hall, who found time in a very busy schedule to give careful guidance to our project.

Jimmie Gilbert
James Spencer
Linda Gilbert

Fundamentals

1-1
Sets
In this chapter, we review the basic properties of some of the number systems which are of primary importance in the study of algebra. The real number system is of particular importance, and it is given special emphasis in later sections.

In any mathematical system, there must be some terms which are undefined. These terms allow one to formulate the initial assumptions made about the system.

The word "set" is one of our undefined terms, but when we use the word, we have in mind a *collection* or *family* of objects which is such that we can determine whether or not a certain object is in the set. The individual objects in the set are called the *members* of the set, or the *elements* of the set.

The simplest way in which a set may be described is by listing the members of the set. For example, we would write

$$T = \{2, 4, 6, 8\}$$

to indicate that T is the set which contains the numbers 2, 4, 6, and 8 and which has no other members.

The use of braces to indicate a set is commonly accepted as standard notation, even when all the elements cannot be listed. Thus the notation

$$E = \{2, 4, 6, 8, \ldots\}$$

would indicate that E is the set of all positive even integers. Another standard way to describe a set is to use braces and indicate the property which is the qualification for membership in the set. In this manner, the same set E would be indicated as

$$E = \{x \mid x \text{ is an even positive integer}\}.$$

This notation[1] is read as "E is the set of all x such that x is an even positive integer." The vertical slash is taken as shorthand for "such that."

There are other shorthand symbolisms that are convenient to use in connection with sets. We write $x \in S$ as shorthand for the phrase "x is an element of S," or "x is in S." If x is not an element of S, we write $x \notin S$. With T as above, we would write $6 \in T$ and $10 \notin T$.

If it happens that every element of the set A is an element of the set B, then A is called a *subset* of B, and we write $A \subseteq B$. With our sets T and E above, $T \subseteq E$. We write $A \nsubseteq B$ to indicate that A is not a subset of B. If T and E are the same as before and C is given by $C = \{4, 8, 12\}$, then $C \nsubseteq T$ and $C \subseteq E$.

The notation $B \supseteq A$ is used interchangeably with $A \subseteq B$. We would read "B contains A" for $B \supseteq A$ and "A is contained in B" for $A \subseteq B$. Similarly, $B \nsupseteq A$ and $A \nsubseteq B$ are interchangeable symbolisms. The symbol "$\in$" is reserved for elements, and "$\subseteq$" is reserved for subsets. For example, one would write $4 \in \{2, 4, 6, 8\}$ and $\{2, 4\} \subseteq \{2, 4, 6, 8\}$. One would *not* write $4 \subseteq \{2, 4, 6, 8\}$ or $\{2, 4\} \in \{2, 4, 6, 8\}$. The subset which consists of the element 4 only would be written $\{4\}$, and $\{4\} \subseteq \{2, 4, 6, 8\}$ would be correct use of the notation.

If $A \subseteq B$ and $A \neq B$, A is called a *proper subset* of B.

We have already used the equality sign in connection with sets several times, under the assumption that the meaning was clear: $A = B$ means that A and B are composed of exactly the same elements. That is, each member of A is a member of B, and each member of B is a member of A.

There are two operations on sets that are frequently used in mathematics. One of these operations is that of forming the union of two sets. If A and B are sets, the *union* of A and B is the set $A \cup B$ given by

$$A \cup B = \{x \mid x \in A \quad \text{or} \quad x \in B\}.$$

---

[1]This notation is frequently referred to as the "set-builder" notation.

That is, $A \cup B$ consists of all those elements x which are either an element of A, or an element of B, or an element of both A and B.

EXAMPLE 1   (a) If $A = \{2, 4, 6\}$ and $B = \{1, 3, 6\}$, then $A \cup B = \{1, 2, 3, 4, 6\}$.
(b) If $A = \{1, 2, 3, 4\}$ and $B = \{2, 4\}$, then $A \cup B = \{1, 2, 3, 4\} = A$.
(c) $\{1, 2, 3\} \cup \{4, 5, 6\} = \{1, 2, 3, 4, 5, 6\}$.

The other operation frequently used is that of forming the intersection of sets. For sets A and B, the *intersection* of A and B is the set $A \cap B$ given by

$$A \cap B = \{x \mid x \in A \quad \text{and} \quad x \in B\}.$$

Thus $A \cap B$ consists of those elements which are in both A and B.

EXAMPLE 2   (a) If $A = \{2, 4, 6, 8\}$ and $B = \{1, 3, 6, 2\}$, then $A \cap B = \{2, 6\}$.
(b) If $A = \{1, 2, 3, 4\}$ and $B = \{2, 4\}$, then $A \cap B = \{2, 4\} = B$.

One can see that it might happen that A and B would have no elements at all in common. That is, there would be no element x such that x is an element of A and also an element of B. In other words, the set $A \cap B$ may be empty. It makes sense, then, to introduce the empty set, the set which has no members. The *empty set* is denoted by $\varnothing$, or $\{\ \}$, and $\varnothing$ is regarded as being a subset of every set. Two sets A and B are called *disjoint* if $A \cap B = \varnothing$.

The operations of union and intersection can be applied in repetition. As an instance, one might form the intersection of A and B, obtaining $A \cap B$, and then form the union of this set with a set C. The resulting set would be denoted by $(A \cap B) \cup C$, where the parentheses indicate the order in which the operations are to be performed.

EXAMPLE 3   Let $A = \{1, 2, 3, 4\}$, $B = \{3, 4, 5, 6\}$, and $C = \{2, 4, 6\}$. Then
$$(A \cap B) \cup C = \{3, 4\} \cup \{2, 4, 6\} = \{2, 3, 4, 6\}$$
and
$$A \cap (B \cup C) = \{1, 2, 3, 4\} \cap \{2, 3, 4, 5, 6\} = \{2, 3, 4\}.$$
This illustrates that the placing of symbols of grouping in an expression may affect the resulting set, for we have $(A \cap B) \cup C \neq A \cap (B \cup C)$.

When working with sets, we make an agreement or assumption that all the sets we are dealing with are subsets of some *universal set*. Having made this assumption, performing the operations of union and intersection on some of our sets results once again in one of the sets in our collection. Often, it is helpful to draw a picture that diagrams the various sets involved. For example, in Figure 1.1 we have indicated two subsets A and B of a universal set U. The shaded circles representing A and B lead to a crosshatching for the intersection of the regions A and B. Pictorial representations such as we have in Figure 1.1 are called *Venn diagrams*.

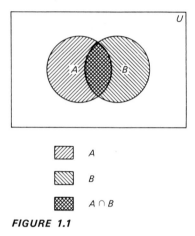

FIGURE 1.1

In some situations, it is useful to have additional notation for special subsets of the universal set. For arbitrary subsets A and B of U, the set of all elements of A which are not elements of B is denoted by $A - B$. That is,

$$A - B = \{x \in U \,|\, x \in A \quad \text{and} \quad x \notin B\}.$$

This set is called the *complement* of B in A. In Figure 1.2, we have represented sets A and B by circles as before. These circles naturally separate the points representing U into four regions, labeled 1, 2, 3, and 4. The set $A - B$ is represented by the region numbered 3.

The set $U - A$ is given the special notation $A' = U - A$. Thus,

$$A' = \{x \in U \,|\, x \notin A\}.$$

In Figure 1.2, A' is represented by the union of the regions numbered 1 and 4. We simply speak of the *complement of A* when referring to A', rather than the complement of A in U. There are some interesting relations which exist between complements and unions or intersections. For example, it is true that $(A \cup B)' = A' \cap B'$. This statement is known as *De Morgan's Theorem*. The proof of this fact and similar results can be found in textbooks for abstract algebra or other more advanced courses.

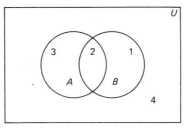

FIGURE 1.2

EXAMPLE 4 Let $U = \{1, 2, 3, 4, 5, 6, 7, 8, 9\}$, $A = \{1, 2, 3, 4, 5, 6\}$, and $B = \{1, 3, 5, 7, 9\}$. The operations and notations introduced in this section are illustrated below with several examples.

$$A - B = \{2, 4, 6\},$$
$$B - A = \{7, 9\},$$
$$A' = \{7, 8, 9\},$$
$$B' = \{2, 4, 6, 8\},$$
$$A \cap B' = \{2, 4, 6\},$$
$$(A' \cup B)' = \{1, 3, 5, 7, 8, 9\}' = \{2, 4, 6\},$$
$$A \cup A' = U,$$
$$(A \cup A')' = \varnothing.$$

There are several types of numbers which are given special designations. Those that we are concerned with now are listed below.

1. The set of *natural numbers*

$$N = \{1, 2, 3, \ldots\}.$$

This set is also referred to as the set of *counting numbers*, or as the set of *positive integers*.

2. The set of *integers*

$$I = \{\ldots, -3, -2, -1, 0, 1, 2, 3, \ldots\}.$$

An *even integer* is an integer which is a multiple of 2. An odd integer is an integer which is not even. Note that 0 is an even integer, since $0 = 0 \cdot 2$.

3. The set of *rational numbers*

$$Q = \left\{ \frac{a}{b} \,\middle|\, a \in I, \quad b \in I, \quad \text{and} \quad b \neq 0 \right\}.$$

Thus a rational number is a quotient of integers with a nonzero denominator. The rational numbers are also characterized by the fact that they have decimal representations which are terminating

(such as $\frac{1}{4} = 0.25$) or repeating (such as $\frac{70}{33} = 2.121212\cdots$).
Note that $I \subseteq Q$.

4. The set of *irrational numbers*, which consists of those numbers which have a decimal representation that is nonterminating and nonrepeating. The numbers $\sqrt{2}$ and π are examples of irrational numbers.

5. The set of *real numbers*, which is the union of the set of all rational numbers and the set of all irrational numbers.

EXERCISES 1-1[2]

1. Let $A = \{0, 1, 2, 3, 4, 5, 6, 7, 8, 9\}$. Exhibit the set $B = \{x \mid x \in A$ and x is even$\}$ by listing the members of B.

2. Describe the set A in Exercise 1 by indicating a property which is qualification for membership in A.

3. Describe the set $A = \{1, 4, 9, 16, 25\}$ by indicating a property which is qualification for membership in A.

4. Write out $A \cup B$ and $A \cap B$ if $A = \{3, 5, 7, 8, 9\}$ and $B = \{7, 8, 9, 13, 15\}$.

5. Let $U = \{0, 1, 2, 3, \ldots, 9, 10\}$, $A = \{0, 2, 3, 4, 6, 9\}$, $B = \{0, 2, 4, 6, 8, 10\}$, and $C = \{3, 4, 5, 6\}$. Find
   (a) $(A' \cup B)'$ (b) $A' \cap B'$
   (c) $A \cap (B \cup C)$ (d) $(A \cup B) \cap (A \cup C)$
   (e) $(A \cap B) \cup C$ (f) $(A \cup B) \cap C$
   (g) $C \cap A$ (h) $C \cap B$
   (i) $(C \cap A) \cup (C \cap B)$ (j) $C \cap (A \cup B)$
   (k) $A - B$ (l) $A - C$
   (m) $B - A$ (n) $C - A$
   (o) $A - (B - C)$ (p) $(A - B) - C$
   (q) $(A - B) \cap (A - C)$ (r) $(A - B) \cup (A - C)$
   (s) $(A - B) \cap (B - C)$ (t) $(A - B) \cup (B - C)$

6. Determine whether each of the following statements is true or false.
   (a) $\{a, c\} = \{a, c, c\}$ (b) $\{a, c\} \subseteq \{c, a\}$ (c) $a \in \{a, b, c\}$
   (d) $0 \in \emptyset$ (e) $\emptyset \subseteq \{a, b, c\}$ (f) $a \subseteq \{a, b, c\}$
   (g) $\emptyset \in \{a, b, c\}$ (h) $\emptyset \in \{\emptyset\}$

7. Determine whether or not each statement is true for all sets A, B, and C.
   (a) $A \cap A' = \emptyset$
   (b) $A \cup \emptyset = \emptyset$
   (c) $A \cap (B \cup C) = A \cup (B \cap C)$
   (d) $A \cup B = B \cap A$
   (e) $A \cap (B \cup C)' = A \cap B' \cap C'$
   (f) $A \cup \emptyset = A \cap \emptyset$

8. Let A be an arbitrary set and U the universal set. Complete the following statements.

[2]Starred exercises are either specialized or more difficult.

(a) $A \cup A' =$

(b) $A \cap A' =$

(c) $A \cup \emptyset =$

(d) $A \cap \emptyset =$

(e) $A \cup U =$

(f) $A \cap U =$

(g) $A \cap A =$

(h) $A \cup A =$

(i) $U \cup \emptyset =$

(j) $\emptyset \cup \emptyset =$

(k) $U \cap \emptyset =$

(l) $\emptyset \cap \emptyset =$

(m) $U' =$

(n) $A \cap \emptyset' =$

(o) $\emptyset' =$

(p) $(A \cap \emptyset)' =$

9. List all the subsets of $\{a, b\}$.

10. List all the subsets of $\{a, b, c\}$.

*11. State the most general conditions on the sets A and B under which the given equality holds.

(a) $A \cap B' = \emptyset$

(b) $A \cup B = \emptyset$

(c) $A \cap B = U$

(d) $A \cap B = B$

(e) $A \cup B = B$

12. Let A, B, C, and U be as indicated in the following diagram. Shade the regions corresponding to these sets: (a) $A \cup B$; (b) $A \cap B$; (c) $A' \cap B'$; (d) $A \cap B \cap C$; (e) $A' \cup (B \cap C)$.

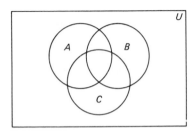

13. Let N be the set of natural numbers, I the set of integers, Q the set of rational numbers, R the set of real numbers, and I_r the set of irrational numbers. In which of these sets does each of the following belong?

(a) 1

(b) $\frac{2}{3}$

(c) $-\pi$

(d) 1.3

(e) $-\sqrt{2}$

(f) 0

(g) $0.6666\cdots$

(h) $-\frac{135}{222}$

(i) $\frac{22}{7}$

(j) $\sqrt{-1}$

14. Let N, I, Q, R, and I_r be as described in problem 13. Determine the following.

(a) $N \cap I$

(b) $Q \cap I_r$

(c) $Q \cup I_r$

(d) $R \cup I_r$

(e) $I \cap Q$

(f) $I \cap I_r$

1-2
The Field
Properties
of the
Real Numbers

In this section, we briefly review some of the basic properties of the real numbers. For the remainder of the book, the set of all real numbers will be denoted by $\Re$.

One of the fundamental properties of the real numbers is that there is a relation of equality defined in $\Re$. The basic properties of equality are these:

1. *Reflexive Property:* $a = a$, for all $a \in \mathfrak{R}$.

2. *Symmetric Property:* If $a = b$, then $b = a$.

3. *Transitive Property:* If $a = b$ and $b = c$, then $a = c$.

4. *Substitution Property:* If $a = b$, then $a + c = b + c$ and $ac = bc$ for all $c \in \mathfrak{R}$.

We are familiar with the operations of addition, subtraction, and multiplication on real numbers. These are examples of *binary operations*. When we speak of a binary operation on a set, we have in mind a process which combines two elements of the set to produce a third element of the set. This third element, the result of the operation on the first two, must be unique. That is, there must be one and only one result from the combination. Also, it must always be possible to combine the two elements, no matter which two are chosen. The operations of union and intersection of subsets of U are examples of binary operations. The two fundamental binary operations on the real numbers are addition and multiplication (subtraction and division are defined in terms of these). We adopt the usual notations for these operations.

Concerning the operations of addition and multiplication of real numbers, the basic properties are given in the list below. This list of properties, or postulates, is called the *field properties*. Any set in which equality, addition, and multiplication are defined so that the field properties hold is called a *field*. There are many fields other than the real numbers. One example is the set of all rational numbers.

Field Properties for the Real Numbers

1. *Addition Postulates.* The operation of addition defined in $\mathfrak{R}$ has the following properties.
   (a) *Closure Property.* $a + b \in \mathfrak{R}$ for all $a \in \mathfrak{R}$ and $b \in \mathfrak{R}$.
   (b) *Associative Property.* $a + (b + c) = (a + b) + c$ for all $a, b, c \in \mathfrak{R}$.
   (c) *Additive Identity.* There is a real number 0 such that $a + 0 = 0 + a = a$ for all $a \in \mathfrak{R}$.
   (d) *Additive Inverses.* For each $a \in \mathfrak{R}$, there is an element $-a \in \mathfrak{R}$ such that $a + (-a) = (-a) + a = 0$.
   (e) *Commutative Property.* $a + b = b + a$ for all a and $b \in \mathfrak{R}$.

2. *Multiplication Postulates.* The operation of multiplication defined in $\mathfrak{R}$ has the following properties.
   (a) *Closure Property.* $a \cdot b \in \mathfrak{R}$ for all $a \in \mathfrak{R}$ and $b \in \mathfrak{R}$.

(b) *Associative Property.* $a \cdot (b \cdot c) = (a \cdot b) \cdot c$ for all $a, b, c \in \mathcal{R}$.

(c) *Multiplicative Identity.* There is a real number 1 such that $1 \neq 0$ and $a \cdot 1 = 1 \cdot a = a$ for all $a \in \mathcal{R}$.

(d) *Multiplicative Inverses.* For each *nonzero* $a \in \mathcal{R}$, there is an element $\dfrac{1}{a} \in \mathcal{R}$ such that $a\left(\dfrac{1}{a}\right) = \left(\dfrac{1}{a}\right)a = 1$.

(e) *Commutative Property.* $a \cdot b = b \cdot a$ for all $a, b \in \mathcal{R}$.

3. *Distributive Property.* $a \cdot (b + c) = a \cdot b + a \cdot c$ for all $a, b, c \in \mathcal{R}$.

It is worth noting that this list of field properties is a list of *assumptions* concerning the real numbers. It is a list of basic properties from which the other properties of addition and multiplication of real numbers can be derived. We derive some of these in the next section.

EXAMPLE 1   Each of the statements below is a true statement concerning addition or multiplication of real numbers. To the right of each statement, field properties which justify the statement are given. Letters represent arbitrary real numbers.

(a) $2 + (3 + \pi) = 5 + \pi$ (Associative property, $+$)

(b) $5(\frac{1}{5}) = 1$ (Multiplicative inverse)

(c) $a(x + 1) = ax + a \cdot 1$ (Distributive property)

           $= ax + a$ (Multiplicative identity)

(d) $(2 + 3) + (-3) = 2 + [3 + (-3)]$ (Associative property, $+$)

              $= 2 + 0$ (Additive inverse)

              $= 2$ (Additive identity)

(e) $ax + ay = ay + ax$ (Commutative property, $+$)

(f) $ax + ay = ax + ya$ (Commutative property, $\cdot$)

(g) $x + y = 1 \cdot x + 1 \cdot y$ (Multiplicative identity)

(h) $2 + \dfrac{\pi}{3} \in \mathcal{R}$ (Closure property, $+$)

The *difference*, $a - b$, of real numbers a and b is defined by

$$a - b = a + (-b).$$

The operation which combines a and b to yield $a - b$ is called *subtraction*.

We note the two different uses of the minus sign "$-$" in the equation $a - b = a + (-b)$. In $a - b$, it represents the operation of subtraction, and in $a + (-b)$, the term $-b$ denotes the real number which yields 0 when added to b.

Just as the operation of addition is used to define subtraction, the operation of multiplication is used to define division. Let a and b be real numbers with $b \neq 0$. The quotient $\dfrac{a}{b}$ (or $a \div b$, or a/b) is defined by

$$\frac{a}{b} = a \cdot \left(\frac{1}{b}\right), \qquad \text{where } b \neq 0.$$

The operation which combines a and b to yield a/b when $b \neq 0$ is called *division*.

We note that division by 0 is excluded. We do not go into the details here, but any attempt to extend the operation of division so as to include division by zero leads invariably to a conflict with the field properties.

The operations of addition, multiplication, subtraction, and division are frequently referred to as the *four fundamental operations*.

In many instances, it is necessary to indicate a sequence of operations which are to be performed. Such a situation could give rise to ambiguity. For example, $5 \cdot 6 + 3$ could mean $(5)(9) = 45$, or it could mean $30 + 3 = 33$. Rules concerning the hierarchy of operations and the use of symbols of grouping have been established to govern these situations. These are stated below.

Hierarchy of Operations and Symbols of Grouping

1. In an expression involving several of the four fundamental operations, all *multiplications and divisions are performed first*, from left to right, then followed by the additions and subtractions, from left to right.

2. Parentheses, ( ), brackets, [ ], or braces, { }, may be used to indicate the order of operations, with the understanding that the *innermost symbols of grouping are removed first*. That is, the innermost operations are performed first.

EXAMPLE 2   The rules given above are illustrated in the following examples.

(a) $48 + 16 \div 8 - 3 \cdot 2 = 48 + 2 - 6$
$$= 44$$

(b) $(48 + 16) \div 8 - 3 \cdot 2 = 64 \div 8 - 3 \cdot 2$
$$= 8 - 6$$
$$= 2$$

(c) $(48 + 16) \div (8 - 3 \cdot 2) = 64 \div (8 - 6)$
$$= 64 \div 2$$
$$= 32$$

(d) $\{[(48 + 16) \div 8] - 3\} \cdot 2 = \{[64 \div 8] - 3\} \cdot 2$
$$= \{8 - 3\} \cdot 2$$
$$= 5 \cdot 2$$
$$= 10$$

The use of expressions such as those in (a), (b), and (c) is rare. It is preferable to avoid confusion by the use of parentheses. For example, in part (a), one would write $48 + (16 \div 8) - (3 \cdot 2)$ instead of $48 + 16 \div 8 - 3 \cdot 2$.

A set is *closed* under an operation if the result of performing the operation on two elements of the set, distinct or not, is always again a member of the set.

EXAMPLE 3 The closure property is illustrated in the following examples.

(a) The set of all positive integers is *closed* under addition since the sum of two positive integers is always a positive integer.

(b) The set of all positive integers is *not closed* under subtraction. The numbers 3 and 7 are positive integers, but $3 - 7 = -4$ is *not* a positive integer.

(c) The set $\{0, 1\}$ is *not closed* under addition since $1 + 1 = 2$, and $2 \notin \{0, 1\}$.

EXERCISES 1-2

In each of problems 1-10, a true statement is made concerning addition or multiplication of real numbers. For each statement, identify the field properties which justify that statement.

1. $3 + [4 + (-4)] = 3 + 0$

2. $4 + (-4) = 0$

3. $ax + ay = (x + y)a$

4. $ax + ay = ay + ax$

5. $(1)(3) + 4 = 3 + 4$

6. $(1)(3) + (1)(3) = (1)(3 + 3)$

7. $0 + (1)(5) = (1)(5)$

8. $0 + (1)(5) = 0 + 5$

9. $(-\sqrt{2})\left(\dfrac{1}{-\sqrt{2}}\right) = 1$

10. $3 + (4 + 5) = 7 + 5$

In problems 11-16, decide whether or not the given set is closed under the stated operation. If the set is not closed, give an example to illustrate this fact.

11. The set $\{0, 1\}$ under multiplication.

12. The set of all integers under addition.

13. The set of all even integers under multiplication.

14. The set of all odd integers under multiplication.

15. The set of all odd integers under addition.

16. The set of all negative integers under multiplication.

17. Does subtraction of real numbers have the associative property? That is, does $a - (b - c) = (a - b) - c$ for all real numbers a, b, and c? Explain why, or why not.

18. Does division of nonzero real numbers have the associative property? Why, or why not?

19. Does subtraction of real numbers have the commutative property? Illustrate your answer.

20. Does division of nonzero real numbers have the commutative property? Illustrate your answer.

21. If $x = 4$ and $y = 5$, find the value of
$$[17x - x(2 + y)] \div (3 + y).$$

22. If $x = 3$ and $y = 5$, find the value of
$$\{8x[3y + 2 - 2(2y - 7)] - 4[12x - 2(2y)]\} \div [y + 2 + x(14 - x)].$$

23. If $a = 3$, $b = 5$, and $c = 2$, find the value of
$$[2b(a - b) + 4ab(a - 2b + 3c)] \div [bc(2ab - c)].$$

24. If $a = -2$, $b = 3$, and $c = 5$, find the value of
$$\{a[b - c(a - b)] + b[a - 2c(3b + c)]\} \div (2ab).$$

1-3
Properties
of Real Numbers

The field properties and their consequences are the basis for many of the techniques developed in elementary algebra. In this section, we study some of the important properties of the real numbers which are consequences of the field properties.

The behavior of 0 with respect to the operation of multiplication is of primary importance. The following two theorems describe this behavior.

THEOREM 1-1   If either $a = 0$ or $b = 0$, then $a \cdot b = 0$.

PROOF   We note first that the theorem states that if either factor in a product is zero, then the product must be zero.

Suppose that $b = 0$. We shall reduce $a \cdot 0$ to 0 by using various field properties, as indicated.

$a \cdot b = a \cdot 0$

$\qquad = a \cdot 0 + 0$ (Additive identity)

$\qquad = a \cdot 0 + \{a \cdot 0 + [-(a \cdot 0)]\}$ (Additive inverse)

$\qquad = \{a \cdot 0 + a \cdot 0\} + [-(a \cdot 0)]$ (Assoc. property, $+$)

$\qquad = \{a \cdot (0 + 0)\} + [-(a \cdot 0)]$ (Distributive property)

$\qquad = a \cdot 0 + [-(a \cdot 0)]$ (Additive identity)

$\qquad = 0$ (Additive inverse)

In the case where $a = 0$, we have $0 \cdot b = b \cdot 0 = 0$, using the commutative property of multiplication.

In mathematics, an "either-or" phrase is used in the inclusive sense, so that the "both" possibility is always included. In Theorem 1-1, for instance, the possibility that both $a = 0$ and $b = 0$ is included, and $0 \cdot 0 = 0$.

The converse of Theorem 1-1 is also true. That is, if a product is zero, then at least one of the factors is zero.

THEOREM 1-2   If a and b are real numbers such that $ab = 0$, then either $a = 0$ or $b = 0$.

PROOF   Let a and b be real numbers such that $ab = 0$. We need to show that at least one of the numbers a, b is zero. If $a = 0$, the conclusion of the theorem is satisfied. Suppose, then, that $a \neq 0$. By the multiplicative inverse postulate, there is a real number $1/a$ such that $1/a \cdot a = 1$. The field properties can be used to show that $b = 0$ as follows:

$$b = 1 \cdot b \qquad \text{(Multiplicative identity)}$$

$$= \left(\frac{1}{a} \cdot a\right)b \qquad \text{(Multiplicative inverse)}$$

$$= \frac{1}{a}(ab) \qquad \text{(Associative property, } \cdot)$$

$$= \frac{1}{a} \cdot 0 \qquad \text{(Since } ab = 0)$$

$$= 0 \qquad \text{(Theorem 1-1)}$$

This completes the proof of the theorem.

One important consequence of Theorem 1-1 should be noted: if $a \neq 0$ and $b \neq 0$, then $ab \neq 0$. That is, a product of nonzero factors is not zero.

The following theorem is known as the *cancellation law for addition*. It is one of the most important properties mentioned in this section, since it is extremely useful in simplifying equations.

THEOREM 1-3 (CANCELLATION LAW FOR ADDITION)   If a, b, and c are real numbers and $a + c = b + c$, then $a = b$.

PROOF   Suppose that $a + c = b + c$. Now c has an additive inverse $-c$, and adding $-c$ to both sides of this equation yields

$$(a + c) + (-c) = (b + c) + (-c)$$

and

$$a + [c + (-c)] = b + [c + (-c)],$$

by the associative property of addition. Since $c + (-c) = 0$, we have

$$a + 0 = b + 0,$$

and therefore

$$a = b.$$

There is a property for multiplication which is similar to Theorem 1-3, and it is known as the *cancellation law for multiplication*. Instead of simplifying an equation by removing a *term* in a *sum*, it simplifies an equation by removing a *factor* in a *product*. There is the important condition, however, that the factor removed *must be different from zero*.

THEOREM 1-4 (CANCELLATION LAW FOR MULTIPLICATION) If $a, b,$ and c are real numbers such that

$$ab = ac \quad \text{with} \quad a \neq 0,$$

then

$$b = c.$$

PROOF   Let $a, b,$ and $c \in \Re$ be such that $ab = ac$, and $a \neq 0$. The condition $a \neq 0$ means that $1/a$ exists, and multiplying both sides of

$$ab = ac$$

by $1/a$ yields

$$\frac{1}{a}(ab) = \frac{1}{a}(ac).$$

Hence

$$\left(\frac{1}{a} \cdot a\right)b = \left(\frac{1}{a} \cdot a\right)c$$

or

$$1 \cdot b = 1 \cdot c.$$

Therefore, $b = c$, and the theorem is proven.

The next theorem states several familiar properties which can be proved by using the field properties and the preceding theorems. There is a great deal of similarity in the proofs of the various parts of the theorem, so a proof is provided for only one part. The proofs of the other parts are left as exercises.

We recall from the definition of division that $a/b = a(1/b)$ when $b \neq 0$.

THEOREM 1-5   Let $a, b, c,$ and d be real numbers with $b \neq 0$ and $d \neq 0$. Then

(a) $\dfrac{a}{b} = \dfrac{c}{d}$ if and only if[3] $ad = bc$;

(b) $\dfrac{ad}{bd} = \dfrac{a}{b}$;

(c) $\dfrac{1}{b} \cdot \dfrac{1}{d} = \dfrac{1}{bd}$;

(d) $\dfrac{a}{b} \cdot \dfrac{c}{d} = \dfrac{ac}{bd}$;

(e) $\dfrac{a}{b} + \dfrac{c}{d} = \dfrac{ad + bc}{bd}$.

PROOF   As mentioned before the theorem, we shall prove only one part of the theorem. Consider part (c).

Since $b \neq 0$ and $d \neq 0$, we have $bd \neq 0$ by Theorem 1-2. Thus $1/bd$ exists and

$$(bd)\left(\frac{1}{bd}\right) = 1.$$

Using the field properties for multiplication, we also have

$$(bd)\left(\frac{1}{b} \cdot \frac{1}{d}\right) = \left[(bd) \cdot \frac{1}{b}\right] \cdot \frac{1}{d} \qquad \text{(Associative property)}$$

$$= \left[\frac{1}{b} \cdot (bd)\right] \cdot \frac{1}{d} \qquad \text{(Commutative property)}$$

$$= \left[\left(\frac{1}{b} \cdot b\right)d\right] \cdot \frac{1}{d} \qquad \text{(Associative property)}$$

$$= (1 \cdot d) \cdot \frac{1}{d} \qquad \text{(Multiplicative inverse)}$$

$$= d \cdot \frac{1}{d} \qquad \text{(Multiplicative identity)}$$

$$= 1 \qquad \text{(Multiplicative inverse)}$$

Thus we have

$$(bd)\left(\frac{1}{b} \cdot \frac{1}{d}\right) = (bd)\left(\frac{1}{bd}\right),$$

since each side of this equation equals 1. It follows from Theorem 1-4 that

$$\frac{1}{b} \cdot \frac{1}{d} = \frac{1}{bd}.$$

A special case of part (e) of Theorem 1-5 is worth noting:

$$\frac{a}{d} + \frac{c}{d} = \frac{a + c}{d}.$$

(See problem 4.)

[3]That is, $a/b = c/d$ if $ad = bc$, and $ad = bc$ if $a/b = c/d$.

The next theorem gives two of the relationships involving multiplication and additive inverses. Other similar results may be found in the exercises.

THEOREM 1-6 Let a and b be real numbers. Then

> (a) $(-a)b = -(ab)$;
> (b) $(-a)(-b) = ab$.

PROOF OF PART (a) From the definition of $-(ab)$, we have $-(ab) + ab = 0$. Using the field properties and Theorem 1-1, we can also show that $(-a)b + ab = 0$:

$$
\begin{aligned}
(-a)b + ab &= b(-a) + ba & \text{(Commutative property, }\cdot\text{)} \\
&= b(-a + a) & \text{(Distributive property)} \\
&= b \cdot 0 & \text{(Additive inverse)} \\
&= 0 & \text{(Theorem 1-1)}
\end{aligned}
$$

Thus

$$(-a)b + ab = -(ab) + ab,$$

and it follows from the cancellation law for addition that

$$(-a)b = -(ab).$$

The proof of part (b) is left as an exercise (see problem 7).

EXERCISES 1-3

Prove each statement by using the field properties, definitions, or previously proved results. a, b, c, and d are real numbers with $b \neq 0$ and $d \neq 0$. (See Theorem 1-5.)

1. $\dfrac{a}{b} = \dfrac{c}{d}$ if and only if $ad = bc$

2. $\dfrac{ad}{bd} = \dfrac{a}{b}$

3. $\dfrac{a}{b} \cdot \dfrac{c}{d} = \dfrac{ac}{bd}$

4. (a) $\dfrac{a}{b} + \dfrac{c}{b} = \dfrac{a + c}{b}$ (b) $\dfrac{a}{b} + \dfrac{c}{d} = \dfrac{ad + bc}{bd}$

Use the field properties, definitions, or previously stated results to prove each statement. a, b, and c are arbitrary real numbers.

5. $a + (b - c) = (a + b) - c$

6. $b + (a - b) = a$

7. $(-a)(-b) = ab$

8. $-(-a) = a$

9. $a - (-b) = a + b$

10. $(-a) + (-b) = -(a + b)$

11. If $b \neq 0$, then $\dfrac{-a}{b} = \dfrac{a}{-b} = -\dfrac{a}{b}$.

12. $\dfrac{-a}{-b} = \dfrac{a}{b}$

13. If $b \neq 0$, then $\dfrac{a}{b} - \dfrac{c}{b} = \dfrac{a - c}{b}$.

14. $-a = (-1) \cdot a$

15. $a(b - c) = ab - ac$

16. $(b - c) \cdot a = ba - ca$

17. $(a - b) + (b - c) = a - c$

18. Let $b \neq 0$. Prove that $a/b = 0$ if and only if $a = 0$. (That is, $a/b = 0$ if $a = 0$, and $a = 0$ if $a/b = 0$.)

*19. Let $a \neq 0$ and $b \neq 0$. Prove that

$$\frac{1}{a/b} = \frac{b}{a}.$$

*20. Let $a \in \mathcal{R}$ be arbitrary, and let b, c, and d be nonzero real numbers. Prove that

$$\frac{a/b}{c/d} = \frac{ad}{bc}.$$

***1-4**
*Ordering
of the
Real Numbers*

One of the most useful aids in working with real numbers is the association between real numbers and points on a straight line. We start with a horizontal line, on which a point is chosen and labeled with 0. This point is referred to as the *origin*. A unit of measure is chosen, and points successively one unit apart are located on the line in both directions from the origin. It is conventional to label the points so located to the right of 0 in succession with the positive integers $1, 2, 3, \ldots$, and those to the left of 0 are labeled successively with the negative integers $-1, -2, -3, \ldots$. The resulting configuration is pictured in Figure 1.3. The arrows at the end indicate that the line extends indefinitely in each direction.

Rational numbers are then located on the line by using appropriate portions of the chosen unit of measure. For example, $5/2 = 2.5$ is located two and one-half units to the right of 0, $-7/4 = -1.75$ is located one and three-fourths units to the left of 0, and so on (see Figure 1.4). Irrational numbers can be located to any desired degree of accuracy by using their

FIGURE 1.3

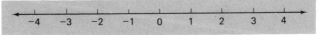

decimal approximations. In Figure 1.4, $\sqrt{2}$ is located using $\sqrt{2} \approx 1.41$, and $-\pi$ is located using $-\pi \approx -3.14$. (The symbol $\approx$ is used to mean "approximately equals.")

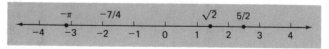

FIGURE 1.4

The association that has thus been established between real numbers and the points on the line is a one-to-one correspondence: each real number corresponds to exactly one point on the line, and each point on the line corresponds to one real number. The property that every point on the line corresponds to a real number is frequently called the *completeness property* of the real numbers. This relates to the fact that there are no gaps or jumps between the numbers when they are represented by points on the line. After the correspondence between real numbers and points on the line has been made, the line is commonly referred to as "the number line."

We recall the following fundamental properties of the positive real numbers.

1. The sum of two positive real numbers is a positive real number.
2. The product of two positive real numbers is a positive real number.
3. For any real number a, exactly one of the following statements is true:
   (i) a is positive;
   (ii) $a = 0$;
   (iii) $-a$ is positive.

In connection with statement (iii), we recall that $-a$ represents the additive inverse of a, and does not necessarily represent a negative number. When a is negative, then $-a$ is positive. For example, if $a = -4$, then $-a = 4$, a positive number.

DEFINITION 1-7   The symbol ">" is read as "greater than."

$a > b$ means $a - b$ is a positive number.

The graphical representation of $a > b$ on the number line is that *a lies to the right of b* (see Figure 1.5). In particular, $a > 0$ means that a lies to the right of 0, or that *a is positive*.

The properties of positive numbers referred to just prior to Definition 1-7 translate readily into statements about the relation ">." Specifically, we have the following theorem.

FIGURE 1.5

THEOREM 1-8   The relation ">" has the properties listed below, where a, b, and c represent arbitrary real numbers.

(a) *Transitive Property.* If $a > b$ and $b > c$, then $a > c$.
(b) *Closure Property.* If $a > 0$ and $b > 0$, then $a + b > 0$ and $ab > 0$.
(c) *Trichotomy Property.* Exactly one of the following is true:
$$a > b, \qquad a = b, \qquad b > a.$$

PROOF   To prove part (a), suppose that $a > b$ and $b > c$. Then $a - b$ is positive, and $b - c$ is positive. Since the sum of two positive numbers is positive, this means that
$$(a - b) + (b - c) = a - c$$
is positive. That is, $a > c$.

The proofs of parts (b) and (c) are left as exercises.

Ordering between real numbers is often indicated by using the relation "less than" instead of "greater than."

DEFINITION 1-9   The symbol "<" is read as "less than."

$a < b$ means $b - a$ is a positive number.

Equivalently, $a < b$ means $b > a$.

On the number line, $a < b$ means that *a lies to the left of b.* In particular, $a < 0$ means that a lies to the left of 0, or that *a is negative.* Whether we write $a < b$ or $b > a$ is often a matter of emphasis or of personal preference.

The symbols "<" and ">" are often used in combination with equality to form compound statements. For example, either of the symbolisms $x \leqq a$ or $x \leq a$ means that x is less than or equal to a. In this text, the symbol $x \leq a$ is used for this relation. Similarly, $x \geq a$ indicates that x is greater than or equal to a.

EXAMPLE 1   Some uses of "$\leq$" and "$\geq$" are illustrated by the statements below.

$-9 \leq -3$, since $-9 < -3$.

$8 \geq 8$, since $8 = 8$.

$x \leq 2$ means that x lies at 2 or to the left of 2 on the number line.

$x \geq 3$ means that x lies at 3 or to the right of 3 on the number line.

A graphical representation of $x \leq 2$ is obtained by placing a large dot at 2 and indicating a bold arrow to the left on the number line. This is illustrated in Figure 1.6. Similarly, $x \geq 3$ is represented as shown in Figure 1.7. The large dot at the end of the arrow is to indicate that the end point is included. If the end point is not to be included, we indicate this by an open circle at the end of the arrow. As an illustration, $x > 1$ is represented in Figure 1.8.

$x \leq 2$

FIGURE 1.6

$x \geq 3$

FIGURE 1.7

$x > 1$

FIGURE 1.8

Statements using the symbols "$<$," "$>$," "$\leq$," or "$\geq$" are called *inequalities*. In later work, we will be concerned with obtaining the solution sets for inequalities. At that time, the following theorem will be invaluable.

THEOREM 1-10   For any real numbers a, b, and c:

> (a) if $a > b$, then $a + c > b + c$;
> (b) if $a > b$ and $c > 0$, then $ac > bc$;
> (c) if $a > b$ and $c < 0$, then $ac < bc$.

PROOF   To prove part (a), suppose that $a > b$. Then $a - b$ is positive. Since

$$(a + c) - (b + c) = a + c - b - c$$
$$= a - b,$$

this means that $(a + c) - (b + c)$ is positive, so

$$a + c > b + c.$$

The proofs of parts (b) and (c) are left as exercises. It should be noted carefully in part (c) that the direction of the inequality reverses when both sides are multiplied by a negative number, whereas it does not change in part (b).

At times it is convenient to use inequalities which involve two of the symbols "$<$," "$>$," "$\leq$," "$\geq$." As an instance, suppose it is known that x is located either to the right of 2 or to the left of -1 on the number line. This could be indicated by

$$x > 2 \quad \text{or} \quad x < -1.$$

Graphically, this would be represented as shown in Figure 1.9, where the open circles again indicate that the end points are not included.

$x > 2 \quad \text{or} \quad x < -1$

FIGURE 1.9

As another illustration, suppose that we wish to indicate that x lies between 2 and 4 on the number line. This could be indicated by

$$x > 2 \quad \text{and} \quad x < 4$$

or by

$$2 < x \quad \text{and} \quad x < 4.$$

Graphically, this is represented by a line segment from 2 to 4, not including the end points. Notice that in Figure 1.10 we have combined the compound statement $2 < x$ and $x < 4$ into the compact form $2 < x < 4$. This is conventional notation to indicate that x lies between 2 and 4. There are two important rules for use of this type of compact notation:

$2 < x < 4$

FIGURE 1.10

1. Compact forms such as $a < x < b$ or $a > x > b$ are used only with compound statements using "and."

2. The two inequality symbols involved must be of the same kind (both "$<$," or both "$>$," etc.).

EXAMPLE 2   Figures 1.11 to 1.13 present the graphical representations of several compound statements involving inequalities.

(a) *Statement:* $-1 < x$ and $x \le 2$
Compact Form: $-1 < x \le 2$
Representation:

$-1 < x \le 2$

FIGURE 1.11

(b) *Statement:* $-2 \le x$ and $x \le 3$
Compact Form: $-2 \le x \le 3$
Representation:

$-2 \le x \le 3$

FIGURE 1.12

(c) *Statement:* $x \le 0$ or $x > 2$
No Compact Form: Not an "and" statement.
Representation:

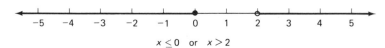

$x \le 0$ or $x > 2$

FIGURE 1.13

EXAMPLE 3   Name the part of Theorem 1-10 which justifies the given statement.

(a) If $-3x - 4 > 8$, then $-3x > 12$.
(b) If $-3x > 12$, then $x < -4$.

SOLUTION

(a) Theorem 1-10(a) allows us to add 4 to both sides of $-3x - 4 > 8$ to obtain $-3x > 12$.
(b) Since $-\frac{1}{3} < 0$, Theorem 1-10(c) allows us to multiply both sides of $-3x > 12$ by $-\frac{1}{3}$, and obtain $x < -4$.

The combined parts of Example 3 show how Theorem 1-10 can be used in some cases to isolate the variable x. Other illustrations are given in the next example.

EXAMPLE 4 Insert appropriate numbers in the blank spaces so as to make a true statement.

(a) If $4x - 5 \geq 7$, then $x \geq$ ___ .

(b) If $-9 < -2x + 3 \leq 7$, then ___ $> x \geq$ ___ .

SOLUTION

(a) In order to isolate x, we add 5 to both sides of the given inequality, and then multiply both sides by $\frac{1}{4}$. [See Theorem 1-10(a) and (b).]

$$4x - 5 \geq 7,$$
$$4x \geq 12,$$
$$x \geq \underline{3}.$$

(b) Using parts (a) and (c) of Theorem 1-10, we first add -3 to both sides of the given inequality, and then multiply both sides by $-\frac{1}{2}$:

$$-9 < -2x + 3 \leq 7,$$
$$-12 < -2x \leq 4,$$
$$\underline{6} > x \geq \underline{-2}.$$

The appropriate numbers for the blanks are underlined in the last step.

EXERCISES 1-4

In problems 1–10, name the part of Theorem 1-8 or Theorem 1-10 which justifies the given statement.

1. If $2x - 3 > 7$, then $2x > 10$.

2. If $2x > 10$, then $x > 5$.

3. If $x > 2$ and $2 > y$, then $x > y$.

4. If $x > 0$, then $x + 3 > 0$.

5. Either $x > 5$, $x = 5$, or $5 > x$.

6. If $x > 0$, then $5x > 0$.

7. If $5x > 0$, then $x > 0$.

8. If $-3x > 12$, then $x < -4$.

9. If $2x + 3 > 7$, then $2x > 4$.

10. If $x > 0$ and $y < 0$, then $xy < 0$.

In problems 11–22, insert appropriate numbers in the blank spaces so as to make a true statement.

11. If $x - 2 < 5$, then $x <$ ___ .

12. If $x + 3 < 2$, then $x <$ ___ .

13. If $x - 3 \leq -4$, then $x \leq$ ___ .

14. If $x + 2 \geq 6$, then $x \geq$ ____.

15. If $\frac{2}{3}x < 12$, then $x <$ ____.

16. If $\frac{4}{3}x \leq 48$, then $x \leq$ ____.

17. If $-2x \geq 6$, then $x \leq$ ____.

18. If $-3x \leq 18$, then $x \geq$ ____.

19. If $-1 \leq x - 2 < 4$, then ____ $\leq x <$ ____.

20. If $1 > x + 2 > -3$, then ____ $> x >$ ____.

21. If $-3 < -2x + 3 \leq 5$, then ____ $> x \geq$ ____.

22. If $10 > 4 - 3x \geq -2$, then ____ $< x \leq$ ____.

In problems 23–38, draw graphical representations of the given statements (see Figures 1.11, 1.12, 1.13).

23. $-2 > x \geq -7$

24. $-1 \leq x \leq 3$

25. $-4 < x \leq 0$

26. $-2 < x < 2$

27. $3 \geq x > 1$

28. $2 \geq x \geq 0$

29. $x > 1$ or $x < -1$

30. $x \leq 1$ or $x > 3$

31. $x > -1$ or $x < 1$

32. $x > 2$ or $x \leq 0$

33. $-2 < x \leq 0$ or $x > 1$

34. $3 > x \geq 1$ or $x < 0$

35. $2 \geq x > -1$ or $x < -3$

36. $3 > x > 1$ or $x > 5$

37. $-1 \leq x \leq 2$ or $x < -4$

38. $1 \leq x \leq 3$ or $-3 < x < -1$

*39. Prove that if a and b are real numbers with $a > 0$ and $b > 0$, then $a + b > 0$ and $ab > 0$.

*40. Prove that, for any real numbers a and b, exactly one of the following statements is true: $a > b$, $a = b$, $b > a$.

*41. Let $a, b,$ and c be real numbers. Prove that if $a > b$ and $c > 0$, then $ac > bc$.

*42. Let $a, b,$ and c be real numbers. Prove that if $a > b$ and $c < 0$, then $ac < bc$.

*43. Prove that if $a < b$, then $a < \dfrac{a + b}{2} < b$.

*44. Prove that if $0 < a < b$, then $a^2 < b^2$. ($a^2 = a \cdot a$, and $b^2 = b \cdot b$.)

1-5
Absolute Value

In describing distances between points on the number line, the concept of absolute value is essential.

DEFINITION 1-11   The absolute value of a real number a is denoted by $|a|$. It is defined by

$$|a| = \begin{cases} a & \text{if } a \geq 0, \\ -a & \text{if } a < 0. \end{cases}$$

We note that $|a|$ is *never negative*. If $a \neq 0$, then $|a|$ is a *positive number.* It is also worth noting that in the case where $|a| = -a$, the value $-a$ signifies a positive number. That is, when a is a *negative number*, then $-a$ is a *positive number*. For example, if $a = -3$, then $-a = -(-3) = 3$, and $|a| = 3$.

The graphical interpretation of $|a|$ on the number line is that $|a|$ denotes the distance between 0 and a.

EXAMPLE 1   (a) $|2| = 2$, and 2 is located two units from 0.
(b) $|-2| = 2$, and -2 is located two units from 0.
(c) $|a| = |-a|$ for any real number a, since both a and $-a$ are located at distance $|a|$ from 0.

When one thinks of absolute value in terms of distance on the number line, the following theorem seems fairly evident. A proof using the definition of absolute value involves consideration of different cases, and is omitted in this book.

THEOREM 1-12   Let x and d be real numbers with $d > 0$. Then

(a) $-|x| \leq x \leq |x|$;
(b) $|x| < d$ if and only if $-d < x < d$;
(c) $|x| > d$ if and only if either $x > d$ or $x < -d$.

Graphical representations of parts (b) and (c) of Theorem 1-12 are given in Figures 1.14 and 1.15.

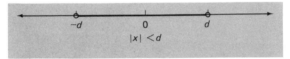

FIGURE 1.14

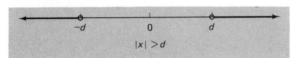

FIGURE 1.15

The following theorem can be proved by using the definition of absolute value and considering the various cases.

THEOREM 1-13   Let a and b be real numbers. Then

(a) $|a| \cdot |b| = |ab|$;
(b) $\left| \dfrac{a}{b} \right| = \dfrac{|a|}{|b|}$ if $b \neq 0$;
(c) $|a + b| \leq |a| + |b|$.

The inequality $|a + b| \leq |a| + |b|$ is frequently referred to as the *triangular inequality*.

There is one more connection to be made between absolute value and distance. Let d represent a positive real number, and consider

$$|x - a| < d.$$

By Theorem 1-12(b), $|x - a| < d$ if and only if

$$-d < x - a < d.$$

Adding a to all members of this compound inequality, we have

$$a - d < x < a + d.$$

This means that x lies between $a - d$ and $a + d$ (see Figure 1.16). Looking at it another way, $|x - a|$ represents the distance between x and a on the number line, and $|x - a| < d$ means that the distance between x and a is less than d.

Similarly, $|x - a| > d$ means that the distance between x and a is greater than d. This is represented in Figure 1.17.

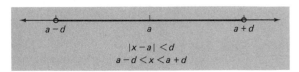

FIGURE 1.16

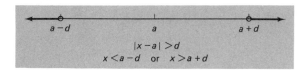

FIGURE 1.17

EXAMPLE 2   The foregoing discussion is illustrated in the following examples.

(a) $|x - 2| < 3$ means that the distance between x and 2 is less than 3. Thus x cannot lie as far to the right of 2 as $2 + 3 = 5$, and x cannot lie as far to the left of 2 as $2 - 3 = -1$. This is represented in Figure 1.18.

FIGURE 1.18

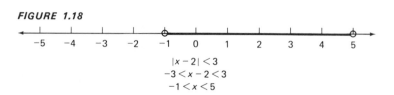

(b) $|x + 1| < 4$ means that the distance between x and -1 is less than 4, since $|x + 1| = |x - (-1)|$. This means that x cannot lie as far to the right of -1 as $-1 + 4 = 3$, or as far to the left as $-1 - 4 = -5$. The situation is pictured in Figure 1.19.

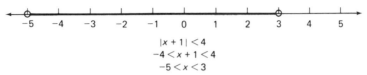

$$|x + 1| < 4$$
$$-4 < x + 1 < 4$$
$$-5 < x < 3$$

FIGURE 1.19

(c) $|x + 2| > 1$ means that the distance between x and -2 is greater than 1, since $|x + 2| = |x - (-2)|$. On the number line, x is either to the right of $-2 + 1 = -1$, or x is to the left of $-2 - 1 = -3$. See Figure 1.20.

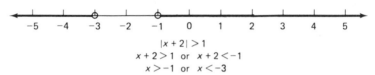

$$|x + 2| > 1$$
$$x + 2 > 1 \quad \text{or} \quad x + 2 < -1$$
$$x > -1 \quad \text{or} \quad x < -3$$

FIGURE 1.20

EXERCISES 1-5

Arrange each set of numbers in order from smallest to largest.

1. $-3, \pi, |-2|, |-3|, -\pi, -2$
2. $|-2|, -1, \sqrt{2}, -\sqrt{2}, -2, 1$
3. $\left|\dfrac{-22}{7}\right|, \pi, |-3|, -\left|\dfrac{22}{7}\right|, 0, -\pi$
4. $\sqrt{80}, -9, |-9|, -\sqrt{80}, \sqrt{80} - 9, |\sqrt{80} - 9|$
5. $-|8 - 3|, -8 - |3|, |8 - 3|, |8| + 3, 0$
6. $\left|-\dfrac{11}{7}\right|, -\left|\dfrac{11}{7}\right|, \left|-\dfrac{9}{7}\right|, -\left|\dfrac{9}{7}\right|, \left|-\dfrac{7}{11}\right|, -\left|\dfrac{7}{11}\right|, \left|\dfrac{-7}{9}\right|, -\left|\dfrac{7}{9}\right|$

Express each of the following without using absolute value symbols.

7. $|-7|$

8. $\left|-\dfrac{3}{2}\right|$

9. $|-7 + 3|$

10. $\left|\dfrac{2}{3}\right|$

11. $|-7| + 3$

12. $\left|\dfrac{2}{7}\right| + \left|\dfrac{4}{-7}\right|$

13. $|9 - 7|$

14. $|9| - |7|$

15. $|9| + |-7|$

16. $|7 - 9|$

17. $|7| - |9|$

18. $|x - 5|$, if $x > 5$

19. $|y - 4|$, if $y > 4$

20. $|x - 5|$, if $x < 5$

21. $|y - 4|$, if $y < 4$

22. $|7 - \sqrt{50}|$

23. $|9 - \sqrt{80}|$

24. $\left|\dfrac{22}{7} - \pi\right|$

25. $|2x - 10|$, if $x < 5$

26. $|3x - 12|$, if $x < 4$

*27. $|a - b|$, if $a < b$

*28. $|2a - b|$, if $a > b/2$

*29. $|2a - b|$, if $a < b/2$

*30. $|a - 2b|$, if $a < 2b$

31. $\left|\dfrac{x}{5}\right|$, if $x < 0$

32. $\dfrac{|x|}{5}$, if $x > 0$

33. $|(-1)^2|$

34. $|(-4)|^2$

35. $|-a^2|$, where $a^2 = a \cdot a$

36. $|a^2|$, where $a^2 = a \cdot a$

37. $|-5 - 2a|$, if $a < -3$

38. $|3a - 14|$, if $a < 4$

39. $|3a - 14|$, if $a > 6$

40. $|15 - 6a|$, if $a > 3$

Give the part of Theorem 1-12 or Theorem 1-13 that justifies each statement.

41. $|x| \cdot |-5| = |-5x|$

42. $\left|\dfrac{x}{5}\right| = \dfrac{|x|}{5}$

43. $|-3x| = 3|x|$

44. $|-3 + x| \le 3 + |x|$

45. $|x - y| \le |x| + |-y|$

46. $|x| > 5$ if $x > 5$

47. $|x| > 4$ if $x < -4$

48. $|x| < 3$ if $-3 < x < 3$

49. $\dfrac{x + 1}{y - 1} \le \left|\dfrac{x + 1}{y - 1}\right|$ if $y \ne 1$

50. $|x + 3| - 3 \le |x|$

Draw a graphical representation for each inequality. (See Figures 1.14 to 1.20.)

51. $|x| < 2$

52. $|x| < 4$

53. $|x| > 3$

54. $|x| > 1$

55. $|x| \le 4$

56. $|x| \le 1$

57. $|x| \ge 2$

58. $|x| \ge 3$

59. $|x - 2| < 1$

60. $|x - 1| < 3$

61. $|x - 4| \le 6$

62. $|x - 5| \le 2$

63. $|x + 1| < 2$

64. $|x + 3| < 1$

65. $|x + 6| \le 4$

66. $|x + 2| \le 5$

67. $|x - 3| \ge 4$

68. $|x - 1| \ge 5$

69. $|x - 2| > 1$

70. $|x - 4| > 3$

71. $|x + 4| > 5$

72. $|x + 2| > 3$

73. $|x + 6| \ge 2$

74. $|x + 3| \ge 1$

PRACTICE TEST for Chapter 1

1. Let $U = \{0, 1, 2, 3, 4, 5, 6, 7, 8, 9\}$, $A = \{1, 3, 5, 6, 7, 8\}$, and $B = \{2, 4, 6, 8, 9\}$. Find
   (a) $A \cap B$ (b) $A \cup B$ (c) A'
   (d) $A - B$ (e) $A \cap (B' \cup A)$

2. Let N be the set of natural numbers, I the set of integers, Q the set of rational numbers, $\mathfrak{R}$ the set of real numbers, and I_r the set of irrational numbers. In which of these sets does each of the following belong?
   (a) -4 (b) $-\frac{3}{2}$ (c) $\sqrt{3}$
   (d) $-0.7666 \cdots$ (e) π

3. Determine if the following sets are closed under the given operation. If the set is not closed, give an example to illustrate this fact.
   (a) The set of integers under division.
   (b) The set of even integers under multiplication.

4. If $x = 2$ and $y = 3$, find the value of
$$[2x(3 - y)] - (7 - y)(2x - 3y).$$

5. Let a and b be real numbers. Use the field properties and the fact that $(-a)(b) = -(ab)$ to prove that $(a)(-b) = -(ab)$.

6. Insert appropriate numbers in the blank spaces so as to make a true statement.
   (a) If $-7x + 4 < 2x - 1$, then $x > $ ____.
   (b) If $-3 < 2x - 7 \leq -4$, then ____ $< x \leq$ ____.

7. Draw a graphical representation of the following statement.
$$-2 < x \leq 1 \quad \text{or} \quad x > 3$$

8. Draw a graphical representation for each inequality.
   (a) $|x - 3| < 4$ (b) $|x + 2| \geq 3$

In problems 9 and 10, find the value of each of the given expressions.

9. (a) $|6 - \sqrt{37}|$ (b) $\left|\dfrac{x}{3}\right|$, if $x < 0$

10. (a) $|y + 3|$, if $y > -3$ (b) $|3x - 5|$, if $x < 1$

Polynomials and Rational Integral Expressions

2-1
Integral Exponents

A letter which is used to refer to an arbitrary element in a given set of numbers is called a *variable*. In Chapter 1, we have used letters such as a, b, x, and y as variables. In this chapter, we shall be concerned with certain types of algebraic expressions. An *algebraic expression* is a statement involving numbers and variables which is obtained by performing the operations of addition, multiplication, subtraction, division, or extraction of roots. In this chapter, the numbers will be real numbers, and the types of algebraic expressions that we deal with are called *polynomials* and *rational integral expressions* (these terms are defined in Sections 2-2 and 2-5). A great deal of elementary mathematics requires that one be able to manipulate these types of expressions to change their forms, to combine, expand, simplify, and factor.

We begin with the definition of a positive integral exponent.

DEFINITION 2-1 Let x denote a real number. If n is a positive integer, we define x^n as follows:

$$x^n = \underbrace{x \cdot x \cdot x \cdots x.}_{n \text{ factors } x}$$

EXAMPLE 1 (a) $3^2 = 3 \cdot 3 = 9$
(b) $4^3 = 4 \cdot 4 \cdot 4 = 64$
(c) $(\frac{2}{3})^4 = \frac{2}{3} \cdot \frac{2}{3} \cdot \frac{2}{3} \cdot \frac{2}{3} = \frac{16}{81}$

In the expression x^n, x is called the *base* and n is called the *exponent*. Note that $3x^4$ is $3 \cdot x \cdot x \cdot x \cdot x$, while $(3x)^4$ is $(3x)(3x)(3x)(3x)$. Also, $-3x^4$ means $(-1)(3x^4)$ and *not* $(-3x)^4$.

There are five basic rules for changing the forms of expressions which involve exponents. These rules are stated in the following theorem.

THEOREM 2-2 If x, y are any real numbers and m, n are any positive integers, then the *laws of exponents* are as follows:

(a) $x^m \cdot x^n = x^{m+n}$;
(b) $(x^m)^n = x^{mn}$;
(c) $(xy)^n = x^n \cdot y^n$;
(d) if $y \neq 0$, $\left(\dfrac{x}{y}\right)^n = \dfrac{x^n}{y^n}$;
(e) if $x \neq 0$,

$$\frac{x^m}{x^n} = \begin{cases} x^{m-n} & \text{if } m > n, \\ \dfrac{1}{x^{n-m}} & \text{if } m < n, \\ 1 & \text{if } m = n. \end{cases}$$

A rigorous proof of this theorem requires techniques that are too involved for us at this point. However, we can get an intuitive feeling for these laws. For example, to see (a), we consider

$$x^m \cdot x^n = \underbrace{(x \cdot x \cdots x)}_{m \text{ factors } x} \cdot \underbrace{(x \cdot x \cdots x)}_{n \text{ factors } x}$$
$$= \underbrace{x \cdot x \cdots x}_{m + n \text{ factors } x}$$
$$= x^{m+n}, \qquad \text{by definition of } x^{m+n}.$$

Likewise, to see (b), consider

$$(x^m)^n = \underbrace{x^m \cdot x^m \cdots x^m.}_{n \text{ factors } x^m}$$

Counting the number of factors x on the right side, and recalling that each x^m contains m factors x, we can agree that the total number of factors x on the right side is $m \cdot n$ since there are n factors x^m. Thus

$$(x^m)^n = x^{mn}.$$

Laws (c) and (d) can similarly be obtained in this fashion. As for (e), if $m > n$, then $m - n$ is a positive integer and we can write

$$\frac{x^m}{x^n} = \frac{x^n \cdot x^{m-n}}{x^n}, \qquad \text{since } x^n \cdot x^{m-n} = x^m \text{ by (a)},$$

$$= \frac{x^n}{x^n} \cdot x^{m-n}, \qquad \text{by rearrangement},$$

$$= 1 \cdot x^{m-n}, \qquad \text{since } \frac{x^n}{x^n} = 1,$$

$$= x^{m-n}.$$

Thus, if $m > n$, $x^m/x^n = x^{m-n}$. Similarly, if $m < n$, then $n - m$ is a positive integer and $x^m/x^n = 1/x^{n-m}$. If $m = n$, then $x^m/x^n = 1$.

These laws can be extended to more than two factors. For example,

$$x^m \cdot x^n \cdot x^p = x^{m+n+p}$$

and

$$(x \cdot y \cdot z)^n = x^n \cdot y^n \cdot z^n.$$

EXAMPLE 2 Simplify each of the following expressions.

(a) $(3x^4y^2) \cdot (5x^2y^3) = 3 \cdot 5 \cdot x^4 \cdot x^2 \cdot y^2 \cdot y^3$
$= 15x^6y^5, \qquad$ by Theorem 2-2(a).

(b) $(2x^2y^3)^4 = (2)^4(x^2)^4(y^3)^4, \qquad$ by Theorem 2-2(c),
$= 16x^8y^{12}, \qquad$ by Theorem 2-2(b).

(c) $\left(\dfrac{2x^2}{y^3}\right)^2 \cdot \left(\dfrac{y}{4x}\right)^3 = \dfrac{2^2(x^2)^2}{(y^3)^2} \cdot \dfrac{y^3}{4^3x^3}, \qquad$ by Theorem 2-2(d) and 2-2(c),

$= \dfrac{4x^4}{y^6} \cdot \dfrac{y^3}{64x^3}, \qquad$ by Theorem 2-2(b),

$= \dfrac{4}{64} \cdot \dfrac{x^4}{x^3} \cdot \dfrac{y^3}{y^6}, \qquad$ by rearrangement,

$= \dfrac{1}{16} \cdot x \cdot \dfrac{1}{y^3}, \qquad$ by Theorem 2-2(e),

$= \dfrac{x}{16y^3}, \qquad$ by rearrangement.

We note that part (e) of Theorem 2-2 says that $x^m/x^m = 1$. Now if we use the first part of (e) and ignore the condition that $m > n$, we get $x^m/x^m = x^{m-m} = x^0$. It is natural, then, for us to define x^0 as 1.

DEFINITION 2-3   If x is any nonzero real number, then we define

$$x^0 = 1.$$

Again, if we ignore the conditions on m and n in Theorem 2-2(e), we could write

$$\frac{x^m}{x^n} = x^{m-n} = \frac{1}{x^{n-m}} \qquad \text{if } x \neq 0.$$

It would be desirable that this property hold without restriction. In order to make this happen, we make the following definition of x^{-n}, where n is a positive integer.

DEFINITION 2-4   If n is a positive integer and x is any nonzero real number, then

$$x^{-1} = \frac{1}{x} \quad \text{and} \quad x^{-n} = (x^n)^{-1} = \frac{1}{x^n}.$$

With this definition, we can rewrite the last part of Theorem 2-2 as

$$(\text{e}') \text{ if } x \neq 0, \frac{x^m}{x^n} = x^{m-n} = \frac{1}{x^{n-m}}$$

for any positive integers m and n. All parts of Theorem 2-2 hold for zero and negative exponents.

EXAMPLE 3   Eliminate zero and negative exponents, and simplify each of the following expressions.

(a) $(x^3y^{-2})^{-2} = (x^3)^{-2} \cdot (y^{-2})^{-2},$ by Theorem 2-2(c),

$\qquad\qquad\quad = x^{-6} \cdot y^4,$ by Theorem 2-2(b),

$\qquad\qquad\quad = \dfrac{1}{x^6} \cdot y^4,$ by Definition 2-4,

$\qquad\qquad\quad = \dfrac{y^4}{x^6}.$

(b) $\dfrac{9x^2y^{-5}}{3x^{-2}y^3} = \dfrac{9}{3} \cdot \dfrac{x^2}{x^{-2}} \cdot \dfrac{y^{-5}}{y^3},$ by rearrangement,

$\qquad\qquad = 3x^{2-(-2)} \cdot y^{-5-3},$ by Theorem 2-2(e'),

$\qquad\qquad = 3x^4 \cdot y^{-8},$

$\qquad\qquad = \dfrac{3x^4}{y^8},$ by Definition 2-4.

(c) $\dfrac{x^{-2}+y^{-1}}{(xy)^{-2}} = \dfrac{\dfrac{1}{x^2}+\dfrac{1}{y}}{\dfrac{1}{(xy)^2}},$ by Definition 2-4,

$$= \dfrac{\dfrac{y+x^2}{x^2y}}{\dfrac{1}{x^2y^2}},$$ by Theorem 1-5(e) and Theorem 2-2(c),

$$= \dfrac{y+x^2}{x^2y} \cdot \dfrac{x^2y^2}{1},$$ by Theorem 1-5(b),

$$= \dfrac{y+x^2}{1} \cdot \dfrac{x^2}{x^2} \cdot \dfrac{y^2}{y},$$ by rearrangement,

$$= y(y+x^2),$$ by Theorem 2-2(e').

EXERCISES 2-1

Compute the value when possible; otherwise, simplify by removing all negative and zero exponents and by using the laws of exponents.

1. 2^3

2. $(-3)^3$

3. $\left(\dfrac{4}{5}\right)^4$

4. $\left(-\dfrac{2}{3}\right)^4$

5. $\left(\dfrac{3y}{5}\right)^2$

6. $\left(\dfrac{c}{2x}\right)^4$

7. $(3x^2y)^3$

8. $(-2x^2)^3$

9. $(2^3)^2$

10. $2^3 \cdot 2^5$

11. $\left(\dfrac{3x^n}{a^h}\right)^2$

12. $\left(\dfrac{a^kb^h}{c^2}\right)^2$

13. $(3)^{-2}$

14. $(3)^{-3}$

15. $(\tfrac{2}{3})^{-4}$

16. $(-\tfrac{1}{2})^3$

17. x^3y^{-4}

18. $5y^{-3}$

19. $2xy^{-4}$

20. $2^{-4}z^{-3}$

21. $\dfrac{(3x^4)(4x^5)}{2x^6}$

22. $\dfrac{(yz)^2(yz^2)}{y^2z^2}$

23. $[(xy^{-1})^2]^{-3}$

24. $[(a^{-1}b^{-1})^{-2}]^{-1}$

25. $\left(\dfrac{4x^2y^3}{3xy^{-4}}\right)^0$

26. $\dfrac{(3x^2)^3}{(4x^3)^0}$

27. $(5x^2y)^{-2} \cdot (x^{-3}y^2)^2$

28. $(2a^2b^{-3})^{-1}(a^3b)^{-2}$

29. $\dfrac{x^{2n-3}}{x^{n-1}}$

30. $\dfrac{x^{n+4}}{x^{2n-1}} \cdot \dfrac{x^n}{x^2}$

31. $\dfrac{3x^{-2}}{(3y)^{-2}}$

32. $\dfrac{cx^{-3}}{3a^{-2}}$

33. $\dfrac{3^{-2}x^{-2}}{4^{-2}x^{-3}}$

34. $\dfrac{2y^{-3}x^2}{3^{-1}y^2x^{-3}}$

35. $2^{-3}xy^{-2}$

36. $3y^{-2}z$

37. $\dfrac{2^{-3}x^{-2}}{5^{-2}x^{-2}}$

38. $\dfrac{3^{-2}a^2y^2}{a^{-3}y^{-3}}$

39. $\dfrac{2a^{-2}y^{-3}}{6^{-1}ay^2}$

40. $\dfrac{3x^{-3}y^2}{2^{-2}x^2y^{-3}}$

41. $x^{-3}y$

42. $\dfrac{a^{-4}}{2y^{-2}}$

43. $\dfrac{x^{-2}+y^{-2}}{(xy)^{-2}}$

44. $(x^{-1}+y^{-1})^{-1}$

Write each expression without a denominator (or with a denominator of 1) by use of negative exponents.

45. $\dfrac{3}{x^3}$

46. $\dfrac{x^3}{4}$

47. $\dfrac{3y^2}{5x^2}$

48. $\dfrac{4xy^2}{3x^2y}$

49. $\dfrac{5z^2}{y^4}$

50. $\dfrac{2x^2yz}{3y^{-1}z^2}$

2-2 Polynomials

One of the simplest forms of algebraic expressions is called a polynomial.

DEFINITION 2-5 A *polynomial* in the variable x is an algebraic expression of the form

$$a_nx^n + a_{n-1}x^{n-1} + \cdots + a_1x + a_0,$$

in which n is a nonnegative integer and the coefficients $a_0, a_1, a_2, \ldots, a_n$ are real numbers.

Note that a polynomial in x has as exponents only positive integers or zero. In a polynomial, the expressions separated by plus signs are called *terms*. The *degree* of a nonzero term is the exponent which appears on x. The *degree* of a nonzero polynomial is the same as that of the term of highest degree appearing in the polynomial. Nonzero real numbers are polynomials of degree zero, and the real number 0 is considered as a polynomial which has no degree. A *monic* polynomial is a polynomial in which the coefficient a_n in the highest degree term is 1.

EXAMPLE 1 Determine which of the following are polynomials, and if they are polynomials, find the degree of the polynomial.

(a) $3x^4 - 5x^2 + 3x$
(b) $4x^2 - 3x + 4 - x^{-1}$
(c) 5
(d) $x^2 - 7$

(a) The expression $3x^4 - 5x^2 + 3x$ is a polynomial of degree 4. The terms are $3x^4$, $-5x^2$, and $3x$.

(b) The expression $4x^2 - 3x + 4 - x^{-1}$ is not a polynomial because of the term $-x^{-1}$.

(c) The real number 5 is a polynomial of degree 0.

(d) The expression $x^2 - 7$ is a monic polynomial of degree 2. The terms are x^2 and -7.

A polynomial having only one term is called a *monomial*. Hence, a polynomial is a sum of monomials. A *binomial* is a polynomial consisting of exactly two terms, and a *trinomial* is a polynomial consisting of exactly three terms.

Terms are called *like terms* if they have the same variable part. For example, $3x^2$ and $-5x^2$ are like terms since they have the same variable part, namely x^2.

Polynomials may be added, subtracted, or multiplied following the rules below.

---

1. To add or subtract polynomials, simply add or subtract their like terms, as called for by the operation.

2. To form the product of two polynomials, multiply each term of the first polynomial by each term of the second polynomial using the laws of exponents, and then combine like terms.

EXAMPLE 2 Perform the indicated operations.

(a) $(5x^2 - 6x + 7) + (9x^2 + 3x - 3)$

(b) $(3x^4 - 7x^3 - 3x^2) - (4x^2 + 4x^3 - x)$

(c) $(-x + 4 + 3x^2)(x^3 + 7 - 3x)$

(d) $(4x^2 - 3x + 2)(3x - 2)$

SOLUTION (a) Although not necessary, it is often convenient to write one polynomial under the other to facilitate the addition of like terms:

$$\begin{array}{r} 5x^2 - 6x + 7 \\ +\quad 9x^2 + 3x - 3 \\ \hline 14x^2 - 3x + 4 \end{array}$$

(b) Aligning like terms and subtracting, we have

$$\begin{array}{r} 3x^4 - \quad 7x^3 - 3x^2 \\ - \quad 4x^3 - 4x^2 + x \\ \hline 3x^4 - 11x^3 - 7x^2 + x \end{array}$$

(c) A systematic way to follow the rule for multiplication of polynomials is to write one polynomial under the other with each in
descending powers of x, then multiply the top polynomial by each
term in the lower polynomial, and finally combine like terms.

$$
\begin{array}{r}
3x^2 - x + 4 \\
\times \quad x^3 - 3x + 7 \\
\hline
3x^5 - x^4 + 4x^3 \\
- 9x^3 + 3x^2 - 12x \\
21x^2 - 7x + 28 \\
\hline
3x^5 - x^4 - 5x^3 + 24x^2 - 19x + 28
\end{array}
$$

(d) Usually, the rewriting is not essential, and one may systematically
write out all of the products involved. Following this procedure,
we have

$$(4x^2)(3x) + (4x^2)(-2) + (-3x)(3x)$$
$$+ (-3x)(-2) + (2)(3x) + (2)(-2)$$
$$= 12x^3 - 8x^2 - 9x^2 + 6x + 6x - 4$$
$$= 12x^3 - 17x^2 + 12x - 4.$$

To this point, our discussion of polynomials has been in terms of
polynomials in the variable x. It is true, of course, that the variable does
not have to be designated by the letter x. One might just as well have a
polynomial in y, or in z, or any other letter. In some instances, an expression may be a polynomial in more than one variable. For example,
$3x^2 - xy^3 + 5xy^2 + y^3$ is a polynomial in both x and y. The *degree* of
a term in a polynomial in more than one variable is defined as the sum of
the exponents on the variables, and the *degree* of the polynomial is the
same as that of the term of highest degree. The term $-xy^3$ in $3x^2 - xy^3$
$+ 5xy^2 + y^3$ has degree $1 + 3 = 4$, and the polynomial has degree 4 also.

Certain products of polynomials occur often enough to warrant
special attention. These are given in Theorem 2-6 below. A proof of the
theorem is not furnished, but the validity of each statement can be established by simple multiplication.

THEOREM 2-6   The following special product rules are valid, where
a, b, c, d, x, and y represent arbitrary real numbers.

(a) $a(x + y) = ax + ay$;
(b) $(x + y)(x - y) = x^2 - y^2$;
(c) $(ax + b)(cx + d) = acx^2 + (ad + bc)x + bd$;
(d) $(x \pm y)^2 = x^2 \pm 2xy + y^2$;
(e) $(x \pm y)^3 = x^3 \pm 3x^2y + 3xy^2 \pm y^3$;
(f) $(x + y)(x^2 - xy + y^2) = x^3 + y^3$;
(g) $(x - y)(x^2 + xy + y^2) = x^3 - y^3$.

In parts (d) and (e), the symbol $\pm$ is read "+ or $-$," and we use either the top sign on both sides or the bottom sign on both sides.

EXAMPLE 3 Each of the following products illustrates one of the special product rules.

(a) $3x(2x - 7) = 6x^2 - 21x$, by Theorem 2-6(a).
(b) $(2x^2 - y)(2x^2 + y) = 4x^4 - y^2$, by Theorem 2-6(b).
(c) $(2x - 1)(3x + 2) = 6x^2 + x - 2$, by Theorem 2-6(c).
(d) $(2x + y)^2 = 4x^2 + 4xy + y^2$, by Theorem 2-6(d).
(e) $(x - 2y)^3 = x^3 - 6x^2y + 12xy^2 - 8y^3$, by Theorem 2-6(e).
(f) $(2x - y)(4x^2 + 2xy + y^2) = 8x^3 - y^3$, by Theorem 2-6(g).

EXERCISES 2-2

Perform the indicated operations and simplify. State the degree of each result.

1. $(3x^3 - 7x + 2) + (4x - 7x^2 + 11)$
2. $(4x^4 - 3x^2 + 1) - (5x^3 - 7x^2 + 2x)$
3. $(2x - 3)^2$ 4. $(2x - 7)(2x + 7)$
5. $(5x^3 - 7x + 4)(3x^2 - 7x - 2)$ 6. $(2x^2 - x + 1)(3x^2 + x - 2)$
7. $(x^3 + 2x^2 - 1)(x^2 - 3)$ 8. $(x^3 - 2)(2x^3 + x^2 - 4x)$
9. $(4x^3 - 2x^2 + 3)(x^3 - x + 2)$ 10. $(x^3 - 2x + 1)(3x^3 + 3x^2 - 1)$
11. $(4x^3 - 5x^2 + 7x + 2) + (3x^3 - x^4 + 7x + 2)$
12. $(4y^2 - 2y + 7) - (6y - 8)$
13. $(6m^4 - 3m^2 + m) - (2m^3 + 4m^2 + 5m)$
14. $-(7x^3 + 2x - 3) + (4x^3 - 7x^2 + 3x - 2)$
15. $4(2x^2 - 3x + 2) - (4x - 3)(2x - 1)$
16. $x^2 - 3x + 2 - (x - 1)(x + 2)$

Use the special product rules to perform the following multiplications.

17. $(c - k)(c + k)$ 18. $(b + 2c)(b - 2c)$
19. $(x - 2)(x + 2)$ 20. $(5 - 2y)(5 + 2y)$
21. $(3x - 4y)^2$ 22. $(x^2 + 4)^2$
23. $(2x + 3)(4x - 5)$ 24. $(2x - 5)(-2x - 5)$
25. $4x(x^2 - 3)$ 26. $(2x - 4)(3x)$
27. $(3x - 7)(2x + 2)$ 28. $(5x - 4)(3x + 8)$
29. $(3y + 5)(7y - 2)$ 30. $(6 - 5x)(2 - x)$

31. $(x^2 + 2)^2$

32. $(3 + 4ax)^2$

33. $(x - 2yz^2)^2$

34. $\left(\dfrac{x}{2} - \dfrac{y}{3}\right)^2$

35. $(\tfrac{1}{2}a - \tfrac{1}{3}b)(\tfrac{1}{2}a + \tfrac{1}{3}b)$

36. $(\tfrac{3}{4}x - 1)(\tfrac{3}{4}x + 1)$

37. $(a - b)(a^2 + ab + b^2)$

38. $(x + 2y)(x^2 - 2xy + 4y^2)$

39. $(3x - 1)(9x^2 + 3x + 1)$

40. $(x - 2y)(x^2 + 2xy + 4y^2)$

41. $(x - 2y)^3$

42. $(x + y)^3$

43. $(2a^2 + 5b^2)(a^2 - 3b^2)$

44. $(3y^2 - 5)(y^2 + 2)$

45. $(5u^2 - 7v^2)(3u^2 - 2v^2)$

46. $(x^3 + 3)(3x^3 - 4)$

47. $(2x - 3y)^3$

48. $(x + 2y)^3$

49. $(3x + 4)(9x^2 - 12x + 16)$

50. $(2x - 3y)(4x^2 + 6xy + 9y^2)$

51. $[2(5x + 3y)]^2$

52. $[5(3c - 4d)]^2$

2-3
Factoring
Polynomials

If a given polynomial is written as a product of two or more other polynomials, the other polynomials are called *factors* of the given polynomial. Since $x^2 - y^2 = (x + y)(x - y)$, the polynomials $x + y$ and $x - y$ are factors of $x^2 - y^2$.

When we speak of factoring a polynomial, we mean that we wish to write the polynomial as a product of two or more factors. The ability to factor polynomials is absolutely essential in order to progress into higher mathematics and its applications. In our work here, we shall be concerned only with factoring polynomials which have integers as coefficients, and we shall require that all factors be polynomials which have only integers as coefficients. This means that we do *not* allow factorizations such as $x^2 - y^2 = (2x + 2y)(\tfrac{1}{2}x - \tfrac{1}{2}y)$ or $x^2 - y^2 = (x + y)(\sqrt{x} + \sqrt{y})(\sqrt{x} - \sqrt{y})$.

The simplest method of factoring is to use the distributive property to factor out a common factor from each term. This is illustrated in the example below.

EXAMPLE 1

(a) The terms in $4m^2 - 2m^3 - 8m$ have $2m$ as a common factor, so we can write

$$4m^2 - 2m^3 - 8m = 2m(2m - m^2 - 4).$$

(b) The terms in $3x^2y - 4xy - 7xy^2$ have xy as a common factor, and

$$3x^2y - 4xy - 7xy^2 = xy(3x - 4 - 7y).$$

Each special product in Theorem 2-6 gives us a technique for factoring. It is of primary importance in later work to possess reasonable skill in using these techniques. With emphasis now on the special products rules as patterns for factoring, we have the following formulas.

(a) $ax + ay = a(x + y)$.
(b) $x^2 - y^2 = (x + y)(x - y)$.
(c) $acx^2 + (ad + bc)x + bd = (ax + b)(cx + d)$.
(d) $x^2 + 2xy + y^2 = (x + y)^2$ and
 $x^2 - 2xy + y^2 = (x - y)^2$.
(e) $x^3 + 3x^2y + 3xy^2 + y^3 = (x + y)^3$ and
 $x^3 - 3x^2y + 3xy^2 - y^3 = (x - y)^3$.
(f) $x^3 + y^3 = (x + y)(x^2 - xy + y^2)$.
(g) $x^3 - y^3 = (x - y)(x^2 + xy + y^2)$.

EXAMPLE 2 The use of these formulas in factoring polynomials is illustrated in the following examples.

(a) Consider the problem of factoring $9x^2 - 25y^2$. From the fact that $9x^2 = (3x)^2$ and $25y^2 = (5y)^2$, we see from formula (b) that

$$9x^2 - 25y^2 = (3x + 5y)(3x - 5y).$$

(b) In order to factor $4x^2 + 12xy + 9y^2$, we observe that $4x^2 = (2x)^2$, $9y^2 = (3y)^2$, and $12xy = 2(2x)(3y)$. Thus formula (d) gives

$$4x^2 + 12xy + 9y^2 = (2x + 3y)^2.$$

(c) In attempting to factor $8x^3 - 12x^2y + 6xy^2 - y^3$, we note that the four terms suggest one of the formulas in (e). Since $8x^3 = (2x)^3$, $-12x^2y = 3(2x)^2(-y)$, and $6xy^2 = 3(2x)(-y)^2$, formula (e) gives

$$8x^3 - 12x^2y + 6xy^2 - y^3 = (2x - y)^3.$$

(d) Consider the problem of factoring $x^2 - 2x - 8$. It soon becomes apparent that this polynomial can only fit the pattern in formula (c), and we must use "trial and error" to find integers b and d such that $b + d = -2$ and $b \cdot d = -8$. A choice that works is $b = -4$ and $d = 2$. Hence we have

$$x^2 - 2x - 8 = (x - 4)(x + 2).$$

(e) We desire to factor $8w^2 + 14wz - 15z^2$. The only promising formula is a slight modification of (c), where we look for factors of the form $(aw + bz)(cw + dz)$ with $ac = 8$, $bd = -15$, and $ac + bd = 14$. As a first guess, we might try $a = 4$, $c = 2$, $b = 3$, $d = -5$. This selection does not work, however, for $(4w + 3z)(2w - 5z) = 8w^2 - 14wz - 15z^2$. But if we choose $a = 4$, $c = 2$, $b = -3$, and $d = 5$, then we have

$$8w^2 + 14wz - 15z^2 = (4w - 3z)(2w + 5z).$$

The last example indicates that one may sometimes make several guesses before being successful in factoring a polynomial, and this is indeed the case. In some instances, a factorization of the type we are considering may not be possible. The polynomial $x^2 - 2$ cannot be factored as a product of polynomials with integers as coefficients (except trivial factorizations, with one factor 1 or -1). The polynomial $x^2 + y^2$ furnishes another example. A polynomial with integral coefficients which cannot be factored as a product of polynomials, both different from 1 and -1, and both with integral coefficients, is called a *prime polynomial*, or an *irreducible polynomial*.[1]

Another method of factoring which is used in combination with other methods is called *factoring by grouping*. This is illustrated in the next example.

EXAMPLE 3   Factor $4hx - 4bh - 8cx + 8bc$.

SOLUTION   First we group the first two terms together and factor from them the common factor $4h$. Then we group the last two terms together and remove the common factor $-8c$, so that we have

$$4h(x - b) - 8c(x - b).$$

We can now factor out $4(x - b)$, and we have

$$4hx - 4bh - 8cx + 8bc = 4(x - b)(h - 2c).$$

EXERCISES 2-3

Factor each expression completely.

1. $12x - 4$ 2. $8m^2n^3 - 4m^3n^2 + 2m^2n^2$

3. $bx + x + c^2x$ 4. $bx + b^2 + c^2b$

5. $xy - x - y + 1$ 6. $ab - bc + ac - c^2$

7. $25y^2 - 10y + 1$ 8. $49x^2 - 14x + 1$

9. $16y^2 - 1$ 10. $49 - v^2$

11. $ax^2 - 9ay^4$ 12. $x^4 - 81y^4$

13. $u^2 - 8u + 16$ 14. $4x^4 - 28x^2 + 49$

15. $x^2 + 2x - 24$ 16. $x^2 + 8x + 15$

17. $x^2 + 10x + 21$ 18. $12 - 7y + y^2$

19. $4 - 3x - x^2$ 20. $45x^2 - 8xy - 4y^2$

21. $3x^2 + 7ax - 6a^2$ 22. $5a^2 - 12ab + 7b^2$

23. $8w^6 - 6w^3 - 9$ 24. $2x^4 - 3x^2 - 20$

25. $7(x + 2y) - 3(x + 2y)$ 26. $-2x(a + h) - 3y(a + h)$

27. $3ac + 3bc + ad + bd$ 28. $5ax + 2bx - 10ad - 4bd$

29. $ax^2 + bx^2 + ad^2 + bd^2$ 30. $a(x - y) - b(y - x)$

[1]In Chapter 9, other factorizations of polynomials are considered.

31. $(4x - 3y)^2 - 25$

32. $(2x + y)^2 - 9z^2$

33. $x^3 - z^3$

34. $1000 - x^3$

35. $8 + u^3$

36. $8x^3 + 27$

37. $216x^3 - 8b^3$

38. $64x^3 + 27y^3$

39. $2 + 4x - 10x^4 - 5x^3$

40. $9x^2 - y^2 + 2yz - z^2$

41. $(4a - b)^2 - (2x - z)^2$

42. $x^6 - y^6$

2-4
Division
of Polynomials

At this point, we have studied the operations of addition, subtraction, and multiplication of polynomials. It is natural to now consider division of polynomials.

In dealing with real numbers in arithmetic, we have the usual equalities

$$\frac{\text{dividend}}{\text{divisor}} = \text{quotient} + \frac{\text{remainder}}{\text{divisor}},$$

or, when written in an alternative way,

$$\text{dividend} = (\text{quotient})(\text{divisor}) + \text{remainder},$$

where the remainder is required to be nonnegative and less than the absolute value of the divisor.

EXAMPLE 1

When 237 is divided by 11, the quotient is 21 and the remainder is 6. Writing this in the forms above, it appears as

$$\frac{237}{11} = 21 + \frac{6}{11},$$

or

$$237 = (21)(11) + 6.$$

In division of polynomials, we desire the same relations between divisor, dividend, quotient, and remainder as given above for real numbers. The analogous condition on the remainder in the case of polynomials is that the degree of the remainder is required to be less than the degree of the divisor.

The simplest type of division of polynomials is where the divisor is a monomial. In this case, the procedure for division is exceedingly simple:

to divide a polynomial by a monomial, divide each term of the polynomial by the monomial.

EXAMPLE 2 To divide $x^3 - 6x^2 + 3x + 5$ by x^2, we write

$$\frac{x^3 - 6x^2 + 3x + 5}{x^2} = \frac{x^3}{x^2} + \frac{-6x^2}{x^2} + \frac{3x}{x^2} + \frac{5}{x^2}$$

$$= x - 6 + \frac{3x + 5}{x^2}.$$

The reader should check that when we multiply the answer by the divisor, x^2, we obtain the dividend.

The general procedure is more complicated, but it is straightforward if we follow the steps below.

Procedure for Division of Polynomials

1. Arrange both polynomials in descending powers of the variable, from left to right.

2. Divide the first term of the dividend by the first term of the divisor, and write this as part of the quotient.

3. Multiply the whole divisor by the part of the quotient obtained in step 2, and subtract this product from the dividend.

4. Consider any remainder now left as the new dividend, and repeat the process outlined in steps 2 and 3.

5. The procedure terminates when the degree of the remainder obtained in step 3 is less than the degree of the divisor.

We illustrate this procedure with an example.

EXAMPLE 3 Divide $8y^3 - 18y^2 - 6 + 11y$ by $-3y + 2 + 4y^2$.

SOLUTION Following step 1 above, we first write the dividend and divisor in descending powers of y, and record them in this form:

$$4y^2 - 3y + 2 \,\overline{\big)\, 8y^3 - 18y^2 + 11y - 6}.$$

Performing steps 2 and 3, we have $8y^3 \div 4y^2 = 2y$, and

$$
\begin{array}{r}
2y \\
4y^2 - 3y + 2 \,\overline{\big)\, 8y^3 - 18y^2 + 11y - 6} \\
8y^3 - 6y^2 + 4y \\
\hline
-12y^2 + 7y - 6
\end{array}
$$

Following step 4, we have $-12y^2 \div 4y^2 = -3$, and

$$
\begin{array}{r}
2y - 3 \quad \text{(quotient)} \\
\text{(divisor)} \quad 4y^2 - 3y + 2\,\overline{\big)\,8y^3 - 18y^2 + 11y - 6} \quad \text{(dividend)} \\
8y^3 - 6y^2 + 4y \\
\hline
-12y^2 + 7y - 6 \\
-12y^2 + 9y - 6 \\
\hline
-2y \quad \text{(remainder)}.
\end{array}
$$

Since the remainder, $-2y$, has degree less than the degree of the divisor, $4y^2 - 3y + 2$, the procedure terminates here, and we have

$$
\frac{8y^3 - 18y^2 + 11y - 6}{4y^2 - 3y + 2} = 2y - 3 + \frac{-2y}{4y^2 - 3y + 2},
$$

or

$$
8y^3 - 18y^2 + 11y - 6 = (2y - 3)(4y^2 - 3y + 2) - 2y.
$$

Quite often we need to quickly divide a polynomial by a binomial of the form $x - c$, where c is a constant. Consider the following example.

EXAMPLE 4   Divide $x^3 - 3x^2 - x + 4$ by $x - 2$.

SOLUTION

$$
\begin{array}{r}
x^2 - x - 3 \quad \text{(quotient)} \\
\text{(divisor)} \quad x - 2\,\overline{\big)\,x^3 - 3x^2 - x + 4} \quad \text{(dividend)} \\
x^3 - 2x^2 \\
\hline
-x^2 - x + 4 \\
-x^2 + 2x \\
\hline
-3x + 4 \\
-3x + 6 \\
\hline
-2 \quad \text{(remainder)}.
\end{array}
$$

Thus

$$
\frac{x^3 - 3x^2 - x + 4}{x - 2} = x^2 - x - 3 + \frac{-2}{x - 2},
$$

or

$$
x^3 - 3x^2 - x + 4 = (x^2 - x - 3)(x - 2) - 2.
$$

Note that in this example the remainder is a constant, and it is true that when any polynomial is divided by $x - c$, the remainder will be a constant.

Many people prefer a shortened form which allows more rapid division of a polynomial by a binomial of the form $x - c$. This procedure is called *synthetic division* or *detached coefficients*. To understand why the synthetic division procedure works, we shall rework the last example with the goal in mind of streamlining the procedure.

$$
\begin{array}{r}
x^2 - x - 3 \\
x - 2\,\overline{\smash{\big)}\,x^3 - 3x^2 - x + 4} \\
\underline{x^3 - 2x^2} \\
-x^2 - x + 4 \\
\underline{-x^2 + 2x} \\
-3x + 4 \\
\underline{-3x + 6} \\
-2
\end{array}
$$

We eliminate the circled terms and write this more compact form:

$$
\begin{array}{r}
x^2 - x - 3 \\
x - 2\,\overline{\smash{\big)}\,x^3 - 3x^2 - x + 4} \\
-2x^2 + 2x + 6 \\
\underline{} \\
x^3 - x^2 - 3x - 2
\end{array}
$$

If we go a step further and eliminate the variables and omit the quotient, we have

$$
\begin{array}{r}
1 - 2\,\overline{\smash{\big)}\,1 - 3 - 1 \quad 4} \\
-2 \quad 2 \quad 6 \\
\underline{} \\
1 - 1 - 3 - 2
\end{array}
$$

As final refinements, we drop the coefficient of x in the divisor and change the sign of the constant in the divisor $x - 2$ to 2, so that we may add at each stage rather than subtract. This yields the routine known as synthetic division:

$$
\begin{array}{r}
2\,\overline{\smash{\big)}\,1 - 3 - 1 \quad 4} \\
2 - 2 - 6 \\
\underline{} \\
1 - 1 - 3 - 2
\end{array}
$$

$$\text{quotient} = x^2 - x - 3 \qquad \text{remainder} = -2.$$

This more efficient arrangement of the division will be very useful in Chapter 9. An even more efficient form will be used there.

We illustrate the synthetic division procedure with one more example. This example illustrates observance of the following

---

Warning: if any coefficients are 0 in the dividend, be sure to record these 0's in the synthetic division procedure.

EXAMPLE 5   Divide $2x^4 - 12x^2 - 5$ by $x + 3$.

SOLUTION

$$
\begin{array}{r|rrrrr}
-3 & 2 & 0 & -12 & 0 & -5 \\
   &   & -6 & 18 & -18 & 54 \\
\hline
   & 2 & -6 & 6 & -18 & 49
\end{array}
$$

Hence

$$
\frac{2x^4 - 12x^2 - 5}{x + 3} = 2x^3 - 6x^2 + 6x - 18 + \frac{49}{x + 3}.
$$

Note that $x + 3 = x - (-3)$, and -3 was recorded in the divisor position. Also, 0 was recorded as the coefficient of x^3 and of x in the dividend.

EXERCISES 2-4

Perform the following divisions. Identify the quotient and remainder.

1. $\dfrac{4x^3 - 8x^2 + 16x}{2x}$

2. $\dfrac{30y^2 - 12y^3 + 18y^4}{6y^2}$

3. $\dfrac{40x^4 - 15x^3 + 25x^2 + 10x - 1}{5x^2}$

4. $\dfrac{15x^4 - 9x^2 + x - 3}{3x^2}$

5. $\dfrac{14y^5 + 21y^3 - 7y + 1}{-y^2}$

6. $\dfrac{5x^6 - x^4 + 3x^3 - x + 2}{-x^3}$

In problems 7–18, divide and leave the results in the form

$$
\frac{\text{dividend}}{\text{divisor}} = \text{quotient} + \frac{\text{remainder}}{\text{divisor}}.
$$

7. $\dfrac{2x^3 - 11x^2 + 19x - 10}{2x - 5}$

8. $\dfrac{3x^3 - 11x^2 + 5x + 3}{3x + 1}$

9. $\dfrac{x^3 - 4x^2 + 6x - 3}{x - 1}$

10. $\dfrac{22 - 33x + 5x^2 + 2x^3}{2x - 5}$

11. $\dfrac{4x^4 + 5x^2 - 7x + 3}{2x^2 + 3x - 2}$

12. $\dfrac{2x^4 + 9x - 9 - 3x^3 + x^2}{x^2 - 2x + 3}$

13. $\dfrac{4m^3 - 3m - 2}{m + 1}$

14. $\dfrac{-5 + x^3 + 2x^2}{x + 3}$

15. $\dfrac{3x^4 - 10x^3 - 10x^2 - 8x + 3x^5 - 8}{x^2 + 3x + 2}$

16. $\dfrac{3x^5 + 3x^4 - 10x^3 - 10x^2 + 8x + 8}{x^2 + 3x + 2}$

17. $\dfrac{4x^4 - 13ax^3 + 12a^2x^2 - 5a^3x + 2a^4}{4x^2 - ax + a^2}$

18. $\dfrac{18k^2 - 3rk - 10r^2}{6k - 5r}$

In problems 19–28, use synthetic division and leave the results in the form

$$\frac{\text{dividend}}{\text{divisor}} = \text{quotient} + \frac{\text{remainder}}{\text{divisor}}.$$

19. $(x^2 - 9x + 20) \div (x - 4)$ 20. $(z^2 + 3z - 4) \div (z + 4)$

21. $(2x^3 - 2x + 7) \div (x - 2)$ 22. $(2x^3 - 5x^2 + 7) \div (x + 2)$

23. $(-3x^3 + 2x - 75) \div (x + 3)$ 24. $(x^4 - 7x^2 - 6x) \div (x + 2)$

25. $(x^3 - 4x^2 + 7x - 6) \div (x + 4)$

26. $(x^4 - 12x^3 + 46x^2 - 60x + 9) \div (x - 3)$

27. $(x^4 - a^4) \div (x + a)$ 28. $(x^5 - a^5) \div (x - a)$

2-5
Rational Integral
Expressions

A *rational integral expression* is a quotient of polynomials which have integral coefficients, that is, a fraction formed by dividing one polynomial with integral coefficients by another. In all manipulations involving rational integral expressions, it is understood that *all values of the variables which make a zero denominator are excluded.*

We say that a rational integral expression is reduced to *lowest terms* if the only common factors in the numerator and denominator are 1 and -1.

Any rational integral expression can be reduced to lowest terms by following this procedure:

---

factor the numerator and denominator completely, and then divide out all common factors.

In this procedure, it is understood that all values of the variables which make a common factor zero are *excluded* in the resulting fraction as well as in the original fraction.

EXAMPLE 1

Reduce $\dfrac{x^2 - 4xy + 4y^2}{x^2 - 4y^2}$ to lowest terms.

SOLUTION

$$\frac{x^2 - 4xy + 4y^2}{x^2 - 4y^2} = \frac{(x - 2y)^2}{(x + 2y)(x - 2y)} = \frac{x - 2y}{x + 2y}$$

It is understood that all values of x and y are excluded for which $x - 2y = 0$ or $x + 2y = 0$. In other words,

$$\frac{x^2 - 4xy + 4y^2}{x^2 - 4y^2} = \frac{x - 2y}{x + 2y},$$

subject to the conditions that $x \neq 2y$ and $x \neq -2y$.

To add or subtract fractions with the same denominator, we use the pattern

$$\frac{a}{d} - \frac{b}{d} + \frac{c}{d} = \frac{a - b + c}{d}.$$

With different denominators, we first convert the fractions to fractions with the same denominator. The most efficient denominator to use is the least common multiple.

A *least common multiple* (LCM) of two or more polynomials with integral coefficients is a polynomial of least degree which has each polynomial as a factor. We take as LCM the polynomial in which the coefficient of the highest-degree term is the smallest possible positive integer. To find this LCM, we

1. factor each polynomial completely (i.e., into prime factors), and

2. form the product of all of the different prime factors, with each prime factor raised to the highest exponent with which it appears in any of the given polynomials. In the prime factors, the coefficient of the highest degree term is chosen to be positive.

A moment's reflection reveals that this is the same procedure as that used for finding the least common multiple of integers, except that prime polynomials are used in the place of prime integers. Actually, then, the procedure for integers is a special case of the procedure for polynomials with integral coefficients.

EXAMPLE 2 (a) To find the LCM of 6, 9, and 15, we first factor each into prime factors as

$$6 = 2 \cdot 3,$$
$$9 = 3^2,$$
$$15 = 3 \cdot 5.$$

Then we form the product of the prime factors, with each factor raised to the highest exponent with which it appears in any of the factorizations:

$$2 \cdot 3^2 \cdot 5 = 90,$$

and 90 is the LCM of 6, 9, and 15.

(b) To find the LCM of $6xy^2$, $20x^3y$, and $225xy^3$, we first factor each into prime factors as

$$6xy^2 = 2 \cdot 3xy^2,$$
$$20x^3y = 2^2 \cdot 5x^3y,$$
$$225xy^3 = 3^2 \cdot 5^2xy^3.$$

Then we write

$$2^2 \cdot 3^2 \cdot 5^2 \cdot x^3 \cdot y^3 = 900x^3y^3,$$

and $900x^3y^3$ is the LCM of $6xy^2$, $20x^3y$, and $225xy^3$.

(c) To find the LCM of $3x^2 - 12x + 9$ and $x^2 - 2x + 1$, we factor the polynomials as

$$3x^2 - 12x + 9 = 3(x - 1)(x - 3),$$

$$x^2 - 2x + 1 = (x - 1)^2,$$

and the LCM is $3(x - 1)^2(x - 3)$.

A *least common denominator* (LCD) of two or more rational integral expressions is a least common multiple of the denominators of the expressions. In order to form a sum or difference involving two or more rational integral expressions which are in lowest terms, we follow these steps:

1. Find the LCD of the expressions.

2. Change each fraction to a fraction having as denominator the LCD found in step 1.

3. Combine the new numerators, and place the result over the LCD.

EXAMPLE 3 Perform the indicated operations.

(a) $\dfrac{3a}{2b} - \dfrac{2 - 5b}{c^2} + 6$

(b) $\dfrac{2}{6x - 3} - \dfrac{x}{4x^2 - 1}$

SOLUTION

(a) A LCD of the denominators $2b$, c^2, and 1 is $2bc^2$, so we have

$$\frac{3a}{2b} - \frac{2 - 5b}{c^2} + 6 = \frac{3ac^2}{2bc^2} - \frac{(2 - 5b)(2b)}{2bc^2} + \frac{12bc^2}{2bc^2}$$

$$= \frac{3ac^2 - 4b + 10b^2 + 12bc^2}{2bc^2}.$$

(b) Since $6x - 3 = 3(2x - 1)$ and $4x^2 - 1 = (2x + 1)(2x - 1)$, the LCD is $3(2x - 1)(2x + 1) = 3(4x^2 - 1)$. Thus

$$\frac{2}{6x - 3} - \frac{x}{4x^2 - 1} = \frac{2(2x + 1)}{3(4x^2 - 1)} - \frac{3x}{3(4x^2 - 1)}$$

$$= \frac{4x + 2 - 3x}{3(4x^2 - 1)}$$

$$= \frac{x + 2}{3(4x^2 - 1)}.$$

Multiplication and division of rational integral expressions follow the usual rules for products or quotients of real numbers: that is,

$$\frac{a}{b} \cdot \frac{c}{d} = \frac{ac}{bd},$$

and

$$\frac{a}{b} \div \frac{c}{d} = \frac{ad}{bc}.$$

After applying either of these rules, we factor both numerators and denominators completely and reduce the result to lowest terms by dividing out common factors.

EXAMPLE 4 Perform the indicated operations.

(a) $\dfrac{y^2 - 9}{y^2 + 6y + 9} \cdot \dfrac{y^3 - 1}{y^2 - 4y + 3}$

$$= \frac{(y-3)(y+3)}{(y+3)^2} \cdot \frac{(y-1)(y^2 + y + 1)}{(y-1)(y-3)}$$

$$= \frac{y^2 + y + 1}{y + 3}$$

(b) $\dfrac{25 - 9x^2}{x + 3} \div (5x - 3x^2) = \dfrac{25 - 9x^2}{x + 3} \cdot \dfrac{1}{5x - 3x^2}$

$$= \frac{(5 - 3x)(5 + 3x)}{x + 3} \cdot \frac{1}{x(5 - 3x)}$$

$$= \frac{5 + 3x}{x^2 + 3x}$$

(c) $\dfrac{\dfrac{6y^2 + 5y - 4}{2y^2 + 7y + 3}}{\dfrac{2y^2 + 3y - 2}{y^2 - 9}} = \dfrac{6y^2 + 5y - 4}{2y^2 + 7y + 3} \cdot \dfrac{y^2 - 9}{2y^2 + 3y - 2}$

$$= \frac{(3y + 4)(2y - 1)}{(2y + 1)(y + 3)} \cdot \frac{(y - 3)(y + 3)}{(2y - 1)(y + 2)}$$

$$= \frac{(3y + 4)(y - 3)}{(2y + 1)(y + 2)}$$

When a quotient of rational integral expressions is written in fractional form as was done in Example 4(c), it is referred to as a *complex fraction*.

Some operations with rational integral expressions may involve both sums and products or quotients. In such cases, the operations are performed subject to the same rules as those given in Chapter 1 for real numbers.

EXAMPLE 5 Express

$$\left(\frac{2}{y} + \frac{5}{y^2} - \frac{12}{y^3}\right) \div \left(4 - \frac{8}{y} + \frac{3}{y^2}\right)$$

as a single fraction in lowest terms.

SOLUTION

$$\left(\frac{2}{y} + \frac{5}{y^2} - \frac{12}{y^3}\right) \div \left(4 - \frac{8}{y} + \frac{3}{y^2}\right)$$

$$= \left(\frac{2y^2 + 5y - 12}{y^3}\right) \div \left(\frac{4y^2 - 8y + 3}{y^2}\right)$$

$$= \frac{2y^2 + 5y - 12}{y^3} \cdot \frac{y^2}{4y^2 - 8y + 3}$$

$$= \frac{(2y - 3)(y + 4)}{y^3 \, y} \cdot \frac{y^2}{(2y - 3)(2y - 1)}$$

$$= \frac{y + 4}{y(2y - 1)}$$

EXERCISES 2-5

Reduce each expression to lowest terms.

1. $\dfrac{x^2 - y^2}{x - y}$

2. $\dfrac{x^2 - y^2}{x^3 + y^3}$

3. $\dfrac{x^2 - x - 12}{x^2 - 2x - 8}$

4. $\dfrac{2x^2 + 5x - 3}{x^2 + 2x - 3}$

5. $\dfrac{4(a - 1)^2(a + 2)^3}{(a + 2)(a - 1)^3}$

6. $\dfrac{12x^2(x - 3)^2(3x - 1)}{15x(x - 3)(3x - 1)^2}$

7. $\dfrac{(x^3 - 27y^3)(x^2 - 4y^2)}{x^2 - 5xy + 6y^2}$

8. $\dfrac{(x + y)(x^2 + 3xy - 4y^2)}{(x - y)(x^2 + 2xy + y^2)}$

Find a least common multiple of the polynomials given in each problem.

9. $20, 35, 5$

10. $14, 35, 15$

11. $9x^2, 6x, 8x^5$

12. $16y^4, 27y, 12$

13. $x^2 - y^2, y - x$

14. $(x - y)^2, y - x$

15. $2x^2 - 2x - 12, x^2 + 4x + 4$

16. $12x^2 + 12x + 3, 8x^2 - 2$

Perform the indicated operations, and reduce the result to lowest terms.

17. $\dfrac{y}{4x^2} - \dfrac{z}{3x}$

18. $\dfrac{2 - y}{3xy^2} - \dfrac{3 - 4y}{2x^2y}$

19. $\dfrac{2y}{2x + 2y} - \dfrac{3}{x^2 - y^2}$

20. $\dfrac{2}{3x - 6y} - \dfrac{3}{x^2 - 4y^2}$

21. $\dfrac{x + 5}{x^2 + 7x + 10} - \dfrac{x + 5}{2x^2 + x - 6}$

22. $\dfrac{2y + 3}{3y^2 + y - 2} + \dfrac{4 - 3y}{2y^2 - 3y - 5}$

23. $\dfrac{z - 5}{z^2 - 9} + \dfrac{z - 2}{12 - 4z}$

24. $y - \dfrac{2y}{y^2 - 1} + \dfrac{3}{y + 1}$

25. $\dfrac{w-1}{2w^2-13w+15}+\dfrac{w+3}{2w^2-15w+18}$

26. $\dfrac{x+5}{x^2+7x+10}-\dfrac{x-1}{x^2+5x+6}$

27. $\dfrac{2b-3y}{3c+6d}\cdot\dfrac{2c+4d}{4b-6y}$

28. $(x^2-4y^2)\cdot\dfrac{5a}{xy-2y^2}$

29. $\dfrac{3a-ay}{6z-xz}\cdot\dfrac{4z-xz}{3a-ya}$

30. $\dfrac{3x-3y}{4x+2y}\cdot\dfrac{4x^2-y^2}{(x-y)^2}$

31. $\dfrac{y^2-2y-15}{y^2-9}\cdot\dfrac{y^2-6y+9}{12-4y}$

32. $\dfrac{25-4x^2}{x^3-1}\cdot\dfrac{x^3+x^2+x}{15-x-2x^2}$

33. $\dfrac{w^3-4w^2+9w-36}{w^4-81}\cdot\dfrac{2aw^2-aw-15a}{w^3-64}$

34. $\dfrac{2z^2-5z-12}{2z^2+5z-12}\cdot\dfrac{2z^2+3z-9}{2z^2+9z+9}$

35. $\dfrac{x^3+y^3}{2x-3y}\div\dfrac{2x+2y}{4x^2-9y^2}$

36. $(cy+dy)\div\dfrac{d^2-c^2}{3y}$

37. $\dfrac{x^3-27}{2x^2+5}\div(3x-9)$

38. $(9x^2-25y^2)\div\left(1+\dfrac{5y}{3x}\right)$

39. $\dfrac{w+3}{w^2+7w+10}\div\dfrac{w^2+6w+9}{w^2+5w+6}$

40. $\dfrac{y^2+2y-15}{y^2+11y+30}\div\dfrac{y^2-8y+15}{y^2+2y-24}$

41. $\dfrac{3w^2+w-2}{4w^2-w-5}\div\dfrac{3w^2-11w+6}{4w^2-7w-15}$

42. $\left(x-1-\dfrac{6}{x}\right)\div\left(1+\dfrac{2}{x}-\dfrac{15}{x^2}\right)$

43. $\dfrac{3a+2b}{2ya-2xyb}\cdot\dfrac{a^3-x^3b^3}{xa+2xb}\cdot\dfrac{a^2+4ab+4b^2}{9a^2-4b^2}$

44. $\dfrac{2x^2+x-6}{x^3+2x^2+4x}\cdot\dfrac{6x^3+24x}{3x^4-48}\div\dfrac{8x^2-18}{x^3-8}$

45. $\left(\dfrac{2}{x+2}-\dfrac{1}{x-2}\right)\left(\dfrac{x-2}{x+1}\right)$

46. $\left(1-\dfrac{8}{x^3}\right)\div\left(\dfrac{2}{x}-1\right)$

In problems 47–54 express the given complex fraction as a rational integral expression in lowest terms.

47. $\dfrac{1+\dfrac{y}{x}}{1-\dfrac{y}{x}}$

48. $\dfrac{4-\dfrac{3}{x}}{3+\dfrac{5}{x^2}}$

49. $\dfrac{\dfrac{b^2}{a^2}-\dfrac{a}{b}}{\dfrac{1}{2b}-\dfrac{a}{2b^2}}$

50. $\dfrac{\dfrac{2xy}{x+2y}-x}{\dfrac{3xy}{x+3y}+x}$

51. $\dfrac{\dfrac{1}{x^2}-\dfrac{1}{y^2}}{\dfrac{1}{x}+\dfrac{1}{y}}$

52. $\dfrac{\dfrac{1}{a}+\dfrac{1}{b}}{\dfrac{1}{a+b}}$

53. $\dfrac{9x^2-4y^2}{\dfrac{x-y}{y-2x}-1}$

54. $\dfrac{\dfrac{a+3b}{xy}+\dfrac{b}{yz}}{\dfrac{b-a}{x}+\dfrac{b+a}{x}}$

1. Simplify by using the laws of exponents.

   (a) $\left(\dfrac{-2x^2}{y}\right)^3$

   (b) $(x^2y)^2(2xy^3)^2$

2. Simplify by removing all negative and zero exponents, and by using the laws of exponents.

   (a) $(a+b)^0$

   (b) $\dfrac{3a^{-3}y^{-2}}{6ay^{-4}}$

3. Perform the following operations, and simplify if possible.

   (a) $(5x^2 - 3x + 2) + (-7x^3 - 3x^2 + x - 1)$

   (b) $(9x^4 - 3x^2 + 7x) - (5x^3 + 4x^2 - 3x - 2)$

4. Perform the following multiplications, and simplify if possible.

   (a) $(6x^2 - 3x + 1)(5x^3 - 3x)$ (b) $(3x + y)(9x^2 - 3xy + y^2)$

5. Factor completely.

   (a) $36x^2 - 12x + 1$

   (b) $8x^3 - 27$

6. Factor completely.

   (a) $6x^3 - 14x^2 - 12x$

   (b) $x^3 - 4x^2 - 9x + 36$

7. Perform the following divisions, and leave the answer in the form

   $$\dfrac{\text{dividend}}{\text{divisor}} = \text{quotient} + \dfrac{\text{remainder}}{\text{divisor}}.$$

   (a) $\dfrac{4x^5 - 20x^3 + 14x^2 + 3x - 1}{2x^2}$

   (b) $\dfrac{3x^3 - 7x^2 + 3x - 2}{3x - 1}$

8. Perform the following division by using synthetic division, and leave the result in the form described in problem 7.

   $$\dfrac{x^4 - 7x^2 + 3x + 5}{x - 1}$$

9. Perform the indicated operations, and reduce the result to lowest terms.

   (a) $\dfrac{1}{2x - 2} - \dfrac{x}{x^2 - 4x + 3}$

   (b) $\dfrac{(x - y)^2}{2x^2 + 3xy + y^2} \div \dfrac{x^2 - y^2}{x^2 + 2xy + y^2}$

10. Express the following complex fraction as a rational integral expression in lowest terms.

    $$\dfrac{\dfrac{x}{y} - \dfrac{y}{x}}{\dfrac{1}{x} - \dfrac{1}{y}}$$

Radicals and Rational Exponents

3-1
Radicals

If a is a positive real number and n is a positive integer, then there exists a positive number b such that $b^n = a$. The number b is called the *principal nth root of a*. Similarly, if a is a negative real number and n is an *odd* positive integer, then there exists a negative real number b such that $b^n = a$. Again, this number b is called the *principal nth root of a*. In either case, we denote the principal nth root of a by $\sqrt[n]{a}$.

DEFINITION 3-1 If a and b are nonnegative real numbers and n is a positive integer, or if a and b are negative real numbers and n is an odd positive integer, then $\sqrt[n]{a} = b$ if and only if $b^n = a$.

At this point, we do not define $\sqrt[n]{a}$ when n is even and a is negative. This is done later, in the study of complex numbers.

As is customary when $n = 2$, the number $\sqrt[2]{a}$ is called the *square root* of a, and is written simply as $\sqrt{a}$. Similarly, $\sqrt[3]{a}$ is called the *cube root* of a.

EXAMPLE 1 Find the value of each of the following:

(a) $\sqrt[4]{16}$; (b) $\sqrt[3]{-8}$; (c) $\sqrt[5]{32}$;

(d) $\sqrt{\frac{4}{9}}$; (e) $\sqrt[4]{-16}$.

SOLUTION We have

(a) $\sqrt[4]{16} = 2$ since $2^4 = 16$ with both 2 and 16 positive;
(b) $\sqrt[3]{-8} = -2$ since $(-2)^3 = -8$ with both -2 and -8 negative;
(c) $\sqrt[5]{32} = 2$ since $2^5 = 32$ with both 2 and 32 positive;
(d) $\sqrt{\frac{4}{9}} = \frac{2}{3}$ since $(\frac{2}{3})^2 = \frac{4}{9}$ with both $\frac{2}{3}$ and $\frac{4}{9}$ positive;
(e) $\sqrt[4]{-16}$ is not defined since it requires $\sqrt[n]{a}$ with n even and a negative.

We emphasize that the definition of $\sqrt[n]{a}$ defines one and only one real number b such that $b^n = a$. Hence, when we write $\sqrt[n]{a}$ we mean that unique number which when raised to the nth power is a, and whose sign is the same as that of a. Although $(-2)^2 = 4$, when we write $\sqrt{4}$, we mean 2, not -2. Similarly, $-\sqrt{4} = -2$. Note that $\sqrt{a^2}$ designates that *positive number* which yields a^2 when it is squared. Since a may be negative, $\sqrt{a^2}$ *is not always the same as a*. However, we can write, in all cases,

$$\sqrt{a^2} = |a|.$$

The symbol $\sqrt[n]{a}$ is called a *radical*, a is called the *radicand*, and n is called the *index* or *order* of the radical. The symbol $\sqrt{}$ is called the radical sign. The fundamental rules for radicals are stated in the following theorem.

THEOREM 3-2 Let n be a positive integer, and let x and y be real numbers. Then

(a) $(\sqrt[n]{x})^n = x$ if $\sqrt[n]{x}$ exists;
(b) $\sqrt[n]{x} \cdot \sqrt[n]{y} = \sqrt[n]{xy}$ if $\sqrt[n]{x}$ and $\sqrt[n]{y}$ exist;
(c) $\dfrac{\sqrt[n]{x}}{\sqrt[n]{y}} = \sqrt[n]{\dfrac{x}{y}}$ if $\sqrt[n]{x}$ and $\sqrt[n]{y}$ exist;
(d) $\sqrt[n]{x^n} = \begin{cases} x & \text{if } n \text{ is odd,} \\ |x| & \text{if } n \text{ is even;} \end{cases}$
(e) $\sqrt[m]{\sqrt[n]{x}} = \sqrt[mn]{x}$ if m is a positive integer and $\sqrt[mn]{x}$ exists.

EXAMPLE 2 Find the value of the indicated root, or simplify using the properties in Theorem 3-2:

(a) $\sqrt[3]{125}$; (b) $\sqrt[5]{-32a^5}$; (c) $\sqrt[4]{16a^4}$;

(d) $\sqrt[3]{\dfrac{-64}{a^3b^6}}$.

SOLUTION   We have

(a) $\sqrt[3]{125} = \sqrt[3]{(5)^3} = 5$;

(b) $\sqrt[5]{-32a^5} = \sqrt[5]{(-2)^5a^5} = \sqrt[5]{(-2)^5}\sqrt[5]{a^5} = -2a$;

(c) $\sqrt[4]{16a^4} = \sqrt[4]{(2)^4}\sqrt[4]{a^4} = 2|a|$;

(d) $\sqrt[3]{\dfrac{-64}{a^3b^6}} = \dfrac{\sqrt[3]{(-4)^3}}{\sqrt[3]{a^3}\sqrt[3]{(b^2)^3}} = \dfrac{-4}{ab^2}$.

EXERCISES 3-1

Find the principal square root of each of the following numbers.

1. 25

2. 64

3. $\frac{9}{16}$

4. $\frac{36}{256}$

5. $\frac{4}{81}$

6. $\frac{100}{49}$

7. $\frac{144}{25}$

8. 0.04

9. 1.21

10. 2.25

11. 0.0036

12. 0.0169

Find the principal cube root of each of the following numbers.

13. 64

14. -216

15. $-\frac{27}{125}$

16. $-\frac{64}{216}$

17. $\frac{8}{1000}$

18. -0.008

19. $\frac{64}{27}$

20. 0.027

21. -0.064

Find the value of each expression.

22. $(\sqrt{29})^2$

23. $(\sqrt[5]{73})^5$

24. $\sqrt[3]{-\dfrac{1}{8}}$

25. $\sqrt[4]{\dfrac{256}{81}}$

26. $\sqrt[3]{-1}$

27. $\sqrt[4]{10,000}$

28. $\sqrt[5]{0.00032}$

29. $\sqrt{49}$

30. $\sqrt[3]{\sqrt{64}}$

31. $\sqrt{\sqrt[5]{1024}}$

32. $\sqrt[4]{(\sqrt[3]{16})^3}$

33. $\sqrt{\sqrt{81}}$

Find the value of the indicated root, or simplify using the properties in Theorem 3-2. Assume that letter symbols represent positive real numbers.

34. $\sqrt{a^2}$

35. $\sqrt[3]{8b^6}$

36. $\sqrt[4]{a^{12}}$

37. $\sqrt{y^4w^6}$

38. $\sqrt[3]{0.125x^3}$

39. $\sqrt{0.16x^4}$

40. $\sqrt[3]{-8x^3y^6}$

41. $\sqrt[4]{\dfrac{16x^4}{81y^8}}$

42. $\sqrt[5]{\dfrac{-243x^{10}}{32y^5}}$

43. $\sqrt[7]{-1}$

44. $\sqrt{\dfrac{y^4}{w^6}}$

45. $\sqrt[3]{\dfrac{216x^3}{125}}$

46. $\sqrt[5]{\dfrac{-1}{x^{15}}}$

47. $(\sqrt[4]{19y^5})^4$

48. $(\sqrt[3]{-2x})^3$

49. $\sqrt{\sqrt[3]{x^6y^{12}}}$

50. $\sqrt[4]{\sqrt[5]{\dfrac{1024}{y^{40}}}}$

51. $\sqrt[3]{\sqrt[3]{\dfrac{x^{18}}{z^6y^0}}}$

52. $\sqrt[4]{x^8y^{-12}}$

53. $\sqrt{\left(\dfrac{81x^4}{27z^6}\right)^{-1}}$

54. $\sqrt[3]{\left(\dfrac{-8y^0}{x^6}\right)^{-2}}$

**3-2
Rational
Exponents**

In Chapter 2, we defined what was meant by a^n when n was an integer. We wish now to extend this definition to a^n, where n is a rational number. At the same time, we wish to retain the properties that hold for integral exponents. This guides us in making our new definitions.

If rational exponents are to satisfy property (b) of Theorem 2-2, for example, we must have

$$(x^{1/3})^3 = x.$$

That is, $x^{1/3}$ must be a cube root of x. This leads us to define $x^{1/n}$ as follows:

$$x^{1/n} = \sqrt[n]{x},$$

where n is a positive integer.

It is a consequence of this definition that $x^{1/n}$ is not defined if n is even and x is negative.

Again using part (b) of Theorem 2-2 as a guide, we would need $x^{5/3}$ to satisfy

$$(x^{5/3})^3 = x^5.$$

That is, $x^{5/3}$ must be a cube root of x^5,

$$x^{5/3} = \sqrt[3]{x^5}.$$

Since $(\sqrt[3]{x})^5$ is also a cube root of x^5, we have

$$x^{5/3} = \sqrt[3]{x^5} = (\sqrt[3]{x})^5.$$

This discussion of our particular example should make the following definition plausible.

DEFINITION 3-3   If m is an integer and n is a positive integer such that the fraction m/n is in lowest terms, then we define $x^{m/n}$ by

$$x^{m/n} = \sqrt[n]{x^m},$$

and this is equivalent to

$$x^{m/n} = (\sqrt[n]{x})^m.$$

Under the conditions we have placed on m and n, each of the expressions $\sqrt[n]{x^m}$ and $(\sqrt[n]{x})^m$ is defined when the other is, and they are always equal.

EXAMPLE 1   We illustrate Definition 3-3 by evaluating several expressions involving rational exponents.

(a) $(16)^{1/4} = \sqrt[4]{16} = 2$

(b) $\left(\dfrac{a^2}{16}\right)^{3/2} = \left(\sqrt{\dfrac{a^2}{16}}\right)^3 = \left(\dfrac{|a|}{4}\right)^3 = \dfrac{|a|^3}{64}$

(c) $\left(-\dfrac{8}{27}\right)^{2/3} = \left(\sqrt[3]{-\dfrac{8}{27}}\right)^2 = \left(-\dfrac{2}{3}\right)^2 = \dfrac{4}{9}$

(d) $\left(\dfrac{-8a^6}{b^3}\right)^{4/3} = \left(\sqrt[3]{\dfrac{-8a^6}{b^3}}\right)^4 = \left(\dfrac{-2a^2}{b}\right)^4 = \dfrac{16a^8}{b^4}$

(e) $(64)^{-5/6} = (\sqrt[6]{64})^{-5} = (2)^{-5} = \dfrac{1}{32}$

The properties stated in Theorem 2-2 for integral exponents remain valid for rational exponents, provided only that the roots involved are defined.

Since rational exponents are defined in terms of radicals, any expression involving one of them can be changed to an expression involving the other. This is illustrated in the following example. The form $a^{m/n}$ involving a rational exponent is referred to as *exponential form*.

EXAMPLE 2 For each expression below, change from radical form to exponential form, or from exponential form to radical form, whichever is appropriate.

(a) $\sqrt[3]{a^5} = a^{5/3}$

(b) $\left(\sqrt[4]{\dfrac{ab}{c^2}}\right)^3 = \left(\dfrac{ab}{c^2}\right)^{3/4}$

(c) $\left(\dfrac{x^2y}{8}\right)^{2/5} = \sqrt[5]{\left(\dfrac{x^2y}{8}\right)^2}$ or $\left(\sqrt[5]{\dfrac{x^2y}{8}}\right)^2$

(d) $a^{7/5} = \sqrt[5]{a^7}$ or $(\sqrt[5]{a})^7$

EXERCISES 3-2

For each expression, change from radical form to exponential form, or from exponential form to radical form, whichever is appropriate.

1. $a^{1/5}$

2. $x^{5/7}$

3. $3a^{1/4}$

4. $bx^{4/5}$

5. $\sqrt[4]{a^5}$

6. $\sqrt[3]{x^8}$

7. $\sqrt[9]{x^{10}}$

8. $\sqrt[5]{x^{10}}$

9. $\sqrt[3]{a^6}$

10. $(3b)^{1/4}$

11. $(2xy^2)^{2/3}$

12. $(5a^3)^{1/5}$

13. $\sqrt[7]{bx^2}$

14. $\sqrt{a^8}$

15. $(xy)^{5/3}$

Evaluate each of the following expressions. Assume that letter symbols represent positive real numbers.

16. $(4)^{1/2}$

17. $(16)^{1/4}$

18. $(-27)^{1/3}$

19. $(-64a^6)^{1/6}$

20. $(25)^{-3/2}$

21. $(64)^{-4/3}$

22. $(-4a^4)^{-3/2}$

23. $(-4)^{-1/2}$

24. $(0.04)^{1/2}$

25. $(-216)^{-2/3}$

26. $(0.0144)^{1/2}$

27. $\left(\frac{1}{32}\right)^{4/5}$

28. $(16)^{5/4}$

29. $(81)^{3/4}$

30. $\left(\frac{1}{16}\right)^{-3/4}$

3-3
Elementary
Operations
on Radicals

In this section, we consider five ways in which an expression involving radicals or rational exponents may be changed. Frequently, these changes lead to a simpler form of an expression. We shall discuss and illustrate each of these five ways.

I. *Removal of factors from the radicand.* A radical may be simplified by using these facts: $\sqrt[n]{ab} = \sqrt[n]{a}\sqrt[n]{b}$ when $\sqrt[n]{a}$ and $\sqrt[n]{b}$ are defined; $\sqrt[n]{a^n} = a$ if n is odd; $\sqrt[n]{a^n} = |a|$ if n is even.

EXAMPLE 1 Simplify the following radicals by removing as many factors as possible from the radicand:

(a) $\sqrt{32a^2}$;

(b) $\sqrt[3]{32a^4b^5}$.

SOLUTION We have

(a) $\sqrt{32a^2} = \sqrt{(4)^2(2)(a^2)} = \sqrt{(4)^2}\sqrt{a^2}\sqrt{2} = 4|a|\sqrt{2}$;

(b) $\sqrt[3]{32a^4b^5} = \sqrt[3]{(2)^3 \cdot (2)^2 \cdot a^3 \cdot a \cdot b^3 \cdot b^2}$

$\qquad = \sqrt[3]{(2)^3}\sqrt[3]{a^3}\sqrt[3]{b^3}\sqrt[3]{(2)^2ab^2}$

$\qquad = 2ab\sqrt[3]{4ab^2}$.

II. *Additions by use of the distributive property.* In a sum involving radicals, we follow the following procedure:
1. first, we simplify the terms by removing all factors possible from under the radical as described in (I) above;
2. then, if two or more radicals have the same index and radicand, they can be combined using the distributive property.

EXAMPLE 2 The following simplifications illustrate the procedure described in (II).

(a) $\sqrt{32} + \sqrt{18} = \sqrt{(16)(2)} + \sqrt{(9)(2)} = 4\sqrt{2} + 3\sqrt{2} = 7\sqrt{2}$.
This simplification illustrates the important fact that $\sqrt{x + y} \neq \sqrt{x} + \sqrt{y}$. For, in this case, we have

$$\sqrt{32 + 18} = \sqrt{50} = \sqrt{(25)(2)} = 5\sqrt{2} \neq 7\sqrt{2}.$$

(b) $\sqrt{8} - b\sqrt{18a^4} = 2\sqrt{2} - 3a^2b\sqrt{2} = (2 - 3a^2b)\sqrt{2}$.

(c) $\sqrt{3} + \sqrt{7} - 5\sqrt{2}$ cannot be simplified.

III. *Performance of multiplications or divisions.* The product or quotient of two radicals of the same index can be expressed as a single radical using the properties $\sqrt[n]{a}\sqrt[n]{b} = \sqrt[n]{ab}$ and $\sqrt[n]{a}/\sqrt[n]{b} = \sqrt[n]{a/b}$, when $\sqrt[n]{a}$ and $\sqrt[n]{b}$ are defined.

EXAMPLE 3 The procedure described in (III) is illustrated by the following examples.

(a) $(2\sqrt{3})(3\sqrt{2}) = 2 \cdot 3 \cdot \sqrt{3} \cdot \sqrt{2} = 6\sqrt{6}$

(b) $\dfrac{\sqrt{4}}{\sqrt{6}} = \sqrt{\dfrac{4}{6}} = \sqrt{\dfrac{2}{3}}$, or $\dfrac{\sqrt{4}}{\sqrt{6}} = \dfrac{\sqrt{2}\sqrt{2}}{\sqrt{2}\sqrt{3}} = \dfrac{\sqrt{2}}{\sqrt{3}} = \sqrt{\dfrac{2}{3}}$

(c) $\dfrac{\sqrt[3]{ab}}{\sqrt[3]{b^2}} = \sqrt[3]{\dfrac{ab}{b^2}} = \sqrt[3]{\dfrac{a}{b}}$

(d) $\sqrt[3]{4a^2} \cdot \sqrt[3]{10a^2b^5} = \sqrt[3]{40a^4b^5}$
$$= \sqrt[3]{8a^3b^3}\sqrt[3]{5ab^2}$$
$$= 2ab\sqrt[3]{5ab^2}$$

IV. *Introduction of factors into the radicand.* A factor c may be moved under the radical sign by using the property that $c = \sqrt[n]{c^n}$, whenever $\sqrt[n]{c}$ exists.

EXAMPLE 4 This procedure is illustrated below.

(a) $2\sqrt{3} = \sqrt{4} \cdot \sqrt{3} = \sqrt{12}$

(b) $3a\sqrt[3]{2} = \sqrt[3]{(3a)^3} \cdot \sqrt[3]{2} = \sqrt[3]{54a^3}$

V. *Reduction of the index of a radical.* It may be possible to reduce the index of a radical if the radicand can be written as a power which has a factor in common with the index.

EXAMPLE 5 The usefulness of rational exponents is brought out in the following illustrations of a reduction of the index.

(a) $\sqrt[4]{4} = (4)^{1/4} = (2^2)^{1/4} = 2^{2/4} = 2^{1/2} = \sqrt{2}$

(b) $\sqrt[9]{8} = (8)^{1/9} = (2^3)^{1/9} = 2^{3/9} = 2^{1/3} = \sqrt[3]{2}$

EXERCISES 3-3

Simplify by removing all possible factors from the radicand. All letters represent positive real numbers.

1. $\sqrt{500}$

2. $\sqrt[3]{-16}$

3. $\sqrt{32a^3b^4}$

4. $\sqrt[3]{108(x + y)^7}$

5. $\sqrt[3]{128x^7}$

6. $\sqrt[3]{8x^8}$

7. $\sqrt[5]{x^{13}}$

8. $\sqrt[5]{-x^6y^7}$

9. $\sqrt[3]{-128a^9}$

10. $\sqrt{4a^6}$

11. $\sqrt[4]{162}$

12. $\sqrt[4]{16z^6}$

Combine whenever possible by use of the distributive property.

13. $\sqrt{2} + \sqrt{3}$

14. $\sqrt{12} + \sqrt{27}$

15. $3\sqrt{50} - 2\sqrt{18}$

16. $3\sqrt{2} - 2\sqrt[3]{2}$

17. $\sqrt{150} - \sqrt{24}$

18. $3a\sqrt[3]{16} - 4b\sqrt[3]{2}$

19. $\sqrt[3]{54a^3} + 2b\sqrt[3]{16}$

20. $\sqrt[3]{16a^3x} + \sqrt[3]{-54a^6x}$

21. $\sqrt[3]{375x^3y^4} + \sqrt[3]{-3x^3y^4}$

Perform the indicated multiplications or divisions, and then remove all factors possible from the radicand. All letters represent positive real numbers.

22. $\sqrt{3} \cdot \sqrt{2}$

23. $\sqrt[3]{5} \cdot \sqrt[3]{50}$

24. $\sqrt[3]{2} \cdot \sqrt[3]{24}$

25. $\dfrac{\sqrt{15}}{\sqrt{3}}$

26. $\dfrac{\sqrt[3]{20}}{\sqrt[3]{5}}$

27. $\dfrac{\sqrt{15x}}{\sqrt{5x}}$

28. $\sqrt{3a} \cdot \sqrt{12a^3}$

29. $\sqrt[3]{4x^2} \cdot \sqrt[3]{6x^2y^4}$

30. $(3 - 2\sqrt{7})(2 + \sqrt{7})$

31. $(2 - \sqrt{3})(3 + 2\sqrt{3})$

32. $(\sqrt{6} - 3\sqrt{2})(\sqrt{6} + 3\sqrt{2})$

33. $(\sqrt{5} - 3\sqrt{2})(\sqrt{10} - 3)$

Replace the coefficient of the radical by an equivalent factor under the radical sign. All letters represent positive real numbers.

34. $3\sqrt{5a}$

35. $2\sqrt{5x}$

36. $2a\sqrt{5a}$

37. $3\sqrt[3]{x}$

38. $2x\sqrt[3]{2x}$

39. $\dfrac{1}{3}\sqrt{\dfrac{3a}{2b}}$

40. $2\sqrt[5]{x^3}$

41. $2\sqrt{\dfrac{2x}{3y}}$

42. $2\sqrt[4]{2x}$

Reduce the radical index and remove any factors possible from underneath the radical sign. All letters represent positive real numbers.

43. $\sqrt[4]{\dfrac{4}{9}}$

44. $\sqrt[12]{16x^4y^8}$

45. $\sqrt[9]{a^6}$

46. $\sqrt[9]{27a^3}$

47. $\sqrt[9]{125}$

48. $\sqrt[12]{x^9}$

49. $\sqrt[6]{0.027x^3}$

50. $\sqrt[4]{0.64}$

51. $\sqrt[6]{u^4}$

3-4
Simplest
Radical Form

In many instances, it is desirable to change a quotient which has radicals in the denominator to an expression which is equal to the original quotient but has no radicals in the denominator. This sort of change is desirable if one wishes to find a decimal value of such a quotient by hand calculations. For example, suppose that one wishes a decimal value for the quotient $3/\sqrt{2}$, using 1.414 as the value of $\sqrt{2}$. Instead of performing the division in the form

$$\frac{3}{\sqrt{2}} \approx \frac{3}{1.414},$$

the quotient can be evaluated much more easily if it is changed to

$$\frac{3}{\sqrt{2}} = \frac{3 \cdot \sqrt{2}}{\sqrt{2} \cdot \sqrt{2}} = \frac{3\sqrt{2}}{2} \approx \frac{3(1.414)}{2}.$$

The process of rewriting a quotient so that the denominator contains no radicals is called *rationalizing the denominator*. This is accomplished by multiplying the numerator and denominator of a fraction by a factor which yields a new denominator which contains no radicals.

EXAMPLE 1 Rationalize the denominators of the following fractions.

(a) $\sqrt{\dfrac{3}{7}}$ (b) $\sqrt[3]{\dfrac{3}{4}}$ (c) $\sqrt[3]{\dfrac{2a}{9b^2}}$

SOLUTION In each case, we multiply by a factor chosen so as to form a power in the denominator which is the same as the index of the radical. It is efficient to make the multiplying factor as small as possible.

(a) $\sqrt{\dfrac{3}{7}} = \sqrt{\dfrac{3 \cdot 7}{7 \cdot 7}} = \sqrt{\dfrac{21}{(7)^2}} = \dfrac{\sqrt{21}}{7}$

(b) $\sqrt[3]{\dfrac{3}{4}} = \sqrt[3]{\dfrac{3 \cdot 2}{4 \cdot 2}} = \sqrt[3]{\dfrac{6}{(2)^3}} = \dfrac{\sqrt[3]{6}}{2}$

(c) $\sqrt[3]{\dfrac{2a}{9b^2}} = \sqrt[3]{\dfrac{(2a)(3b)}{(9b^2)(3b)}} = \sqrt[3]{\dfrac{6ab}{(3)^3b^3}} = \dfrac{\sqrt[3]{6ab}}{3b}$

One frequently encounters quotients with a denominator which contains radicals and is more complicated than those in the example above. Often a denominator is one of the forms $a + \sqrt{b}$, $a - \sqrt{b}$, $a\sqrt{b} + c\sqrt{d}$, or $a\sqrt{b} - c\sqrt{d}$. To rationalize these types of denominators, we multiply the numerator and denominator of the fraction by $a - \sqrt{b}$, $a + \sqrt{b}$, $a\sqrt{b} - c\sqrt{d}$, or $a\sqrt{b} + c\sqrt{d}$, respectively. The expression obtained from a binomial by changing the sign on the second term is called the *conjugate* of the binomial. Thus $a + \sqrt{b}$ and $a - \sqrt{b}$ are conjugates of each other. The numbers $a\sqrt{b} - c\sqrt{d}$ and $a\sqrt{b} +$

$c\sqrt{d}$ are also conjugates of each other. The product of a binomial that contains square roots and its conjugate is always free of radicals:

$$(a\sqrt{b} + c\sqrt{d})(a\sqrt{b} - c\sqrt{d}) = (a\sqrt{b})^2 - (c\sqrt{d})^2 = a^2b - c^2d.$$

EXAMPLE 2 Rationalize the denominators of the following fractions.

(a) $\dfrac{3}{3 - \sqrt{5}}$ 　　　　　　　　　　(b) $\dfrac{2}{4\sqrt{2} + \sqrt{3}}$

SOLUTION In each case, we multiply the numerator and denominator by the conjugate of the denominator.

(a) $\dfrac{3}{3 - \sqrt{5}} = \dfrac{3}{3 - \sqrt{5}} \cdot \dfrac{3 + \sqrt{5}}{3 + \sqrt{5}}$

$= \dfrac{3(3 + \sqrt{5})}{9 - 5}$

$= \dfrac{3(3 + \sqrt{5})}{4}$

(b) $\dfrac{2}{4\sqrt{2} + \sqrt{3}} = \dfrac{2}{4\sqrt{2} + \sqrt{3}} \cdot \dfrac{4\sqrt{2} - \sqrt{3}}{4\sqrt{2} - \sqrt{3}}$

$= \dfrac{2(4\sqrt{2} - \sqrt{3})}{32 - 3}$

$= \dfrac{2(4\sqrt{2} - \sqrt{3})}{29}$

There are some problems in which it is desirable to rationalize the numerator. See problems 64–69.

If a radical expression satisfies a certain list of conditions, it is said to be in *simplest radical form*. These conditions are given in the next definition.

DEFINITION 3-4 A radical expression is in *simplest radical form* if the following conditions are satisfied:

(a) the expression contains no negative or zero exponents;
(b) the radicand contains no factor to a power equal to or greater than the index of the radical;
(c) the denominator contains no radicals;
(d) the index of the radical is as small as possible.

An expression involving radicals can be reduced to simplest form by following the procedure described below.

> *Reduction of a Radical Expression to Simplest Form*
>
> 1. Eliminate all negative or zero exponents.
>
> 2. Express any power or root of a radical, or any product of radicals, as a single radical.
>
> 3. Reduce each radicand to a simple fraction in lowest terms, and rationalize all denominators.
>
> 4. Remove from each radicand any factor which is a perfect nth power, where n is the index of the radical.
>
> 5. Reduce the index of the radical to the lowest possible order.
>
> 6. In a sum, combine any terms with a common radical factor.

EXAMPLE 3 Reduce each of the expressions below to the simplest radical form. Assume that all letters represent positive real numbers.

(a) $\sqrt[6]{\dfrac{a^2}{27c^{10}}}$

(b) $\dfrac{\sqrt[3]{cd^4}}{\sqrt{cd}}$

(c) $\dfrac{3}{2 - 3\sqrt{2}} \cdot \dfrac{\sqrt{2} - 1}{\sqrt{2}}$

SOLUTION Simplifying each of the expressions in turn, we have

(a) $\sqrt[6]{\dfrac{a^2}{27c^{10}}} = \sqrt[6]{\dfrac{a^2}{27c^{10}} \cdot \dfrac{27c^2}{27c^2}}$

$= \sqrt[6]{\dfrac{27a^2c^2}{(3)^6(c^2)^6}}$

$= \dfrac{\sqrt[6]{27a^2c^2}}{3c^2}$

(b) $\dfrac{\sqrt[3]{cd^4}}{\sqrt{cd}} = \dfrac{(cd^4)^{1/3}}{(cd)^{1/2}}$

$= \dfrac{(cd^4)^{2/6}}{(cd)^{3/6}}$

$= \sqrt[6]{\dfrac{(cd^4)^2}{(cd)^3}}$

$= \sqrt[6]{\dfrac{c^2d^8}{c^3d^3}}$

$= \sqrt[6]{\dfrac{d^5}{c} \cdot \dfrac{c^5}{c^5}}$

$= \dfrac{\sqrt[6]{c^5d^5}}{c}$

(c) $\dfrac{3}{2 - 3\sqrt{2}} \cdot \dfrac{\sqrt{2} - 1}{\sqrt{2}} = \dfrac{3(\sqrt{2} - 1)}{2\sqrt{2} - 3\sqrt{4}}$

$= \dfrac{3(\sqrt{2} - 1)}{2\sqrt{2} - 6} \cdot \dfrac{2\sqrt{2} + 6}{2\sqrt{2} + 6}$

$= \dfrac{3(4 + 6\sqrt{2} - 2\sqrt{2} - 6)}{8 - 36}$

$= \dfrac{3(-2 + 4\sqrt{2})}{-28}$

$= \dfrac{3(1 - 2\sqrt{2})}{14}$

EXERCISES 3-4

Rationalize the denominator. Assume that all letters represent positive real numbers.

1. $\sqrt{\dfrac{1}{5}}$

2. $\sqrt{\dfrac{3}{7}}$

3. $\sqrt{\dfrac{5}{8}}$

4. $\dfrac{1}{\sqrt{5}}$

5. $\dfrac{5}{\sqrt{7}}$

6. $\dfrac{2}{\sqrt{11}}$

7. $\sqrt[3]{\dfrac{1}{4}}$

8. $\sqrt[3]{\dfrac{4}{9}}$

9. $\dfrac{3\sqrt{5}}{2\sqrt{7}}$

10. $\dfrac{1}{\sqrt[4]{64}}$

11. $\dfrac{\sqrt[3]{12}}{\sqrt[3]{4}}$

12. $\dfrac{1-\sqrt{5}}{2+\sqrt{5}}$

13. $\dfrac{2\sqrt{7}-3}{\sqrt{7}-2}$

14. $\dfrac{2}{2\sqrt{3}-\sqrt{5}}$

15. $\dfrac{\sqrt{5}-\sqrt{3}}{3\sqrt{3}+\sqrt{2}}$

16. $\dfrac{\sqrt{3}+2\sqrt{2}}{3\sqrt{2}+2\sqrt{3}}$

17. $\sqrt{\dfrac{a}{3}}$

18. $\sqrt{\dfrac{2a}{c}}$

19. $\sqrt{\dfrac{5a}{2b}}$

20. $\sqrt[3]{\dfrac{3d}{4}}$

21. $\sqrt[3]{\dfrac{2cd}{9x^2}}$

22. $\sqrt{\dfrac{x^2}{z^5}}$

23. $\sqrt{\dfrac{1}{5u^5}}$

24. $\sqrt[4]{\dfrac{cu^6}{27}}$

25. $\sqrt[4]{\dfrac{2a^2y^6}{x^3z^5}}$

26. $\sqrt{\dfrac{a}{2}-\dfrac{5}{x}}$

27. $\sqrt{\dfrac{2}{x}+\dfrac{x}{2b}}$

Change to simplest form. Assume that all letters represent positive real numbers.

28. $\sqrt{40u^5}$

29. $\sqrt[5]{32uv^7}$

30. $\sqrt[3]{5x^{-6}}$

31. $\sqrt[4]{36x^2}$

32. $\sqrt[10]{32x^5}$

33. $\sqrt[4]{\dfrac{5}{8}}$

34. $\sqrt[4]{\dfrac{5}{81}}$

35. $a^{1/3}b^{5/2}$

36. $\left(\dfrac{4u^5}{3x^3}\right)^{1/2}$

37. $\left(\dfrac{cx^4}{36u^7}\right)^{1/3}$

38. $2\sqrt{\dfrac{1}{3}} + 6\sqrt{\dfrac{1}{2}} + a\sqrt{27}$

39. $\sqrt{\dfrac{1}{2}} + 7\sqrt{18} - a\sqrt{72}$

40. $(\sqrt{3} + 2\sqrt{b})^2$

41. $(\sqrt{y} - 3\sqrt{x})^2$

42. $\dfrac{3 - \sqrt{y}}{2 - 2\sqrt{y}}$

43. $\dfrac{\sqrt[3]{3c^2v^3}}{\sqrt[3]{15c^2v^2}}$

44. $\sqrt[8]{36}$

45. $\sqrt[6]{9y^4}$

46. $\sqrt{cd} \cdot \sqrt[4]{d^2}$

47. $\sqrt[3]{2x^2} \cdot \sqrt[4]{4x^3}$

48. $\sqrt[5]{xy^2} \cdot \sqrt{xy}$

49. $\sqrt[3]{32y^{-4}} - \sqrt[3]{-4y^2}$

50. $\sqrt{3x^3 + x^2} + \sqrt{12x + 4}$

51. $(3 + \sqrt[3]{2})(3 - \sqrt[3]{2})$

52. $\sqrt[3]{(x - y)^4}$

53. $\sqrt{(a + 2b)^6}$

54. $\dfrac{x}{\sqrt{4x^{-2}}}$

55. $\sqrt{x^9y^{-2}z^0}$

56. $\sqrt{\dfrac{x^{-3}}{xy^{-6}}}$

57. $\sqrt{\dfrac{27(a + b)^2}{(a + b)^{-3}}}$

58. $(\sqrt{a} - \sqrt{b})^{-1}$

59. $(1 + \sqrt{2})^{-2}$

60. $\sqrt{12\sqrt{9}}$

61. $\sqrt[3]{-2\sqrt{16x^8}}$

62. $\sqrt{x\sqrt[3]{x^6}}$

63. $\sqrt{x^3\sqrt[3]{x^{-3}}}$

Rationalize the numerator. Assume that all letters represent positive real numbers.

*64. $\dfrac{\sqrt{x} - 2}{x}$

*65. $\dfrac{\sqrt{x + h} - \sqrt{x}}{h}$

*66. $\dfrac{\sqrt{x} - \sqrt{3}}{x - 3}$

*67. $\dfrac{\sqrt{a} + \sqrt{b}}{a - b}$

*68. $\dfrac{\sqrt[3]{x} - 1}{x - 1}$

*69. $\dfrac{\sqrt[3]{x + h} - \sqrt[3]{x}}{h}$

PRACTICE TEST for Chapter 3

Assume that all letter symbols represent positive real numbers.

1. Find the value of each of the following.

    (a) $\sqrt[3]{-\dfrac{125}{8}}$

    (b) $\sqrt{\sqrt[3]{64}}$

2. Simplify the radical.

    (a) $\sqrt[3]{\dfrac{216x^6}{64y^3}}$

    (b) $\sqrt[4]{\dfrac{16x^8}{81y^{12}}}$

3. For each expression, change from radical form to exponential form, or from exponential form to radical form, whichever is appropriate.

    (a) $(3x)^{-2/5}$

    (b) $\sqrt[4]{y^3}$

4. Evaluate each of the following expressions.

    (a) $(-32)^{3/5}$

    (b) $\left(\dfrac{1}{125}\right)^{-2/3}$

5. Combine whenever possibe by use of the distributive property.

    (a) $3\sqrt{32} - 5\sqrt{50}$

    (b) $\sqrt[3]{81x^4y} - \sqrt[3]{24x^4y}$

6. Perform the indicated operations whenever possible and then remove all factors possible from under the radical.

    (a) $\dfrac{\sqrt[3]{40}}{\sqrt[3]{5a^3}}$

    (b) $(3 - 2\sqrt{5})(2 + 4\sqrt{5})$

7. Replace the coefficient of the radical by an equivalent factor under the radical sign.

    (a) $3a\sqrt{7}$

    (b) $2\sqrt[4]{5x}$

8. Reduce the radical index and remove any factors possible from underneath the radical sign.

    (a) $\sqrt[12]{125}$

    (b) $\sqrt[9]{125x^{12}}$

9. Rationalize the denominator in each of the following.

    (a) $\dfrac{3}{2 - \sqrt{2}}$

    (b) $\sqrt[3]{\dfrac{4x}{3y}}$

10. Change to simplest radical form.

    (a) $\sqrt[3]{9\sqrt{36}}$

    (b) $\sqrt[5]{x^2y^3} \cdot \sqrt{xy}$

Linear
and Quadratic
Equations
in One Variable

4-1
Linear Equations

Many of the problems encountered in mathematics reduce to that of solving a linear equation or a linear inequality. A *linear equation* in the variable x is an equation which can be written in the form

$$ax + b = 0,$$

where $a \neq 0$. The set of all values of the variable x which satisfy the linear equation is called the *solution set* for the linear equation. It is easily seen that $x = -b/a$ satisfies the equation $ax + b = 0$, $a \neq 0$, and that this is the only value of x which satisfies the equation. Thus $\{-b/a\}$ is the solution set for this equation.

Two equations are said to be *equivalent* if they have the same solution set. For example, the equation

$$3x - 6 = 0$$

is equivalent to the equation

$$x = 2$$

since $\{2\}$ is the solution set for both equations.

The method we shall use to solve a linear equation will be to employ the following theorem in order to find an equivalent equation which has an obvious solution set.

THEOREM 4-1   Let $R(x)$, $S(x)$, and $T(x)$ be algebraic expressions[1] in the variable x. Then

$$R(x) = S(x)$$

is equivalent to

$$R(x) + T(x) = S(x) + T(x) \tag{4.1a}$$

and to

$$R(x)T(x) = S(x)T(x) \qquad \text{for } T(x) \neq 0. \tag{4.1b}$$

EXAMPLE 1   In order to determine the solution set for the equation

$$4x - 3 = 7x + 3 \tag{1}$$

we first add $3 - 7x$ to both sides of the equation. This results in the equivalent equation

$$-3x = 6. \tag{2}$$

Next, we multiply both sides of equation (2) by $-\frac{1}{3}$ and obtain the equivalent equation

$$x = -2,$$

whose solution set is obviously $\{-2\}$.

Notice in (4.1b) that the multiplier $T(x)$ is required to be different from zero in order to have an equivalent equation. Consider the next example, in which it happens that $T(x) = 0$ for certain values of the variable x.

EXAMPLE 2   Solve

$$\frac{x + 4}{x - 1} = \frac{x - 3}{x - 2} + \frac{5x - 9}{(x - 1)(x - 2)}. \tag{3}$$

---

[1]The notation $R(x)$ is read "R of x." It does *not* indicate the product of R and x.

SOLUTION If we multiply both sides of (3) by the least common denominator, $(x - 1)(x - 2)$, we obtain

$$(x - 2)(x + 4) = (x - 3)(x - 1) + 5x - 9,$$

or

$$x^2 + 2x - 8 = x^2 - 4x + 3 + 5x - 9.$$

Using Theorem 4-1, this simplifies to

$$2x - 8 = x - 6,$$

and then to

$$x = 2.$$

But we note that the multiplier $(x - 1)(x - 2)$ is equal to zero whenever $x = 2$ or $x = 1$, and that these values yield undefined quantities in the original equation (3). This means that we must restrict these values from our solution set. Hence the solution set for (3) is the empty set.

We recall that the definition of the absolute value of a variable x is given by

$$|x| = \begin{cases} x & \text{if } x \geq 0, \\ -x & \text{if } x < 0. \end{cases}$$

The next three examples illustrate the solution of certain types of equations which involve absolute values.

EXAMPLE 3 Solve

$$|x - 5| = 4.$$

SOLUTION We consider cases corresponding to the definition of absolute value. First, if $x - 5 \geq 0$, we have $|x - 5| = x - 5$, and we must find the solution set of the linear equation

$$x - 5 = 4.$$

Clearly, this solution set is $\{9\}$. But, if $x - 5 < 0$, then $|x - 5| = -(x - 5)$, and we must solve

$$-(x - 5) = 4. \tag{4}$$

The solution set of (4) is $\{1\}$. Thus

$$\{x \,||\, x - 5| = 4\} = \{1, 9\}.$$

EXAMPLE 4 Determine the solution set of

$$|x| + 2 = 3x - 1. \tag{5}$$

SOLUTION If $x \geq 0$, then $|x| = x$, and $x = \frac{3}{2}$ is the only solution to
$$x + 2 = 3x - 1.$$
Whenever $x < 0$, then $|x| = -x$, and $x = \frac{3}{4}$ is the only solution to
$$-x + 2 = 3x - 1.$$
But $x = \frac{3}{4}$ does not satisfy the condition that $x < 0$. Thus the solution set for (5) is $\{\frac{3}{2}\}$.

Suppose that $R(x)$ and $S(x)$ are algebraic expressions in the variable x, and we wish to determine the solution set of

$$|R(x)| = |S(x)|. \tag{6}$$

Using the definition of absolute value, we have four cases to consider.

Case 1 If $R(x) \geq 0$ and $S(x) \geq 0$, then $|R(x)| = R(x)$ and $|S(x)| = S(x)$, so that equation (6) becomes

$$R(x) = S(x), \tag{7}$$

Case 2 If $R(x) < 0$ and $S(x) < 0$, then $|R(x)| = -R(x)$ and $|S(x)| = -S(x)$. Equation (6) becomes

$$-R(x) = -S(x),$$

and this is equivalent to

$$R(x) = S(x).$$

Case 3 If $R(x) \geq 0$ and $S(x) < 0$, then $|R(x)| = R(x)$ and $|S(x)| = -S(x)$. In this case, equation (6) appears as

$$R(x) = -S(x). \tag{8}$$

Case 4 If $R(x) < 0$ and $S(x) \geq 0$, then $|R(x)| = -R(x)$ and $|S(x)| = S(x)$. Equation (6) becomes

$$-R(x) = S(x),$$

and this is equivalent to

$$R(x) = -S(x).$$

The results of these four cases can be combined and simplified by the following: if $|R(x)| = |S(x)|$, then either $R(x) = S(x)$ or $R(x) = -S(x)$, and the solution set of equation (6) is the union of the solution sets of equations (7) and (8). This result is stated in Theorem 4-2.

THEOREM 4-2 If $R(x)$ and $S(x)$ are algebraic expressions in the variable x, then $|R(x)| = |S(x)|$ if and only if either $R(x) = S(x)$ or $R(x) = -S(x)$.
If $|S(x)|$ is replaced by a positive number a, we have the following corollary.

COROLLARY 4-3   If $a > 0, |R(x)| = a$ if and only if either $R(x) = a$ or $R(x) = -a$.

EXAMPLE 5   To determine the solution set for

$$|x - 2| = |3x + 1|,$$

we consider the two possible situations:

$$\text{If } x - 2 = 3x + 1, \text{ then } -2x = 3, \text{ and } x = -\tfrac{3}{2}.$$
$$\text{If } x - 2 = -(3x + 1), \text{ then } 4x = 1, \text{ and } x = \tfrac{1}{4}.$$

Thus

$$\{x \mid |x - 2| = |3x + 1|\} = \{-\tfrac{3}{2}, \tfrac{1}{4}\}.$$

EXAMPLE 6   Solve

$$|x + 3| = 7.$$

SOLUTION   We have $|x + 3| = 7$ if and only if either $x + 3 = 7$ or $x + 3 = -7$. Thus the solution set is $\{4, -10\}$.

EXERCISES 4-1

Determine the solution set for each of the following equations.

1. $9x + 5 = 32$

2. $3x + 2(1 + x) = x + 2$

3. $\frac{10}{3} - \frac{1}{2}x = \frac{3}{4}x$

4. $12x + \frac{3}{2} = \frac{3}{5}x + 3$

5. $(5x - 1)(4x + 2) = (10x - 1)(2x + 3) - 2(x + 13)$

6. $3x - [4x - 2(x - 5)] - 1 = 2(3 - x) - 17$

7. $3.6 - 0.2x = -1.4x$

8. $-1.21 + 0.27x = 0.32x + 0.79$

9. $\dfrac{x - 2}{2} = \dfrac{3x - 1}{4}$

10. $\dfrac{7x - 4}{3} = x - \dfrac{2}{5}$

11. $\dfrac{1}{1 - x} + \dfrac{3}{x} = -\dfrac{1}{x}$

12. $\dfrac{-2}{x + 1} + 3 = \dfrac{7x}{3(x + 1)}$

13. $\dfrac{1}{2 - x} = \dfrac{-3}{2 + x}$

14. $\dfrac{-1}{2x + 3} - \dfrac{2}{x - 2} = \dfrac{3x + 4}{(2x + 3)(x - 2)}$

15. $\dfrac{1}{x + 1} = -\dfrac{x}{x + 1}$

16. $\dfrac{3x + 5}{x} = 4 + \dfrac{5}{x}$

Solve the equation for the variable indicated.

17. $ax + by = c$; for x.

18. $\dfrac{1}{x} + \dfrac{1}{y} = 1$; for y.

19. $A = \frac{1}{2}bh$; for h.

20. $ax + by = bx + ay$; for x.

21. $S = \dfrac{n}{2}(a + t)$; for t.

22. $S = \dfrac{n}{2}[2a + (n - 1)d]$; for a.

Solve the following equations involving absolute values.

23. $|x| = 3$

24. $|x - 2| = 4$

25. $|2x - 7| = 0$

26. $|3x + 5| = -2$

27. $|x| + 3x - 9 = 0$

28. $|x + 5| = 3x - 2$

29. $|3x + 8| = |2 - x|$

30. $|8x + 5| = |x + 4|$

31. $\dfrac{|2x - 5|}{|x + 2|} = 1$

32. $\dfrac{|3 - 2x|}{|x + 9|} = 0$

**4-2
Applications
(Optional)** In many phases of our everyday life, we are presented with problems that can be represented by a linear equation in one variable. To solve such a problem, we first decide what our unknown is in that particular situation. Next, we use the given information to write an equation involving that unknown. Finally, we solve the equation by using the technique described in Section 4-1.

EXAMPLE 1 Find two consecutive even integers such that the difference in their squares is 52.

SOLUTION Let x and $x + 2$ represent the two consecutive even integers. Then we have the equation

$$(x + 2)^2 - x^2 = 52.$$

Expanding and simplifying, we have

$$x^2 + 4x + 4 - x^2 = 52,$$

and

$$4x + 4 = 52.$$

This gives $x = 12$ and $x + 2 = 14$. That is, the two consecutive even integers are 12 and 14.

EXAMPLE 2 Suppose that a chemist has 10 liters of a mixture which is 10% acid. How much pure acid should she add to make a mixture which is 50% acid?

The amount of acid in the final mixture is the sum of the amount of acid in the original mixture and the amount of acid added. Let x represent the number of liters of pure acid to be added. The amount of acid in the original mixture is $(0.10)(10) = 1$ liter, and the amount of acid in the final mixture is $(0.50)(10 + x)$ liters. Thus we have the equation

$$(0.10)(10) + x = (0.50)(10 + x),$$

or

$$1 + x = 5 + 0.5x.$$

Solving for x, we have

$$0.5x = 4,$$

and

$$x = 8.$$

The chemist must add 8 liters of pure acid in order to obtain a mixture which is 50% acid.

EXAMPLE 3 The sum of the digits in a two-digit number is 5. If the digits are reversed, the new number is 9 more than the original number. What is the original number?

SOLUTION Suppose that x is the units digit in the original number. Then the tens digit is $5 - x$, and the original number expressed in terms of x has a value of $10(5 - x) + x$. Similarly, in the new number, x is the tens digit and $5 - x$ is the units digit, so that its value is $10x + (5 - x)$. Thus we have

$$10x + (5 - x) = 10(5 - x) + x + 9.$$

This simplifies to

$$9x + 5 = -9x + 59,$$

and

$$18x = 54.$$

We then obtain

$$x = 3,$$

and

$$5 - x = 2.$$

This means that the original number is 23.

EXAMPLE 4 If it takes Herman 5 days to dig a ditch, and it takes Toby 2 days to dig a ditch of the same size, how long would it take for both Herman and Toby to dig the ditch if they worked together?

Let x represent the number of days it takes both to dig the ditch working together. Then $1/x$ represents the portion of the ditch that they could dig together in 1 day. Similarly, since it takes Herman 5 days to dig the ditch, he would dig $\frac{1}{5}$ of the ditch in 1 day. Toby would dig $\frac{1}{2}$ of the ditch in 1 day, and, working together, they would dig $\frac{1}{5} + \frac{1}{2}$ of the ditch in 1 day. Thus

$$\frac{1}{5} + \frac{1}{2} = \frac{1}{x}.$$

Multiplying by $10x$, we obtain

$$2x + 5x = 10.$$

Thus

$$7x = 10,$$

and

$$x = 1\tfrac{3}{7}.$$

Working together, Herman and Toby can dig the ditch in $1\tfrac{3}{7}$ days.

If an object moves at a constant rate of speed, r, for a time, t, then it will have moved a distance, d, which is equal to the product of the rate and the time:

$$\boxed{d = rt.}$$

This formula is useful in the following example.

EXAMPLE 5 A certain plane travels 600 km/hr in still air. The plane flies north with the wind for 5 hours, and makes the return trip against the wind in 5 hours and 40 minutes. What is the rate of the wind?

SOLUTION Let r represent the rate of the wind. Then the plane's ground speed is $(600 + r)$ km/hr with the wind and $(600 - r)$ km/hr against the wind. Since the distances traveled on both trips are the same, we have

$$(600 + r)(5) = (600 - r)(\tfrac{17}{3}),$$

or

$$3000 + 5r = 3400 - \tfrac{17}{3}r.$$

Thus

$$\tfrac{32}{3}r = 400,$$

and

$$r = 37\tfrac{1}{2}.$$

That is, the rate of the wind is $37\tfrac{1}{2}$ km/hr.

Linear and Quadratic
Equations in One
Variable

1. Find three consecutive integers whose sum is 147.

2. Find two consecutive even integers whose sum is 90.

3. The sum of the digits in a two-digit number is 11, and the tens digit is 1 more than 4 times the units digit. Find the number.

4. The sum of the digits in a two-digit number is 12, and the tens digit is 3 times the units digit. Find the number.

5. In a two-digit number, the units digit is one less than twice the tens digit. If the digits are reversed, the new number is 20 less than twice the original number. Find the original number.

6. In a three-digit number, the tens digit is twice the hundreds digit and half the units digit. If the digits are reversed, the new number is 49 more than three times the original number. Find the original number.

7. Suppose that 12 dimes and nickels are worth $0.95. How many dimes and how many nickels are there?

8. Suppose that 25 nickels, dimes, and quarters are worth $2.75, and there are twice as many dimes as nickels. How many of each denomination are there?

9. Mrs. Phillips went to the grocery store to buy milk and eggs. Suppose that a carton of milk cost 93 cents and a carton of eggs cost 75 cents. How many cartons each of milk and eggs did she buy if she spent $6.54 (excluding tax) and bought 8 items?

10. There are two types of tickets available for a rock concert. The reserved-seat tickets cost $8.00 each, and the cheap-seat tickets cost $2.50 each. How many tickets were sold if three times as many cheap-seat tickets were sold as reserved-seat tickets, and the total proceeds were $192,200?

11. The perimeter of a rectangular garden is 112 meters, and the length is 4 meters less than twice the width. Find the length and width of the garden.

12. The sum of the three angles in a triangle is 180°. Find the measures of the angles if the largest is four times as large as the smallest and equal to the sum of the two smaller.

13. Suppose that we wish to mix peanuts worth 80 cents per pound with cashews worth 95 cents per pound to obtain 12 pounds of a mixture worth 90 cents per pound. How many pounds of each type of nuts should we use?

14. How many pounds of coffee worth $1.50 per pound must be mixed with 15 pounds of coffee worth $2.00 per pound to obtain a blend worth $1.80 per pound?

15. Suppose that we wish to produce a 40-gram bar of metal that is 15% pure silver by melting together parts of one bar of metal which is 20% pure silver and another which is 12% pure silver. How many grams of each should be used?

16. How much of an 85% solution of acid must be added to 16 liters of a 65% solution to obtain a solution that is 70% acid?

17. Suppose that a chemist wishes to dilute 40 liters of a solution which is 80% acid to one which is 50% acid. How much water should be added to the acidic solution?

18. If 15 liters of an acidic solution mixed with 20 liters of a 55% acidic solution produces a 40% acidic solution, what is the strength of the first solution?

19. Suppose that Dan and Fran live 450 kilometers apart, and at the same time they begin driving toward each other with Dan traveling an average rate of 50 km/hr and with Fran's average rate 55 km/hr. How long will it be before they meet?

20. In problem 19, suppose that Dan leaves at 12 noon and Fran leaves at 1 P.M. At what time will they meet?

21. Suppose that a boat travels 1 hour downstream and then returns in 1 hour 20 minutes. If the speed of the current is 3 km/hr, find the speed of the boat and the distance it traveled downstream.

22. It takes a plane as long to fly 400 kilometers against the wind as it does to fly 450 kilometers with the same wind. If the speed of the wind is 20 km/hr, find the speed of the plane in still air.

23. Two cars leave a city at the same time traveling in opposite directions. At the end of 5 hours, they are 725 kilometers apart. How fast was each car traveling if one traveled 5 km/hr faster than the other?

24. One car traveling 77 km/hr overtook another car traveling the same highway in 2 hours. If the faster car started 45 minutes after the slower car, find the speed of the slower car.

25. If it takes 8 hours for Terri to mow the grass with her push mower, and it takes Martin 5 hours to mow the grass with the riding mower, how long will it take to mow the grass if both Terri and Martin work together?

26. If it takes 12 hours for Beckie to clean the house, 6 hours for her sister Lisa to clean the same house, and 10 hours for the third sister Donna to clean the same house, how long will it take the three sisters to clean the house if they work together?

27. Suppose that Jim and Ruth can roof a house working together in 20 hours, and Jim can do the job alone in 36 hours. How long would it take Ruth to do the job alone?

28. Suppose that it takes Mary and Sam 3 hours to grade a set of homework papers if they work together, and Sam works twice as fast as Mary. How long would it take Mary working alone?

29. A tank can normally be filled in 10 hours. But after the tank developed a leak, it took 12 hours to fill it. How long would it take the leak to empty the full tank?

30. Suppose that it takes 6 hours to fill an empty swimming pool with one hose, and it takes 5 hours to fill the same pool with a different hose. Also, suppose that it takes 15 hours to completely drain the full pool.

How long would it take to fill the pool if both hoses are used and the drain is left open?

4-3
Linear and Absolute Value Inequalities

In this section we shall consider inequalities of the form $ax + b > 0$, $a \neq 0$, and similar inequalities where the inequality symbol may be either $\geq, <,$ or $\leq$. Any inequality of this form is called a *linear inequality*. The *solution set* for such an inequality is the set of all values of the variable which make the inequality true. Two inequalities are said to be *equivalent* if they have the same solution set.

We shall determine the solution set of a given linear inequality by finding an inequality whose solution set is easily determined and which is equivalent to the original inequality. The following theorem will be useful.

THEOREM 4-4   Let $R(x)$, $S(x)$, and $T(x)$ be algebraic expressions in the variable x. Then $R(x) > S(x)$ is equivalent to

> 1. $R(x) + T(x) > S(x) + T(x)$ for any $T(x)$;
> 2. (a) $R(x)T(x) > S(x)T(x)$ for $T(x) > 0$,
>   (b) $R(x)T(x) < S(x)T(x)$ for $T(x) < 0$.

Note: Similar theorems can be stated using the symbols $\geq, <,$ or $\leq$.

The solution procedure is illustrated in the following example.

EXAMPLE 1   Determine the solution set for

$$x - 5 > 2x + 7.$$

SOLUTION   Adding $5 - 2x$ to both sides yields the equivalent inequality

$$-x > 12.$$

Then multiplying both sides by -1, we have

$$x < -12.$$

Thus the solution set is

$$\{x \mid x < -12\}.$$

In working with inequalities, one frequently encounters what are called compound inequalities. A *compound inequality* is a compound statement which consists of two inequalities connected by the word *and*, or by the word *or*.

When two inequalities in a variable x are connected by the word *and* to form a compound inequality, the solution set of the compound inequality is the set of all values of x which satisfy *both* of the inequalities involved in the compound statement. This is illustrated in the following example.

EXAMPLE 2 Determine the solution set for

$$3x - 2 > 4 - 5x \quad \text{and} \quad 9 - 3x \geq 4x + 2.$$

SOLUTION The solution set is determined by

$$\{x \mid 3x - 2 > 4 - 5x \quad \text{and} \quad 9 - 3x \geq 4x + 2\},$$

which is the same as

$$\{x \mid 3x - 2 > 4 - 5x\} \cap \{x \mid 9 - 3x \geq 4x + 2\}.$$

Thus we must determine the solution set for each inequality separately, and then take the *intersection* of these two sets. We find that

$$\{x \mid 3x - 2 > 4 - 5x\} = \{x \mid x > \tfrac{3}{4}\}$$

and

$$\{x \mid 9 - 3x \geq 4x + 2\} = \{x \mid x \leq 1\}.$$

Thus our solution set is

$$\{x \mid x > \tfrac{3}{4}\} \cap \{x \mid x \leq 1\} = \{x \mid \tfrac{3}{4} < x \leq 1\}.$$

This can be represented on the number line as shown in Figure 4.1.

$$3/4 < x \leq 1$$

FIGURE 4.1

When two inequalities in a variable x are connected by the word *or* to form a compound inequality, the solution set of the compound inequality is the set of all values of x which satisfy either one or the other of the two inequalities involved in the compound statement. This is illustrated in Example 3.

EXAMPLE 3 Determine the solution set for

$$2x - 7 \leq 17 - x \quad \text{or} \quad x > 4x - 6.$$

SOLUTION The solution set is determined by

$$\{x \mid 2x - 7 \leq 17 - x \quad \text{or} \quad x > 4x - 6\}$$

$$= \{x \mid 2x - 7 \leq 17 - x\} \cup \{x \mid x > 4x - 6\}.$$

Again we determine the solution set for each inequality separately, and then take the union of these two sets. This procedure gives the solution set as

$$\{x \mid x \le 8\} \cup \{x \mid x < 2\} = \{x \mid x \le 8\}.$$

This can be represented on the number line as shown in Figure 4.2.

$$x \le 8$$

FIGURE 4.2

We next consider a particular type of inequality involving absolute values. The type we are concerned with is that of the form

$$|R(x)| < d,$$

where $R(x)$ is a linear expression in x and d is a constant.

If d is not positive, the solution set for $|R(x)| < d$ is empty because $|R(x)|$ can never be negative. If d is positive, we recall from Section 1-5 that

$$|R(x)| < d$$

is equivalent to the simultaneous inequalities

$$-d < R(x) < d.$$

Similarly, if $d > 0$, then

$$|R(x)| > d$$

is equivalent to the compound statement

$$R(x) > d \quad \text{or} \quad R(x) < -d.$$

Combining this discussion with Corollary 4-3, we have

> $|R(x)| = d$ means either $R(x) = d$ or $R(x) = -d,$
>
> $|R(x)| < d$ means $-d < R(x) < d,$
>
> $|R(x)| > d$ means either $R(x) > d$ or $R(x) < -d.$

This is represented graphically in Figure 4.3.

FIGURE 4.3

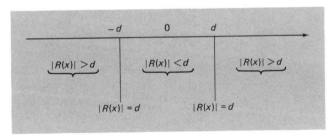

EXAMPLE 4   Solve

$$|3x - 1| < 4.$$

SOLUTION   We must solve the equivalent inequality

$$-4 < 3x - 1 < 4.$$

Adding 1 to all members, we have

$$-3 < 3x < 5.$$

Dividing by 3 then gives

$$-1 < x < \tfrac{5}{3}.$$

The solution set is represented in Figure 4.4.

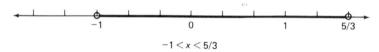

$$-1 < x < 5/3$$

FIGURE 4.4

EXAMPLE 5   Solve

$$|x - 9| > 4.$$

SOLUTION   The solution set is determined by

$$\{x \,|\, x - 9 > 4 \quad \text{or} \quad x - 9 < -4\}.$$

Solving these two inequalities separately, we have

$$\{x \,|\, x > 13 \quad \text{or} \quad x < 5\}$$

as the solution set. The solution set is pictured in Figure 4.5.

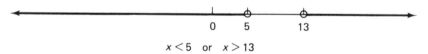

$$x < 5 \quad \text{or} \quad x > 13$$

FIGURE 4.5

EXAMPLE 6   Solve

$$|x - 5| < -1.$$

SOLUTION   Since the absolute value of a real number is either zero or a positive number, $|x - 5| < -1$ is impossible for any x, and the solution set is $\varnothing$.

Determine the solution sets for the following inequalities.

1. $2x < 8$

2. $3x - 5 \geq 1$

3. $25 - 5x < 0$

4. $10 \geq 3y + 4$

5. $2y > \frac{8}{3}y - 4$

6. $-4x \leq -16$

7. $17 - 4x > 2x + 5$

8. $-6x + 3 \geq -5x - 2$

9. $-3x + 2\sqrt{2} < x - 2\sqrt{2}$

10. $3\sqrt{2} - 2x \geq 7\sqrt{2} - 6x$

11. $\frac{1}{2}x \leq \frac{3}{4}x + 9$

12. $\frac{5}{6}x + \frac{1}{3} \geq \frac{2}{3}x$

13. $17 + 5(x + 2) \leq x - 3(x - 2)$

14. $3(x + 2) - 4x > 1 - 2(3x - 2)$

15. $\frac{x}{3} + \frac{x}{2} - \frac{x}{4} \geq \frac{x}{5}$

16. $\frac{1}{2}(x + 2) > \frac{x}{3} - 2$

17. $\frac{7 - 2x}{3} \geq 11$

18. $\frac{3 - x}{4} < \frac{2}{3}$

19. $\frac{2 - x}{3} > \frac{1 - 2x}{5}$

20. $\frac{x - 2}{3} \geq \frac{7x + 1}{2}$

Solve the following compound inequalities.

21. $x - 1 \geq 0$ and $3 \geq x - 1$

22. $x - 1 < 0$ and $3 > x - 1$

23. $7 - 2x \leq 5$ and $1 - 3x \geq 2(2 - x)$

24. $1 - 4x \geq 5$ and $2x - 5 < 3x$

25. $1 - 2x < 3$ and $2(x - 2) < x - 1$

26. $2x + 5 > 1$ and $7x + 6 > 3(x + 2)$

27. $2x + 5 > 1$ or $7x + 6 > 3(x + 2)$

28. $-3x + 2 \geq -1$ or $4x - 5 > 3(x - 1)$

29. $7x - 8 \geq -43$ or $-x > 8 + x$

30. $7 - 2x < 5$ or $1 + 3x < 2x + 3$

31. $\frac{x + 3}{2} < \frac{1 - 2x}{4}$ or $9(x - 4) \leq 3(4 - 7x)$

32. $2x - 1 > x - 1$ or $\frac{4 - 5x}{2} > \frac{6 - x}{3}$

Solve the inequalities below, and represent the solution set on the number line (see Figures 4.1–4.5).

33. $|2x - 5| < 3$

34. $|7 - x| \leq 2$

35. $|4 - 5x| < 0$

36. $|7x - 2| < -3$

37. $|3x - 6| \leq 0$

38. $|2x + 7| \leq 0$

39. $|10x - 3| < 12$

40. $|2 - x| \leq 5$

41. $|2x + 5| > -5$

42. $|5 - x| \geq 0$

43. $|1 - x| > 0$

44. $|3x + 1| > 0$

45. $|3 - 4x| > 2$

46. $|x + 4| \geq 4$

47. $|2x - 4| \geq 3$

48. $|1 - 3x| > 2$

4-4
*Ratio, Proportion,
and Variation
(Optional)*

When we speak of the *ratio* of one quantity a to a second quantity b, we mean the quotient a/b. We write $a:b$ or a/b to indicate the ratio of a to b. In working with ratios, care must be taken in choosing the units of measure to be associated with each of the quantities. If the two measurements are of the same type of quantities, one should use the same units of measure. For example, the ratio of 1 yard to 18 inches could be expressed as:

$$\frac{1}{\frac{1}{2}}, \quad \text{in terms of yards};$$

or

$$\frac{3}{\frac{3}{2}}, \quad \text{in terms of feet};$$

or

$$\frac{36}{18}, \quad \text{in terms of inches}.$$

Notice that each of these ratios is equal to the quotient 2/1, or 2.

If the quantities are not of the same type, then the units of measure must be different. For example, if we consider the ratio of pressure to surface area, our units might be pounds and square inches. That is, the ratio of 3 pounds to 400 square inches would be

$$\frac{3 \text{ lb}}{400 \text{ sq in.}}.$$

This ratio would be read as "3/400 pounds per square inch."

Sometimes it is necessary to express a ratio in units which are different from the ones given. Consider the following example.

EXAMPLE 1   Express 60 miles per hour in terms of feet per second.

SOLUTION   We recall that 5280 feet = 1 mile and 3600 seconds = 1 hour. Now 60 miles per hour can be expressed as the fraction

$$\frac{60 \text{ miles}}{1 \text{ hour}}.$$

Since multiplying numerator and denominator of any fraction by equal quantities does not change the value of the fraction, we can multiply this fraction by $\dfrac{1 \text{ hour}}{3600 \text{ seconds}}$ and $\dfrac{5280 \text{ feet}}{1 \text{ mile}}$ to obtain

$$\frac{60 \text{ miles}}{1 \text{ hour}} = \frac{60 \text{ miles}}{1 \text{ hour}} \cdot \frac{1 \text{ hour}}{3600 \text{ seconds}} \cdot \frac{5280 \text{ feet}}{1 \text{ mile}}$$

$$= \frac{(60)(1)(5280) \text{ feet}}{(1)(3600)(1) \text{ seconds}}$$

$$= \frac{88 \text{ feet}}{1 \text{ second}}.$$

Therefore, 60 miles per hour is equal to 88 feet per second. Notice that we chose the multipliers in the form $\dfrac{1 \text{ hour}}{3600 \text{ seconds}}$ and $\dfrac{5280 \text{ feet}}{1 \text{ mile}}$ $\left(\text{instead}\right.$ of $\dfrac{3600 \text{ seconds}}{1 \text{ hour}}$ and $\left.\dfrac{5280 \text{ feet}}{1 \text{ mile}}\right)$ so that in our product the units "hours" would occur in numerator and denominator, and the units "miles" would occur in numerator and denominator.

An equality of two ratios is called a *proportion*. If we have the ratio of *a* to *b* equal to the ratio of *c* to *d*, we write

$$a : b = c : d,$$

or

$$\frac{a}{b} = \frac{c}{d}.$$

This proportion is sometimes read as "*a* is to *b* as *c* is to *d*." In dealing with proportions, we must use one of the following two rules concerning the units of measure associated with the numerator and denominator of the quotients in the proportion.

1. Both numerators must have the same units, and both denominators must have the same units; or

2. the numerator and denominator of the quotient on the left side of the equality sign must have the same units, and the numerator and denominator of the quotient on the right side of the equality sign must have the same units.

EXAMPLE 2 If a car moves at a uniform rate of speed and travels 400 kilometers in 8 hours, how far does it travel in 6 hours?

SOLUTION Let x represent the number of kilometers traveled in 6 hours. Then a proportion can be set up using the first rule involving units:

$$\frac{400 \text{ km}}{8 \text{ hours}} = \frac{x \text{ km}}{6 \text{ hours}}.$$

This gives a linear equation in one variable to solve:

$$\frac{400}{8} = \frac{x}{6},$$

and we find

$$x = 300 \text{ kilometers}.$$

This same example can be set up using the second rule concerning proportions:

$$\frac{x \text{ km}}{400 \text{ km}} = \frac{6 \text{ hr}}{8 \text{ hr}}.$$

In this case, the linear equation is

$$\frac{x}{400} = \frac{6}{8},$$

and the solution is again $x = 300$ kilometers.

Many of the formulas that occur in mathematics and the other sciences can be described by using the terminology of *variation*. Each variation statement can be translated into a simple mathematical equation.

Direct Variation. The statement "y *varies directly* as x" (or "y is *directly proportional* to x") means that $y = kx$, for some constant k. In direct variation, the word "directly" is often omitted, and the variation statement reads as "y varies as x" or "y is proportional to x."

Inverse Variation. The statement "y *varies inversely* as x" (or "y is inversely proportional to x") means that $y = k/x$, for some constant k.

Joint Variation. The statement "y *varies jointly* as x and z" (or "y is *jointly proportional* to x and z") means that $y = kxz$, for some constant k.

Combined Variation. The statement "y varies directly as x and inversely as z" (or "y is directly proportional to x and inversely proportional to z") means that $y = kx/z$, for some constant k.

In all of the statements of variation, the constant k is called the *constant of variation*, or the *proportionality constant*.

Most problems involving variation can be solved by following the steps listed below:

1. Express the variation statement as an equation with a constant of variation k.

2. Use a known set of values for the variables to find the value of the constant of variation k.

3. Use the equation with the value of the constant k and another set of known values for all but one of the variables to find a corre-ponding value of the unknown variable.

This procedure is illustrated in the following examples.

EXAMPLE 3 The circumference C of a circle varies directly as the radius r. If the circumference is 4π meters whenever the radius is 2 meters, find the circum-ference whenever the radius is 12 meters.

SOLUTION Following the procedure outlined above, we have the steps below.

1. The equation describing the variation is
$$C = kr, \qquad \text{for some constant } k.$$

2. To find the value of the constant k, we use the known set of values of C and r in our variation equation:
$$4\pi = k(2),$$
or
$$k = 2\pi.$$
Thus the equation describing the variation is
$$C = 2\pi r.$$

3. To determine C when $r = 12$ meters, we use $C = 2\pi r$ and find
$$C = 2\pi(12) = 24\pi \text{ meters.}$$

EXAMPLE 4 The height h of a right circular cylinder varies inversely as the square of the radius r. If the height is 9 meters when the radius is 1 meter, find the height of a right circular cylinder whose radius is 2 meters.

SOLUTION 1. The equation describing the variation is
$$h = \frac{k}{r^2}.$$

2. Since $h = 9$ when $r = 1$, we have
$$9 = \frac{k}{(1)^2}$$

and $k = 9$. Thus the variation equation is

$$h = \frac{9}{r^2}. \tag{1}$$

3. To find the value of h when $r = 2$, we substitute $r = 2$ into equation (1) and obtain

$$h = \tfrac{9}{4}.$$

Thus $h = \tfrac{9}{4}$ meters when $r = 2$ meters.

EXAMPLE 5 Suppose that z varies jointly as x and t^2, and inversely as $3w - 1$. If $z = 4$ when $x = -2$, $t = 1$, and $w = 5$, find z when $x = -3$, $t = 4$, and $w = -1$.

SOLUTION

1. The variation equation is

$$z = \frac{kxt^2}{3w - 1}.$$

2. Since $z = 4$ when $x = -2$, $t = 1$, and $w = 5$, we have

$$4 = \frac{k(-2)(1)^2}{3(5) - 1},$$

or

$$4 = \frac{-2k}{14}.$$

This yields $k = -28$, and the variation equation is

$$z = \frac{-28xt^2}{3w - 1}. \tag{2}$$

3. To determine the value of z when $x = -3$, $t = 4$, $w = -1$, we substitute these values into equation (2) and obtain

$$z = \frac{-28(-3)(4)^2}{3(-1) - 1} = -336.$$

EXERCISES 4-4

Express each of the following ratios as a quotient.

1. 3 pounds to 4 pounds
2. 4 feet to 2 inches
3. 12 meters to 150 centimeters
4. $20 to 2 tickets
5. 90 boys to 30 girls
6. 40 miles to $\tfrac{1}{3}$ hour

Perform the following conversions.

7. 44 feet per second to miles per hour
8. 40 pounds per second to tons per hour

9. 48 ounces per square inch to pounds per square foot

10. 55 meters per second to kilometers per hour

11. 20 miles per gallon to miles per quart

12. 90 miles per hour to yards per minute

Solve the following problems by using proportions.

13. If a car can travel 60 miles on 2.5 gallons of gas, how far can it travel on 9 gallons of gas?

14. If a punch recipe calls for 5 parts water for each part syrup, how much syrup must be used to make 10 gallons of punch?

15. If gas must be mixed with oil at a ratio of 50 : 1, how many pints of oil must be used with 6 gallons of gas?

16. Suppose that the ratio of gas to oil in problem 15 is 40 : 1. How many ounces of oil must be used with 1 gallon of gas? (16 ounces = 1 pint.)

17. If two boys divide $40 in a ratio of 3 to 5, how much money will each receive?

18. If a mixture contains 40% alcohol, how many ounces of alcohol are in 6 ounces of the mixture?

Solve the following variation problems.

19. Given that y is directly proportional to z and inversely proportional to w, find y when $z = 1$ and $w = 4$, if $y = \frac{1}{3}$ when $w = 18$ and $z = \frac{1}{6}$.

20. Suppose that w varies directly as the product of x and y, and inversely as the square of z. If $w = 9$ when $x = 6$, $y = 27$, and $z = 3$, find the value of w when $x = 4$, $y = 7$, and $z = 2$.

21. Assume that the square of t varies inversely as the cube of z, and that $t = 4$ when $z = \frac{1}{2}$. Find t when $z = 2$.

22. Suppose that x varies directly as the square root of $y - 1$, and that $x = -8$ when $y = 5$. Find x when $y = 50$.

23. If x varies jointly as y and t^3, and inversely as $4z - 3$, find z when $x = -2$, $y = 7$, and $t = -1$, if $z = 1$ when $x = 2$, $y = -4$, and $t = -2$.

24. Suppose that z varies directly as w, and inversely as $x - 1$ and the square of t. Find z when $x = -1$, $w = -1$, and $t = -1$, if $z = -4$ when $w = 7$, $x = 2$, and $t = -3$.

25. The area of an equilateral triangle varies as the square of its base. If the area is $\sqrt{3}$ square meters when the length of the base is 2 meters, find the area of an equilateral triangle whose base has length 5 meters.

26. The distance traveled by a free-falling body varies directly as the square of the time that it falls. If such a body falls 144 feet in 3 seconds, how far will it fall in 10 seconds?

27. Suppose that the strength S of a beam with constant length and rectangular cross section varies directly as the width x and the square of the breadth y. If a beam 2 inches wide and 4 inches in breadth has strength

40 pounds, how wide must a beam be cut to have the same strength and be 8 inches in breadth?

28. The altitude h of a cone varies directly as its volume and inversely as the square of the radius of its base. If a cone with altitude 32 centimeters and radius of the base 4 centimeters has volume $512\pi/3$ cubic centimeters, find the altitude of a cone with volume 6π cubic centimeters and radius of the base 3 centimeters.

29. The stiffness S of a rectangular beam varies directly as the width x and the cube of the depth d. If $S = 90$ when $x = 2$ and $d = 3$, find the stiffness S for a beam with width 3 and depth 2.

30. The slant height S of a right circular cone varies directly as its lateral surface area and inversely as the radius of its base. If the lateral surface area is 400π square inches when the radius of the base is 20 inches and the slant height is 20 inches, find the slant height of a right circular cone with lateral surface area 12 square inches and radius of the base 2 inches.

31. The distance required to stop a car varies directly as the square of its speed. If the car can stop in 49 feet from a speed of 35 mi/hr, how many feet are required to stop from a speed of 70 mi/hr?

32. The time required to finish a job varies inversely as the number of persons working. If 5 workers take 12 days to finish a job, how many workers are needed to finish the same job in 10 days?

33. Boyle's law states that the pressure P of a gas varies directly as the absolute temperature T and inversely as the volume V. If the pressure is 350 pounds per square inch when the temperature is 70°A and the volume is 2 cubic feet, what is P when the temperature is 35°A and the volume is 4 cubic feet?

*34. If y varies directly as x and y is not constant, show that x varies directly as y.

*35. If y varies inversely as x and y is not constant, show that x varies inversely as y.

4-5
Quadratic
Equations

Equations which can be written in the form $ax^2 + bx + c = 0$, where $a \neq 0$, are called *quadratic equations*. There are at most two distinct values of x which satisfy a quadratic equation. These two values are called the *solutions*, or *roots*, of the equation. In this section, we shall study two methods of solving a quadratic equation which has real coefficients: (1) factoring and (2) completing the square.

The method of factoring relies on Theorems 1-1 and 1-2: *the product of two factors is zero if and only if at least one of the factors is zero.* That is, $xy = 0$ if and only if either $x = 0$ or $y = 0$.

EXAMPLE 1   Solve

$$x^2 + x - 20 = 0.$$

SOLUTION Factoring the quadratic expression on the left, we have

$$(x + 5)(x - 4) = 0.$$

The method described above leads to the problem of solving the two linear equations

$$x + 5 = 0 \quad \text{or} \quad x - 4 = 0.$$

Thus $x = -5$ or $x = 4$ are the two solutions to the quadratic equation $x^2 + x - 20 = 0$, and the solution set is $\{-5, 4\}$.

EXAMPLE 2 Solve

$$x^2 = 9.$$

SOLUTION The given equation is equivalent to

$$x^2 - 9 = 0.$$

Factoring the quadratic expression on the left, we have

$$(x - 3)(x + 3) = 0,$$

and the solutions to the two linear equations are $x = 3$ or $x = -3$.

Example 2 easily generalizes to any equation of the form $x^2 = a$, where a is nonnegative.

THEOREM 4-5 If $a \geq 0$, the solutions to $x^2 = a$ are $x = \sqrt{a}$ and $x = -\sqrt{a}$.

To indicate the solutions of $x^2 = a$, we write $x = \pm\sqrt{a}$ and say that the solutions are obtained by "equating square roots of the sides." Of course, if $a < 0$, the equation $x^2 = a$ has no solutions in the set of real numbers.

EXAMPLE 3 Solve

$$(2x + 3)^2 = 4.$$

SOLUTION Equating square roots as described above, we have

$$2x + 3 = \pm\sqrt{4},$$

or

$$2x + 3 = \pm 2.$$

This is equivalent to

$$2x = -3 \pm 2.$$

Multiplying both sides by $\frac{1}{2}$ gives

$$x = -\tfrac{3}{2} \pm 1.$$

Thus the solution set is $\{-\tfrac{1}{2}, -\tfrac{5}{2}\}$.

The method of factoring is the simplest and most direct approach used to solve a quadratic equation if the quadratic expression can be factored easily using the methods of Section 2-3. But such a factoring is not always possible. For example, suppose that we wish to solve the quadratic equation.

$$2x^2 + 4x - 1 = 0.$$

If we attempt to factor the quadratic expression, we discover that it is impossible to find two linear factors with rational coefficients whose product is $2x^2 + 4x - 1$. However, there is another method which can be used to solve the equation. It is the method of *completing the square*. This is illustrated in the following six steps.

Step 1   Isolate the constant term on one side of the equation.

$$2x^2 + 4x = 1.$$

Step 2   Divide both sides of the equation by the leading coefficient, that is, the coefficient of x^2.

$$x^2 + 2x = \tfrac{1}{2}.$$

We want the left side of the equation to become a perfect square trinomial $x^2 + 2kx + k^2$. In other words, we wish to add some constant to both sides of the equation so that the left side factors as $(x + k)^2$, where k is a constant. This is described in the next step.

Step 3   Compute $\tfrac{1}{2}$ times the coefficient of x, square the result, and add this to both sides of the equation.

$$[\tfrac{1}{2}(2)]^2 = (1)^2 = 1,$$
$$x^2 + 2x + 1 = \tfrac{1}{2} + 1.$$

Step 4   If the constant side of the equation is negative, there is no solution in the real numbers. If it is nonnegative, write each side of the equation as a perfect square.

$$(x + 1)^2 = \left(\sqrt{\tfrac{3}{2}}\right)^2,$$

or

$$(x + 1)^2 = \left(\frac{\sqrt{6}}{2}\right)^2.$$

Step 5   Equate the square roots of the sides.

$$x + 1 = \pm\frac{\sqrt{6}}{2}.$$

Step 6   Solve the resulting linear equations.

$x + 1 = \frac{\sqrt{6}}{2}$ has the solution $x = -1 + \frac{\sqrt{6}}{2}$,

and

$x + 1 = -\frac{\sqrt{6}}{2}$ has the solution $x = -1 - \frac{\sqrt{6}}{2}$.

Thus the two solutions to $2x^2 + 4x - 1 = 0$ are $x = -1 + \sqrt{6}/2$ and $x = -1 - \sqrt{6}/2$.

The method of completing the square, as described above, will either yield the real numbers which are solutions of a quadratic equation, or it will show that there are no solutions in the set of all real numbers.

EXAMPLE 4   Solve

$$3x^2 - 17x + 10 = 0$$

by completing the square.

SOLUTION   We follow the six steps described in the discussion above.

Step 1 $3x^2 - 17x = -10.$

Step 2 $x^2 - \frac{17}{3}x = -\frac{10}{3}.$

Step 3 $x^2 - \frac{17}{3}x + (\frac{17}{6})^2 = -\frac{10}{3} + (\frac{17}{6})^2,$

or

$x^2 - \frac{17}{3}x + (\frac{17}{6})^2 = \frac{169}{36}.$

Step 4 $(x - \frac{17}{6})^2 = (\frac{13}{6})^2.$

Step 5 $x - \frac{17}{6} = \pm\frac{13}{6}.$

Step 6

$x - \frac{17}{6} = \frac{13}{6}$ yields $x = \frac{13}{6} + \frac{17}{6} = 5,$

and

$x - \frac{17}{6} = -\frac{13}{6}$ yields $x = -\frac{13}{6} + \frac{17}{6} = \frac{2}{3}.$

The solution set is $\{5, \frac{2}{3}\}$.

Some equations can be rewritten so as to become quadratic in form by making an appropriate change of variable. This is illustrated in Example 5.

EXAMPLE 5 Solve

$$4z^4 = 13z^2 - 9.$$

SOLUTION With an appropriate change of variables, this equation can be expressed as a quadratic equation. If we let $x = z^2$, then $z^4 = x^2$, and the equation can be written as

$$4x^2 = 13x - 9,$$

or

$$4x^2 - 13x + 9 = 0.$$

This factors as

$$(4x - 9)(x - 1) = 0.$$

Thus $x = \frac{9}{4}$, $x = 1$, are the two solutions of the quadratic equation in x. But we must solve for the variable z. When $x = \frac{9}{4}$, we have $z^2 = \frac{9}{4}$, and $z = \pm\frac{3}{2}$. When $x = 1$, then $z^2 = 1$ and $z = \pm 1$. Thus the solution set for $4z^4 = 13z^2 - 9$ is $\{\frac{3}{2}, -\frac{3}{2}, 1, -1\}$.

EXERCISES 4-5

Solve by factoring.

1. $x^2 + 3x - 10 = 0$
3. $x^2 + 6x + 9 = 0$
5. $4x^2 - 25 = 0$
7. $x^2 + 3x = 0$
9. $49x^3 + 7x^2 = 2x$
11. $x^3 - 4x^2 = 0$

2. $2x^2 - 5x + 2 = 0$
4. $16x^2 = 40x - 25$
6. $x^2 - 36 = 0$
8. $x^3 + 2x^2 = 0$
10. $2x^3 + 5x^2 - 3x = 0$
12. $x^3 - 4x = 0$

Solve by completing the square.

13. $x^2 + x - 1 = 0$
15. $2x^2 + 5x + 1 = 0$
17. $2x^2 + 8x + 6 = 0$
19. $6x^2 + 13x = -6$

14. $x^2 + 2x - 3 = 0$
16. $4x^2 - 10x + 3 = 0$
18. $x^2 + 2x - 15 = 0$
20. $3x^2 = x + 14$

Solve by either factoring or completing the square.

21. $7 - 15x + 2x^2 = 0$
23. $-27y^2 + 3y + 2 = 0$

22. $2y^2 = 6y + 1$
24. $15x^2 - x - 28 = 0$

25. $x^2 + 2x - 1 = 0$

26. $y^2 + 5y + 5 = 0$

27. $x^2 + ax + b = 0$

28. $a^2x^2 - b^2 = 0$

29. $6x^2 + 7x + 2 = 0$

30. $3z^2 = 11z + 4$

Solve by first expressing the equation in quadratic form.

31. $y^4 - 4y^2 + 4 = 0$

32. $y^4 - 2y^2 + 1 = 0$

33. $27z^6 + 215z^3 - 8 = 0$

34. $z^6 + 16z^3 + 64 = 0$

35. $(y + 2)^2 - 5(y + 2) = 14$

36. $(w - 4)^2 + 2(w - 4) = 63$

37. $\left(\dfrac{1}{x}\right)^2 + \dfrac{6}{x} + 5 = 0$

38. $2p^{-2} + 7p^{-1} + 3 = 0$

39. $2y^{2/3} + y^{1/3} - 1 = 0$

40. $y^{2/3} - y^{1/3} - 12 = 0$

41. $x = 8\sqrt{x} - 15$

42. $(x - 3)^{1/2} - 5(x - 3)^{1/4} + 6 = 0$

4-6

The Quadratic Formula

Let us consider again the quadratic equation $ax^2 + bx + c = 0$, $a \neq 0$. As we have already seen, this equation can be solved by either factoring the quadratic expression or by completing the square on the variable x. Suppose that we wish to find, in general, the solutions of the quadratic equation $ax^2 + bx + c = 0$, $a \neq 0$. This can be done by following the steps outlined in the preceding section to complete the square on the variable x.

Step 1 $$ax^2 + bx = -c$$

Step 2 $$x^2 + \frac{b}{a}x = -\frac{c}{a}$$

Step 3 $$x^2 + \frac{b}{a}x + \left[\frac{1}{2}\left(\frac{b}{a}\right)\right]^2 = -\frac{c}{a} + \left[\frac{1}{2}\left(\frac{b}{a}\right)\right]^2$$
$$= \frac{b^2 - 4ac}{4a^2}.$$

At this point, we have the equation

$$\left(x + \frac{b}{2a}\right)^2 = \frac{b^2 - 4ac}{4a^2},$$

and this equation is equivalent to the original $ax^2 + bx + c = 0$. If $b^2 - 4ac < 0$, there is no solution in the set of real numbers, since $(x + b/2a)^2$ is a perfect square, and cannot be negative. Under the condition that $b^2 - 4ac \geq 0$, we can proceed with the method of completing the square.[2]

---

[2] If $b^2 - 4ac < 0$, complex numbers can be used to complete the derivation of the quadratic formula. Complex numbers are treated in detail in Chapter 9.

Step 4
$$\left(x + \frac{b}{2a}\right)^2 = \left(\frac{\sqrt{b^2 - 4ac}}{2a}\right)^2$$

Step 5
$$x + \frac{b}{2a} = \pm\frac{\sqrt{b^2 - 4ac}}{2a}$$

Step 6 $\quad x + \dfrac{b}{2a} = \dfrac{\sqrt{b^2 - 4ac}}{2a} \quad$ yields $\quad x = \dfrac{-b + \sqrt{b^2 - 4ac}}{2a}$

and

$$x + \frac{b}{2a} = \frac{-\sqrt{b^2 - 4ac}}{2a} \quad \text{yields} \quad x = \frac{-b - \sqrt{b^2 - 4ac}}{2a}.$$

Thus the solutions to the quadratic equation $ax^2 + bx + c = 0$, $a \neq 0$, are given by

$$x = \frac{-b \pm \sqrt{b^2 - 4ac}}{2a},$$

provided that $b^2 - 4ac \geq 0$. This expression is known as the *quadratic formula*. It can be used to solve any quadratic equation after it has been written in the form $ax^2 + bx + c = 0$. We simply identify the coefficients of x^2, x, and the constant term as a, b, and c, respectively, and substitute these values into the quadratic formula.

EXAMPLE 1 Solve

$$9x^2 + 7x = 1$$

by using the quadratic formula.

SOLUTION We first write the equation in the form $ax^2 + bx + c = 0$:
$$9x^2 + 7x - 1 = 0.$$

Then we note that $a = 9$, $b = 7$, and $c = -1$. Substituting these values into the quadratic formula yields

$$x = \frac{-7 \pm \sqrt{(7)^2 - 4(9)(-1)}}{2(9)}$$

$$= \frac{-7 \pm \sqrt{85}}{18},$$

and the solution set is $\left\{\dfrac{-7 + \sqrt{85}}{18}, \dfrac{-7 - \sqrt{85}}{18}\right\}$.

EXAMPLE 2 Find two consecutive even integers whose product is 168.

SOLUTION   Let x represent the first even integer and $x + 2$ the next even integer. Then we have

$$x(x + 2) = 168.$$

This equation can be written in the general quadratic form $ax^2 + bx + c = 0$ by performing the multiplication on the left and subtracting 168 from both sides:

$$x^2 + 2x - 168 = 0.$$

Using the quadratic formula with $a = 1$, $b = 2$, and $c = -168$, we have

$$x = \frac{-2 \pm \sqrt{(2)^2 - 4(1)(-168)}}{2(1)}$$

$$= \frac{-2 \pm \sqrt{676}}{2}$$

$$= \frac{-2 \pm 26}{2}$$

$$= -1 \pm 13.$$

This means that $x = 12$ and $x = -14$ are the two solutions to $x(x + 2) = 168$. When $x = 12$, $x + 2 = 14$, and when $x = -14$, $x + 2 = -12$. Thus there are two pairs of consecutive even integers whose product is 168.

The expression, $b^2 - 4ac$, which occurs under the radical in the quadratic formula is called the *discriminant*. The value of $b^2 - 4ac$ will either be positive, negative, or zero. We can classify the roots of any quadratic equation by simply evaluating the discriminant as follows:

1. If $b^2 - 4ac = 0$, then the two roots will be equal real numbers.

2. If $b^2 - 4ac < 0$, then the equation has no real roots.

3. If $b^2 - 4ac > 0$, then the two roots will be distinct (unequal) real numbers. If a, b, and c are rational and $b^2 - 4ac$ is a perfect square, then the roots will be rational numbers. If $b^2 - 4ac$ is not a perfect square, the roots will be irrational.

EXAMPLE 3   Consider the equation

$$9x^2 - 6x + 1 = 0.$$

SOLUTION   The value of the discriminant is

$$b^2 - 4ac = (-6)^2 - 4(9)(1) = 0,$$

which indicates that the equation has two equal roots. Substituting $a = 9$, $b = -6$, $c = 1$ into the quadratic formula, we find the roots to be

$$x = \frac{6 \pm \sqrt{0}}{2(9)} = \frac{1}{3}.$$

EXAMPLE 4 Consider the equation

$$2x^2 - x - 15 = 0.$$

SOLUTION The value of the discriminant is

$$b^2 - 4ac = (-1)^2 - 4(2)(-15) = 121 > 0,$$

which is a perfect square. Thus the two roots of the quadratic equation are distinct rational numbers. Using the quadratic formula, we find these roots to be

$$x = \frac{1 \pm \sqrt{121}}{2(2)} = \frac{1 \pm 11}{4}.$$

That is,

$$x = \frac{1 + 11}{4} = 3 \quad \text{and} \quad x = \frac{1 - 11}{4} = -\frac{5}{2}$$

are the roots.

EXAMPLE 5 Consider the equation

$$x^2 + x + 1 = 0.$$

SOLUTION Since the value of the discriminant is

$$b^2 - 4ac = (1)^2 - 4(1)(1) = -3 < 0,$$

there are no real solutions to $x^2 + x + 1 = 0$.

Suppose now that we have two values, r_1 and r_2, and we want to determine a quadratic equation which has these values as its roots. Now r_1 is a solution to the linear equation $x - r_1 = 0$, and r_2 is a solution to the linear equation $x - r_2 = 0$. Furthermore, the product of $x - r_1$ and $x - r_2$ is zero if and only if one of the factors is zero. Thus

$$(x - r_1)(x - r_2) = 0$$

is a quadratic equation with the desired roots. Expanding and simplifying, we have

$$x^2 - (r_1 + r_2)x + r_1 r_2 = 0.$$

Since multiplying both sides of any equation by the same nonzero constant does not change the solution set, then any constant multiple of this equation also has solutions r_1 and r_2. It is important to observe that, whenever the leading coefficient[3] (the coefficient of x^2) is 1, the constant term is the product of the two roots and the coefficient of x is the negative of the sum of the two roots.

EXAMPLE 6 Determine a monic quadratic equation if the sum of its roots is -2, and the product of its roots is -8.

SOLUTION Suppose that r_1 and r_2 are the two roots. Then

$$r_1 + r_2 = -2$$

and

$$r_1 r_2 = -8.$$

Thus a quadratic equation with the desired roots is

$$x^2 + 2x - 8 = 0.$$

EXAMPLE 7 Find a monic quadratic equation whose roots are 0 and -3.

SOLUTION We can write the desired quadratic expression as a product of its linear factors and obtain

$$(x - 0)[x - (-3)] = 0,$$

or

$$x^2 + 3x = 0.$$

EXERCISES 4-6

Solve the quadratic equations below by using the quadratic formula.

1. $x^2 + 3x - 28 = 0$ 2. $x^2 - 16x + 64 = 0$

3. $4x^2 - 8x + 3 = 0$ 4. $6x^2 - 11x - 35 = 0$

5. $x^2 + x = 0$ 6. $4x^2 = -7x$

7. $-x^2 - 4x + 12 = 0$ 8. $-x^2 - 2x + 1 = 0$

9. $16x^2 - 25 = 0$ 10. $2x^2 = 1$

11. $4x^2 = -7x - 2$ 12. $2x^2 + 7x = 4$

13. $x^2 + 5x + 5 = 0$ 14. $x^2 + 3x + 1 = 0$

15. $3x^2 + 8x + 3 = 0$ 16. $-5x^2 + 2x + 1 = 0$

[3]We recall that such a polynomial is called *monic*. The same terminology is used with equations arising from setting a polynomial equal to zero.

17. $6x^2 - 11x + 2 = 0$

18. $4x^2 - 4x - 19 = 0$

19. $28x^2 = 45 - x$

20. $6x^2 = 29x - 28$

Use the discriminant to determine the *type* of roots of each of the following equations. Do not evaluate the roots.

21. $x^2 - 4 = 0$

22. $4x^2 + 7x - 15 = 0$

23. $2x^2 - 5x + 7 = 0$

24. $-x^2 - x - 2 = 0$

25. $3x^2 + 5x - 1 = 0$

26. $2x^2 + 3x - 1 = 0$

27. $49x^2 + 14x + 1 = 0$

28. $-x^2 + 2x - 1 = 0$

Find a monic quadratic equation which has the given numbers as solutions.

29. $1, -2$

30. $\frac{3}{2}, 4$

31. $-5, 2$

32. $\frac{1}{2}, -\frac{1}{2}$

33. $-2, -2$

34. $0, -2$

Find a monic quadratic equation with solutions that have the sum and product given.

	Sum	Product			Sum	Product
35.	2	1		36.	5	6
37.	-3	-4		38.	0	4
39.	2	0		40.	-11	28

41. Find two consecutive integers whose product is 462.

42. The sum of the squares of four consecutive integers is 174. Find the integers.

43. The length of a rectangular field is twice its width, and the area is 1800 square meters. Find the dimensions of the field.

44. Suppose the perimeter of a rectangular field is 480 meters, and the area is 10,800 square meters. Determine the dimensions of the field.

45. A positive number minus its square is -182. Determine the number.

46. If an object is thrown vertically upward with an initial velocity of v_0 feet per second, then the distance s in feet that the object will be above the earth at time t, in seconds, is given by the equation $s = v_0 t - 16t^2$. (a) Find t if $v_0 = 128$ feet per second and $s = 192$ ft. (b) Find t if $v_0 = 128$ feet per second and $s = 0$ feet.

47. A farmer is plowing a rectangular field which is 100 meters wide and 200 meters long. How wide a strip must she plow around the field so that 52% of the field is plowed?

48. An open box is to be made from a rectangular piece of tin which is 12 cm wide and 14 cm long by cutting equal squares from the four corners and turning up the sides. How large a square must be cut from each corner if the area of the base is to be 60 square centimeters?

49. A car traveled 660 miles at a uniform rate. If the rate had been 5 mi/hr more, the trip would have taken 1 hour less. Find the rate of the car.

50. A plane traveled 800 miles at a uniform rate of speed. If the rate had been 40 mi/hr more, the trip would have taken 1 hour less. Find the rate of the plane.

4-7
Equations
Involving
Radicals

In this section, we shall consider equations in one variable which contain a radical in one or both members. The standard method for solving an equation of this type is to raise both sides of the equation to a power which will eliminate one or more of the radicals, simplify the resulting equation, and solve for the variable by repeating the process, if necessary.

EXAMPLE 1 To solve the equation

$$\sqrt[3]{x + 1} = 2,$$

we cube both sides of the equation and obtain

$$x + 1 = 8.$$

Solving for x then yields $x = 7$.

In some instances, this procedure leads to solutions which do not satisfy the original equation. Such solutions are called *extraneous solutions*, or *extraneous roots*. To determine whether or not a solution is extraneous, one must check the solution in the original equation (or any equivalent equation, before both sides are raised to a power). Consider, for example, the equation $x = -3$. Squaring both sides yields $x^2 = 9$, whose solutions are $x = 3$ and $x = -3$. But $x = 3$ does not satisfy the original equation. That is, $x = 3$ is an extraneous root.

EXAMPLE 2 Solve

$$\sqrt{8x + 1} - 4 = 1 - 2x.$$

SOLUTION Squaring both sides will not eliminate the radical. To eliminate the radical, we must first isolate it on one side of the equation. Adding 4 to both sides yields

$$\sqrt{8x + 1} = 5 - 2x.$$

Squaring both sides gives

$$8x + 1 = 25 - 20x + 4x^2,$$

or

$$0 = 24 - 28x + 4x^2.$$

Dividing both sides by 4, we have

$$0 = 6 - 7x + x^2,$$

and factoring yields

$$0 = (1 - x)(6 - x).$$

The solutions here are $x = 1$ and $x = 6$. Since one or both solutions might be extraneous, they need to be checked in the original equation. We shall use LHS and RHS to indicate the left-hand side and right-hand side, respectively, of the original equation.

Check: $x = 1$

$$\text{LHS} = \sqrt{8(1) + 1} - 4 = \sqrt{9} - 4 = 3 - 4 = -1$$
$$\text{RHS} = 1 - 2(1) = 1 - 2 = -1$$

Since RHS = LHS whenever $x = 1$, the value $x = 1$ is a solution to the original equation.

Check: $x = 6$

$$\text{LHS} = \sqrt{8(6) + 1} - 4 = \sqrt{49} - 4 = 7 - 4 = 3$$
$$\text{RHS} = 1 - 2(6) = 1 - 12 = -11$$

Since LHS $\neq$ RHS whenever $x = 6$, then $x = 6$ is an extraneous solution, and the solution set to the original equation is $\{1\}$.

EXAMPLE 3   Solve

$$\sqrt[3]{7x + 1} = x + 1.$$

SOLUTION   Cubing both sides eliminates the radical:

$$7x + 1 = (x + 1)^3.$$

Expanding gives

$$7x + 1 = x^3 + 3x^2 + 3x + 1,$$

and this simplifies to

$$0 = x^3 + 3x^2 - 4x.$$

This cubic equation can be solved by factoring. We have

$$0 = x(x^2 + 3x - 4),$$

and

$$0 = x(x + 4)(x - 1).$$

Thus the solutions to the cubic equation are $x = 0, -4, 1$. Since one or more of these roots might be extraneous, we must check each in the original equation.

Check: $x = 0$

$$\text{LHS} = \sqrt[3]{7(0) + 1} = \sqrt[3]{1} = 1$$
$$\text{RHS} = 0 + 1 = 1$$

Since LHS = RHS when $x = 0$, then $x = 0$ is a solution to the original equation.

Check: $x = -4$

$$\text{LHS} = \sqrt[3]{7(-4)+1} = \sqrt[3]{-27} = -3$$
$$\text{RHS} = -4+1 = -3$$

So $x = -4$ is also a solution to the original equation.

Check: $x = 1$

$$\text{LHS} = \sqrt[3]{7(1)+1} = \sqrt[3]{8} = 2$$
$$\text{RHS} = 1+1 = 2$$

The root $x = 1$ also checks, and the solution set to the original equation is $\{0, -4, 1\}$.

In some equations involving radicals, it is necessary to apply the procedure of raising both sides of the equation to a power more than once, in order to eliminate all of the radicals. In such cases, it is important that the checking procedure be performed in the original equation (or any equivalent equation, before both sides are raised to a power). Consider the following example.

EXAMPLE 4 Solve

$$\sqrt{3x+1} = \sqrt{x+4}+1.$$

SOLUTION Although squaring both sides will not eliminate both radicals, one of them will be eliminated:

$$3x+1 = x+4+2\sqrt{x+4}+1.$$

Next, we isolate the radical on one side and obtain

$$2x-4 = 2\sqrt{x+4}.$$

Dividing both sides by 2 gives

$$x-2 = \sqrt{x+4}.$$

Squaring both sides leads to the quadratic equation

$$x^2-4x+4 = x+4,$$

which simplifies to

$$x^2-5x = 0,$$

or

$$x(x-5) = 0.$$

The solutions to the quadratic equation are $x = 0$ and $x = 5$. These must be checked in the original equation.

Check: $x = 0$

$$\text{LHS} = \sqrt{3(0)+1} = \sqrt{1} = 1$$
$$\text{RHS} = \sqrt{0+4}+1 = 2+1 = 3.$$

Since LHS $\neq$ RHS when $x = 0$, then $x = 0$ is an extraneous solution.

Check: $x = 5$

$$\text{LHS} = \sqrt{3(5) + 1} = \sqrt{16} = 4$$
$$\text{RHS} = \sqrt{5 + 4} + 1 = 3 + 1 = 4.$$

Therefore, $x = 5$ is a solution, and the solution set for

$$\sqrt{3x + 1} = \sqrt{x + 4} + 1$$

is $\{5\}$.

The procedure illustrated in the preceding examples is summarized in the following steps.

Step 1   Isolate a radical on one side.

Step 2   Raise both sides to the smallest power so that the isolated radical will be removed.

Step 3   Simplify the equation obtained in step 2. If a radical remains, repeat the procedure, starting with step 1. If no radicals remain, proceed to step 4.

Step 4   Solve the equation obtained in step 3.

Step 5   Check the solutions in the original equation.

EXERCISES 4-7

Determine the solution set for each of the following equations.

1. $\sqrt{x + 10} = x - 2$ 　　　2. $4 + \sqrt{x + 2} = x$

3. $\sqrt{x^2} = -4$ 　　　4. $\sqrt{x + 12} = x$

5. $x = \sqrt{8 - 2x}$ 　　　6. $\sqrt{4x + 1} = x - 5$

7. $\sqrt{10 + x} - x = 10$ 　　　8. $\sqrt{x + 8} = 2x + 1$

9. $\sqrt{x + 7} = \sqrt{2x + 1}$ 　　　10. $x - \sqrt{4x - 3} = 2$

11. $\sqrt[3]{x - 1} = 3$ 　　　12. $\sqrt[3]{5 - 20x + 2x^2} + 3 = 0$

13. $\sqrt[3]{3x^2 + 4x - 1} = \sqrt[3]{3x^2 + 7}$ 　　　14. $\sqrt[5]{x^2 + 2x} = \sqrt[5]{3}$

15. $\sqrt[3]{x^2 + 2x} = -1$ 　　　16. $\sqrt[4]{2x - 1} + \sqrt[4]{x} = 0$

17. $\sqrt[3]{3x - 1} + 1 = x$ 　　　18. $\sqrt[5]{7x - 4} + 1 = 0$

19. $\sqrt{9 - x} + \sqrt{x + 8} = 3$ 　　　20. $\sqrt{2x + 3} - \sqrt{2x} = 1$

21. $\sqrt{2x - 3} - \sqrt{x + 2} = 1$ 　　　22. $\sqrt{x - 2} = 3 + \sqrt{x + 2}$

23. $\sqrt{2x + 2} + \sqrt{2x + 6} = 4$ 　　　24. $\sqrt{3 - x} + \sqrt{2 + x} = 3$

25. $\sqrt{11 - x} - \sqrt{x + 6} = 3$ 　　　26. $\sqrt{1 - 5x} + \sqrt{1 - x} = 2$

27. $\sqrt{2x - 3} - \sqrt{x + 7} + 2 = 0$ 　　　28. $\sqrt{x + 3} = \sqrt{2x + 7} - 1$

*29. $\sqrt{5 + \sqrt{x + 1}} + 1 = \sqrt{x + 1}$ 　*30. $\sqrt{x} + \sqrt{3} = \sqrt{x + 3}$

PRACTICE TEST for Chapter 4

In problems 1 and 2, determine the solution set for the given equation.

1. (a) $7x + 4(3 - 2x) = 5 + x$

   (b) $\dfrac{3}{x+1} + 5 = \dfrac{4x}{x+1}$

2. $\dfrac{|3 - 2x|}{|2 - x|} = 2$

3. Ron can mow a yard in 4 hours. If he and his friend Jane mow the yard together with two mowers, it only takes 1 hour. How long would it take Jane to mow the yard if she worked alone?

4. Solve the following inequality, and represent the solution set on the number line.
$$|7 - 4x| < 5$$

5. A concrete mixture calls for 1 part cement, 2 parts sand, and 3 parts gravel by volume. How much concrete can be made from 4 cubic meters of cement?

6. Suppose that z is directly proportional to x and y^2, and inversely proportional to the cube of w. If $z = 12$ when $x = 2$, $y = 6$, and $w = 4$, find z when $x = 3$, $y = 4$, and $w = 2$.

7. Solve by factoring:
$$6x^2 + 7x - 20 = 0.$$

8. Solve the following equation by first expressing the equation in quadratic form.
$$7 - 15\sqrt{x} + 2x = 0$$

9. Solve the following equation by using the quadratic formula.
$$5x^2 - 3x - 1 = 0$$

10. Determine the solution set for the following equation.
$$x = 3 + \sqrt{7 - 3x}$$

Linear and Quadratic Functions and Relations

5-1
Functions and Relations

Much of the study of mathematics involves functions and relations. In this chapter, we define these concepts and study certain types of relations which occur most frequently. More complicated types are discussed later in the book.

By an *ordered pair of real numbers*, we mean a pairing (a, b) of real numbers a and b, where there is a distinction to be made between the pair (a, b) and the pair (b, a), if a and b are not equal. That is, there is to be a first position and a second position in the pair of numbers. It is traditional to use parentheses to indicate ordered pairs, where the number on the left is called the *first entry*, or *first component*, and the number on the right is called the *second entry*, or *second component*. Two ordered pairs (a, b) and (c, d) are *equal* if and only if $a = c$ and $b = d$.

DEFINITION 5-1 A *relation* is a nonempty set of ordered pairs (x, y) of real numbers x and y. The *domain* D of a relation is the set of all first entry elements, or the set of all x-values, which occur in the relation.

The *range R* of a relation is the set of all second entries, or the set of all y-values, which occur in the relation.

It is common practice to designate sets by using a letter as a name for a set. Relations are sets, and we designate them by letters.

EXAMPLE 1 The set

$$f = \{(-2, 4), (1, 1), (2, 4)\}$$

is a relation with domain $D = \{-2, 1, 2\}$ and range $R = \{1, 4\}$. Similarly, the set

$$g = \{(5, -2), (5, 2), (-5, -2), (-5, 2)\}$$

is a relation with domain $D = \{5, -5\}$ and range $R = \{2, -2\}$. As is frequently the case, these two relations can be described by a simple rule:

$(x, y) \in f$ if and only if $y = x^2$ and $x \in \{-2, 1, 2\}$;

$(x, y) \in g$ if and only if $|x| = y^2 + 1$ and $x \in \{5, -5\}$.

In our work here, we have little use for relations with domains which have only two or three elements. Our relations are usually determined by a rule relating the entries x and y of the ordered pairs (x, y), and we make the following convention regarding these rules.

Unless otherwise specified, the domain of a relation described by a certain rule is the set of all real numbers x which yield a real number y in the rule for the relation.

This convention is illustrated in the next example.

EXAMPLE 2 In each relation below, list three sample ordered pairs in the relation, and state the domain D and range R of the relation.

(a) $h = \{(x, y) | y = x^2\}$

(b) $r = \{(x, y) | y = \sqrt{x}\}$

SOLUTION (a) Three ordered pairs in the relation h are $(1, 1)$, $(2, 4)$, and $(-7, 49)$. The domain of h is the set of all real numbers, since any real number x has a square which is a real number. However, the range consists only of nonnegative real numbers since the rule gives $y = x^2$, and the square of a real number is never negative. That is, h has domain[1]

$$D = \{x | x \in \mathcal{R}\}$$

[1] Recall that $\mathcal{R}$ denotes the set of all real numbers.

and range

$$R = \{y \mid y \geq 0\}.$$

(b) Three ordered pairs in r are $(1, 1)$, $(2, \sqrt{2})$, and $(4, 2)$. In this case, the only real numbers that are in the domain are nonnegative real numbers, for they are the only ones which have a square root in the real numbers. The range consists of all nonnegative real numbers, since $\sqrt{x}$ denotes the nonnegative square root of x. That is, r has domain

$$D = \{x \mid x \geq 0\}$$

and range

$$R = \{y \mid y \geq 0\}.$$

A function is a special type of relation, described in the following definition.

DEFINITION 5-2   A *function* is a relation such that for each element in the domain, there is precisely one element in the range.

In other words, a function is a relation which has one and only one y-value for each of its x-values.

EXAMPLE 3   Which of the relations f, g, h, r in the examples above are functions?

SOLUTION   (a) The relation f in Example 1 has one and only one y-value for each value of x, so f is a function.
(b) The relation g in Example 1 has both of the pairs $(5, -2)$ and $(5, 2)$ in it. This means that there are two y-values, -2 and 2, for $x = 5$, so g is not a function.
(c) The relation h in Example 2 has the rule $y = x^2$, which clearly gives exactly one y-value for each x. Thus h is a function.
(d) The relation r in Example 2 is also a function since $y = \sqrt{x}$ gives a unique value when it is defined.

If f is a function and the pair (x, y) belongs to f, then we write $y = f(x)$ and call y the value of f at x. The symbol $f(x)$ *does not* indicate multiplication, and is read "f of x" or "f at x." In Example 2, we could simply write

$$h(x) = x^2$$

and

$$r(x) = \sqrt{x}$$

to describe the functions h and r.

EXAMPLE 4   For $F(x) = x^2 - 1$ and $G(x) = \sqrt{x - 1}$, find the values of $F(2)$, $F(-3)$, $G(1)$, and $G(14)$.

SOLUTION   We have

$$F(2) = (2)^2 - 1 = 3,$$
$$F(-3) = (-3)^2 - 1 = 8,$$
$$G(1) = \sqrt{1 - 1} = \sqrt{0} = 0,$$
$$G(14) = \sqrt{14 - 1} = \sqrt{13}.$$

Using the ordered pair notation, we could give the same information in this form: $(2, 3) \in F$, $(-3, 8) \in F$, $(1, 0) \in G$, $(14, \sqrt{13}) \in G$.

In more advanced mathematics courses such as the calculus, the operation of composition of functions is frequently encountered. This important operation is given in the following definition.

DEFINITION 5-3   If f and g are functions, the composition function $f \circ g$ is defined by

$$(f \circ g)(x) = f(g(x)).$$

The domain of $f \circ g$ is the set of all x in the domain of g such that f is defined at $g(x)$.

That is, the value of $f \circ g$ at x is obtained by computing $g(x)$, and then evaluating f at the number $g(x)$.

EXAMPLE 5   If $g(x) = x^2 - 1$ and $f(x) = \sqrt{x - 1}$, find each of the following:

(a) $(f \circ g)(x)$;   (b) $(f \circ g)(4)$;   (c) $(f \circ g)(0)$;
(d) $(g \circ f)(x)$;   (e) $(g \circ f)(3)$.

SOLUTION
(a) $(f \circ g)(x) = f(g(x)) = f(x^2 - 1) = \sqrt{x^2 - 1 - 1} = \sqrt{x^2 - 2}$
(b) $(f \circ g)(4) = \sqrt{(4)^2 - 2} = \sqrt{14}$
(c) $(f \circ g)(0) = f(g(0)) = f(-1)$, which is undefined since -1 is not in the domain of f.
(d) $(g \circ f)(x) = g(f(x)) = g(\sqrt{x - 1}) = (\sqrt{x - 1})^2 - 1 = x - 2$
(e) $(g \circ f)(3) = g(f(3)) = g(\sqrt{2}) = 1$

The usual rectangular coordinate system in a plane is often very helpful in working with relations and functions. It can be used to provide a "picture" or graph of a relation. With this purpose in mind, we briefly review the construction of a *rectangular coordinate system*[2] in a plane.

[2]The rectangular coordinate system is also frequently referred to as the *Cartesian coordinate system*. This is in honor of René Descartes (1596–1650), the French mathematician who is credited with inventing the system.

To construct a rectangular coordinate system in a plane, we begin with a horizontal line, called the *x-axis*, and a vertical line, called the *y-axis*, which intersect at a point O, called the *origin*. (See Figure 5.1.)

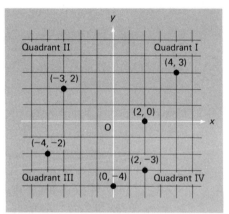

FIGURE 5.1

On each of these lines, we set up a one-to-one correspondence between the points on the line and the real numbers, as was described in Section 1-4. The correspondence on the *x*-axis is set up so that 0 is at the origin with the positive direction to the right, indicated by an arrowhead, and the negative direction to the left. Similarly, the correspondence on the *y*-axis is set up so that 0 is at the origin with the positive direction upward, indicated by an arrowhead, and the negative direction is downward.

We can now set up a one-to-one correspondence between points in the plane and ordered pairs (x, y) of real numbers. To describe this correspondence, let P be a point in the plane. We let x denote the directed horizontal distance from the *y*-axis to the point P, so that x is positive if P is to the right of the *y*-axis, x is negative if P is to the left of the *y*-axis, and x is zero if P is on the *y*-axis. Similarly, we let y denote the directed vertical distance from the *x*-axis to the point P, where y is positive in the upward direction, negative in the downward direction, and zero on the *x*-axis. The ordered pair (x, y) is then assigned to the point P. The first entry, x, is called the *abscissa*, or *x-coordinate*, of P, and the second entry, y, is called the *ordinate*, or *y-coordinate*, of P. The ordered pair (x, y) is referred to as the *coordinates* of P.

Conversely, each ordered pair (x, y) determines a unique point P that has the pair (x, y) as coordinates. The point P is located by simply using x as the directed horizontal distance from the *y*-axis, and y as the directed vertical distance from the *x*-axis. Several points are located by their coordinates in Figure 5.1.

The coordinate axes separate the plane into four regions, called *quadrants*. These quadrants are numbered as shown in Figure 5.1.

As mentioned before, a rectangular coordinate system can be used to obtain a "picture" of a relation. This picture is formed by sketching the graph of the relation.

DEFINITION 5-4   The *graph* of a relation is the set of all points whose coordinates (x, y) are members of the relation.

The sketch of the graph is made by plotting several points on the graph of the relation, and then drawing a curve through these points. This is illustrated in the examples below.

EXAMPLE 6   Sketch the graph of the function $f(x) = x^2$.

SOLUTION   We first assign several sample values to x and compute the corresponding values of $f(x)$. The resulting ordered pairs are recorded in the table of Figure 5.2. We then locate the points corresponding to these coordinates, and sketch the graph as well as possible from these points. The graph provides a picture of the behavior of the function, and it clearly shows that the domain D of f is the set of all real numbers, and the range R of f is the set of all nonnegative real numbers.

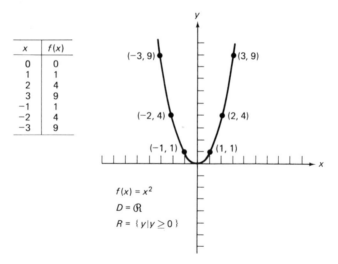

x	$f(x)$
0	0
1	1
2	4
3	9
−1	1
−2	4
−3	9

$f(x) = x^2$

$D = \mathcal{R}$

$R = \{y | y \geq 0\}$

FIGURE 5.2

EXAMPLE 7   Sketch the graph of the relation

$$g = \{(x, y) \mid x + 1 = |y|\}.$$

SOLUTION   Following the same procedure as in Example 6, we obtain the table given in Figure 5.3. It soon becomes evident that g is not a function, since two

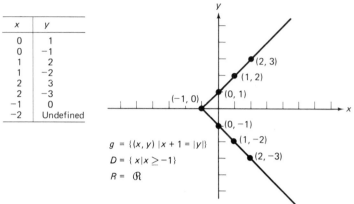

x	y
0	1
0	−1
1	2
1	−2
2	3
2	−3
−1	0
−2	Undefined

$g = \{(x, y) \mid x + 1 = |y|\}$

$D = \{x \mid x \geq -1\}$

$R = \mathcal{R}$

FIGURE 5.3

values of y are sometimes obtained from a given value of x. Also, we discover that if x is assigned a value less than -1, then there is no y which satisfies the equation, since an absolute value cannot be negative. When the points corresponding to the coordinates in the table are plotted, it appears that they lie along two half-lines which have a common end point at $(-1, 0)$, and the graph is drawn accordingly. It is clear that g has domain $D = \{x \mid x \geq -1\}$ and range $R = \mathcal{R}$.

The graph of f in Figure 5.2 makes it easy to see that f is a function, since there is exactly one y-value for each x-value. Similarly, the graph of g in Figure 5.3 shows that there are two y-values for any $x > -1$, so that g is not a function. A graphical description of this situation is that some vertical lines intersect the graph of g at more than one point, and that no vertical line crosses the graph of f at more than one point. This illustrates the *vertical line test* for functions, which is stated below.

Vertical Line Test. If any vertical line intersects the graph of a relation at two or more points, the relation is *not* a function. On the other hand, if no vertical line intersects the graph in more than one point, the relation is a function.

In most cases, the rule defining a relation is simply an equation in x and y which determines the ordered pairs (x, y) that belong to the relation. In such a case, it is common to refer to the graph of the relation as the *graph of the equation*. As examples, the graph of the equation $y = x^2$ is given in Figure 5.2, and the graph of the equation $x + 1 = |y|$ is given in Figure 5.3.

In more advanced work it is sometimes desirable to interchange the components in a relation, and thereby obtain a new relation. This new relation is called the *inverse* of the original relation, and is formally defined as follows.

DEFINITION 5-5 If g is a given relation, the *inverse* of g is the relation g^{-1} defined by

$$g^{-1} = \{(y, x) \,|\, (x, y) \text{ is in } g\}.$$

EXAMPLE 8 Find the inverses of the following relations.

(a) $g = \{(1, 2), (2, 3), (4, 3)\}$
(b) $h = \{(x, y) \,|\, y = 2x - 4\}$

SOLUTION   (a) By the definition,

$$g^{-1} = \{(2, 1), (3, 2), (3, 4)\}.$$

We note that g^{-1} is not a function, even though g is a function.
(b) We have

$$h^{-1} = \{(y, x) \,|\, y = 2x - 4\}.$$

To obtain the more conventional description of h^{-1}, we may interchange the letters x and y:

$$h^{-1} = \{(x, y) \,|\, x = 2y - 4\}$$
$$= \{(x, y) \,|\, y = \tfrac{1}{2}x + 2\}.$$

This illustrates the general fact that *the inverse of a relation g can be obtained by simply interchanging x and y in the defining equation for g.* We note also that both h and h^{-1} are functions in this case.

A more detailed discussion of inverse functions is given in Sections 6-7 and 10-3.

EXERCISES 5-1

Give the domain D and range R of each function.

1. $f(x) = 3x$ 2. $g(x) = 4x - 3$
3. $h(x) = -\sqrt{x - 4}$ 4. $F(x) = \sqrt{x + 2}$
5. $G(x) = x^2 - 3$ 6. $p(x) = 2(x - 1)^2 + 1$
7. $y = \dfrac{1}{x - 1}$ 8. $y = \dfrac{2x}{x - 3}$
9. $y = |x - 1| - 3$ 10. $y = |x|$
11. $y = 4 - x$ 12. $y = 5x - 7$

If $f(x) = 4x - 2$ and $g(x) = x^2$, find each of the following.

13. $f(3)$ 14. $f(-5)$

15. $g(4)$ 16. $g(-7)$

17. $f(b)$ 18. $g(a)$

19. $f(g(4))$ 20. $g[f(-3)]$

21. $(f \circ g)(b)$ 22. $(g \circ f)(a)$

If $f(x) = 2^x$ and $g(x) = x^2 - x$, find each of the following.

23. $f(3)$ 24. $f(-2)$

25. $g(-3)$ 26. $g(2)$

27. $(f \circ g)(2)$ 28. $(g \circ f)(3)$

29. $(f \circ g)(x)$ 30. $(g \circ f)(x)$

31. $f(x + h)$ 32. $g(x + h)$

33. $g(x + h) - g(x)$ 34. $f(x + h) - f(x)$

Sketch the graph of each of the following equations.

35. $y = 3x$ 36. $y = 2x - 1$

37. $2x + 3y = 6$ 38. $y = 4 - x$

39. $|y| = x$ 40. $y = x^2 + 1$

41. $y = |x|$ 42. $y = |x - 1|$

*43. $|y| = |x - 1|$ *44. $|y| = |x|$

Use the vertical line test to determine whether or not the given graph is the graph of a function.

45.

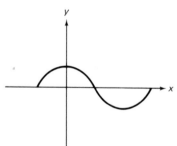

46.

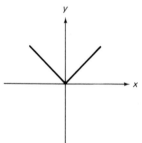

47.

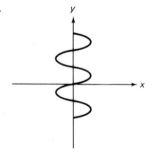

48.

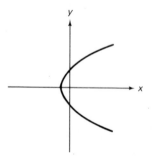

Find the inverse of each of the following relations. Also determine if each relation and/or its inverse is a function.

49. $g = \{(0, 3), (1, 2), (-1, 2)\}$ 50. $g = \{(1, 5), (0, 2), (0, -2)\}$

51. $h = \{(x, y) \mid y = x\}$ 52. $h = \{(x, y) \mid xy = 2\}$

53. $p = \{(x, y) \mid y = |x|\}$ 54. $p = \{(x, y) \mid y = |x + 1|\}$

55. $f = \{(x, y) \mid y = 3x - 2\}$ 56. $f = \{(x, y) \mid x - y = 5\}$

57. $g = \{(x, y) \mid x = 2y^2\}$ 58. $g = \{(x, y) \mid y = 3x^2\}$

5-2
Linear Functions

One of the simplest and most useful types of functions is the linear function. A linear function is a special case of a linear relation, defined as follows.

DEFINITION 5-6   If a, b, and c are real numbers, with not both a and b zero, the set

$$\{(x, y) \mid ax + by = c\}$$

is called a *linear relation*.

EXAMPLE 1   Which of the following sets are linear relations?

(a) $\{(x, y) \mid y = 4x\}$
(b) $\{(x, y) \mid 3x = 4y + 7\}$
(c) $\{(x, y) \mid x^2 + 4y = 7\}$

SOLUTION

(a) The relation $\{(x, y) \mid y = 4x\}$ is a linear relation since it can be written as $\{(x, y) \mid 4x - y = 0\}$, and this fits the form given in Definition 5-6.

(b) The relation $\{(x, y) \mid 3x = 4y + 7\}$ is also a linear relation, since it can be put in the form $\{(x, y) \mid 3x - 4y = 7\}$, and this fits the definition.

(c) The relation $\{(x, y) \mid x^2 + 4y = 7\}$ is not a linear relation. The term, x^2 makes it impossible to write this relation in the form required in Definition 5-6.

A *linear function* is a linear relation which is a function. That is, for each x-value, there is exactly one y-value. Clearly, this will be the case in a linear relation if and only if it is possible to solve for y in the equation $ax + by = c$, and this is possible if and only if $b \neq 0$. Thus we have the following theorem.

THEOREM 5-7   The linear relation

$$\{(x, y) \mid ax + by = c\}$$

is a linear function if and only if $b \neq 0$.

Linear relations get their name from the fact that their graphs are always straight lines. That is, *the graph of any equation of the form*

$$ax + by = c,$$

with not both a and b zero, is a straight line. If there is a point where the graph crosses the y-axis, the y-coordinate of that point is called a *y-intercept* of the graph. To find a y-intercept, we would set $x = 0$ in the equation $ax + by = c$. If there is a solution for y, this y-value would be a y-intercept. If there is not a solution for y when $x = 0$, there is no y-intercept. Similarly, the x-coordinate of a point where the graph crosses the x-axis is called an *x-intercept*. An x-intercept, if there is one, is obtained by setting $y = 0$ in $ax + by = c$, and then solving for x.

EXAMPLE 2 Find the x- and y-intercepts, if they exist, and sketch the graph of the equation

$$2x + 3y = 6.$$

SOLUTION To find the y-intercept, we let $x = 0$ and obtain $3y = 6$. Thus 2 is the y-intercept. Similarly, $y = 0$ gives $2x = 6$, and 3 is the x-intercept. Plotting the points $(0, 2)$ and $(3, 0)$, we draw the graph as in Figure 5.4. The point $(2, \frac{2}{3})$ is included as a check on the line drawn through the other two points.

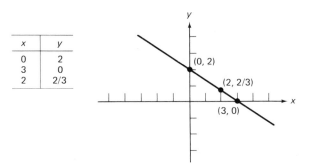

x	y
0	2
3	0
2	2/3

FIGURE 5.4

There are two special cases of the equation $ax + by = c$ which are worth noting. One of these is the equation

$$y = c.$$

The graph of $y = c$ consists of all points with coordinates of the form (x, c), where x may be any real number. Thus the graph of $y = c$, c a constant, is a *horizontal straight line*. Similarly, the graph of the equation

$$x = c,$$

where c is a constant, is a *vertical straight line*. As examples, the graphs of $y = 4$ and $x = -2$ are drawn in Figure 5.5.

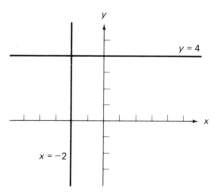

FIGURE 5.5

The concept of the *slope* of a line is extremely useful in the study of the differential calculus, and many other areas as well.

DEFINITION 5-8 The *slope m* of the line through two distinct points (x_1, y_1) and (x_2, y_2) is given by

$$m = \frac{y_2 - y_1}{x_2 - x_1}.$$

There are two special cases that should be noted. If the line is horizontal, then $y_1 = y_2$ and $x_1 \neq x_2$ in the definition, and $m = 0$. That is, *a horizontal line has slope 0*. If the line is vertical, then $x_1 = x_2$ and $y_1 \neq y_2$ in the definition, and $m = (y_2 - y_1)/0$ is undefined, since division by zero is impossible. This means that *the slope of a vertical line is undefined*.

The slope of a line which is not horizontal or vertical is a real number $m \neq 0$, and m is *independent of the choice of the points (x_1, y_1) and (x_2, y_2)*. In Figure 5.6, another pair of points (x_1', y_1') and (x_2', y_2') on the same line is indicated, and we see that

$$\frac{y_2' - y_1'}{x_2' - x_1'} = \frac{y_2 - y_1}{x_2 - x_1}.$$

This follows from the fact that the ratios of corresponding sides of similar triangles are always equal.

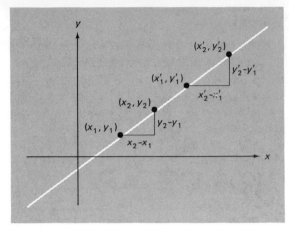

FIGURE 5.6

EXAMPLE 3 Find the slope of the line $2x - y = 4$.

SOLUTION To find the slope, we locate two points on the line and use the formula in Definition 5-8. If we let $x = 1$ and solve for y in the equation of the line, we get $y = -2$. That is, $(1, -2)$ is on the graph. For $x = 3$, we get $y = 2$, and $(3, 2)$ is on the graph. Thus the slope of the line is

$$m = \frac{2 - (-2)}{3 - 1}$$

$$= \frac{4}{2}$$

$$= 2.$$

If the graph of a line slants upward to the right, this indicates that y increases as x increases along the line, and the slope of the line is *positive*. On the other hand, if the graph slants downward to the right, then y decreases as x increases along the line, and the slope of the line is *negative*. These situations are pictured in Figure 5.7, where line l_1 has slope $m_1 > 0$, and line l_2 has slope $m_2 < 0$.

FIGURE 5.7

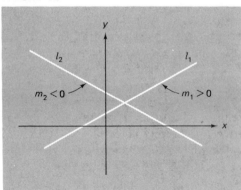

It is possible to find the slope of a line that is not vertical or horizontal directly from its equation $ax + by = c$. To see this, we first note that such a line has both $a \neq 0$ and $b \neq 0$, and the points $(0, c/b)$ and $(c/a, 0)$ are on the line. Using these two points, we find that the slope is

$$m = \frac{0 - \left(\frac{c}{b}\right)}{\frac{c}{a} - 0} = \frac{-\frac{c}{b}}{\frac{c}{a}} = -\frac{a}{b}.$$

This result, together with the fact that a horizontal line has slope 0, shows that

the slope of $ax + by = c$ is $m = -\dfrac{a}{b}$, if it exists.

Notice that, when $ax + by = c$ is solved for y, we get

$$y = -\frac{a}{b}x + \frac{c}{b},$$

and *the coefficient of x in this equation is the slope, $m = -\dfrac{a}{b}$.*

EXAMPLE 4 In Example 3, we used two points to find the slope of $2x - y = 4$. If we solve this equation for y, we get

$$y = 2x - 4.$$

Thus $m = 2$, since 2 is the coefficient of x in this equation.

EXERCISES 5-2

Which of the following equations define linear functions?

1. $2x - y = 6$

2. $y = 7x - 2$

3. $y = |2x - 1|$

4. $y = |x - 2|$

5. $y = \dfrac{3}{x - 2}$

6. $\dfrac{1}{2x - 1} = y$

7. $y = |x| - 2$

8. $y = 3 - |x|$

9. $y = \sqrt{x - 1}$

10. $y = x^2 - 2$

11. $x = 0$

12. $4x = 3$

13. $y^2 = 4x - 3$

14. $x^2 + y^2 = 4$

15. $xy = 3$

16. $y = \dfrac{2}{x}$

Graph each linear relation, and find the x-intercepts and y-intercepts, if they exist.

17. $y = 2x - 3$ 18. $2x - 3y = 6$

19. $4x = 5$ 20. $x = 2$

21. $2x = 3y$ 22. $2y = -3x$

23. $4x - y = 8$ 24. $3x + y = 9$

25. $2x + 5y = 7$ 26. $5x + 2y = 10$

27. $x + y = 3$ 28. $y - x = 5$

Find the slope of the line through the given pair of points, if it exists.

29. $(2, -1)$ and $(-1, 3)$ 30. $(0, 4)$ and $(3, 7)$

31. $(0, 0)$ and $(3, 2)$ 32. $(7, 11)$ and $(11, 7)$

33. $(3, -1)$ and $(3, 3)$ 34. $(2, 7)$ and $(-3, 7)$

35. $(-1, 1)$ and $(7, 5)$ 36. $(-3, -4)$ and $(7, -2)$

If it exists, find the slope of the line which has the given equation.

37. $3x - 4y = 6$ 38. $x = 7$

39. $y = 3$ 40. $4x - 2y = 20$

41. $7x - 5y = 3$ 42. $3x - 7y = 0$

43. $y = 7x - 5$ 44. $y = 5 - 2x$

45. $5x + 2y = 10$ 46. $3x + 5y = 15$

47. $x = 2y + 6$ 48. $x + 8 = 2y$

5-3
The Straight
Line

We have seen in the last section that the graph of any equation of the form

$$ax + by = c, \qquad (1)$$

with not both of a and b zero, is a straight line. For this reason, an equation of this type is called a *linear equation*. It is also true that every line in the plane is the graph of an equation of the form given in (1), and $ax + by = c$ is called the *standard form* of the equation of a straight line.

  There are other forms of the equation of a straight line that are sometimes more useful than the standard form. Two of these are presented in this section.

  The discussion in Section 5-2 shows that $y = mx + b$ is an equation of a line with slope m and y-intercept b. The converse of this statement is also true. To see this, suppose that a line has slope m, and that b is the y-intercept. Then $(0, b)$ is on the line, and any other point (x, y) on the line has coordinates which satisfy

$$\frac{y - b}{x - 0} = m.$$

Solving for y, we have

$$y = mx + b,$$

and this is an equation for the line. This result is stated formally in the following theorem.

THEOREM 5-9   An equation of the straight line which has slope equal to m and y-intercept equal to b is

$$y = mx + b. \tag{2}$$

This form of the equation of the line is called the *slope-intercept form*.

EXAMPLE 1   Find an equation of the straight line whose slope is 2 and whose y-intercept is -3.

SOLUTION   Using $m = 2$ and $b = -3$ in the slope-intercept form, we immediately obtain

$$y = 2x - 3.$$

Another special form for the equation of a straight line is the point-slope form. This is described in the next theorem.

THEOREM 5-10   An equation of the straight line which has slope m and passes through the point (x_1, y_1) is

$$y - y_1 = m(x - x_1). \tag{3}$$

This form of the equation of the line is called the *point-slope form*.

PROOF   Suppose that m is the slope of the line, and (x_1, y_1) is a point on the line. If (x, y) is any other point on the line, then

$$\frac{y - y_1}{x - x_1} = m,$$

by the definition of the slope. When both sides of this equation are multiplied by $x - x_1$, we have

$$y - y_1 = m(x - x_1),$$

and this is the equation stated in the theorem. We note that this last equation is satisfied when $x = x_1$ and $y = y_1$, so every point on the line, including (x_1, y_1), satisfies equation (3).

EXAMPLE 2   Find an equation of the straight line which has slope -2 and passes through the point $(-1, 4)$.

SOLUTION   Using $m = -2$ and $(x_1, y_1) = (-1, 4)$ in the point-slope form of the equation, we have

$$y - 4 = -2(x + 1)$$

as an equation of the line.

EXAMPLE 3   Find an equation of the straight line passing through the points $(-1, 3)$ and $(2, -4)$.

SOLUTION   Here we are given two distinct points, and any two distinct points determine a unique line. We can find the slope of the line by using these two points in the formula from Definition 5-8. We get

$$m = \frac{-4 - 3}{2 - (-1)} = -\frac{7}{3}.$$

Now we can use the point-slope form with this slope and either of the points $(-1, 3)$ or $(2, -4)$ to write an equation of the line. Using $(-1, 3)$, we have

$$y - 3 = -\tfrac{7}{3}(x + 1),$$

and using $(2, -4)$, we have

$$y + 4 = -\tfrac{7}{3}(x - 2).$$

Both of these equations reduce to $7x + 3y = 2$.

It is easy to see that *two nonvertical lines are parallel if and only if they have the same slope.* Slopes can also be used to determine whether or not two lines are perpendicular to each other. If line l_1 has slope m_1 and line l_2 has slope m_2, then l_1 *and* l_2 *are perpendicular to each other if and only if* $m_1 m_2 = -1$. In other words, l_1 *and* l_2 *intersect at right angles if and only if the slope of one line is the negative reciprocal of the slope of the other line.* This fact is proven in the study of analytic geometry.

EXAMPLE 4   Find an equation of the line which satisfies the conditions given in each part below:

(a) parallel to the line $-3x + y = 7$ and passing through the point $(-1, 1)$;

(b) perpendicular to the line $-3x + y = 7$ and having -5 as the y-intercept.

SOLUTION   (a) Solving for y in $-3x + y = 7$, we have

$$y = 3x + 7,$$

so the slope of this line is 3. Thus we are seeking an equation of the line with slope 3 which passes through $(-1, 1)$. Using the point-slope form, we have

$$y - 1 = 3(x + 1)$$

as an equation of the line.

(b) In order for a line to be perpendicular to $-3x + y = 7$, it must have slope equal to $-\frac{1}{3}$. Using $m = -\frac{1}{3}$ and $b = -5$ in the slope-intercept form, we obtain

$$y = -\tfrac{1}{3}x - 5$$

as an equation of the line.

Three special forms of an equation of a straight line have been presented in this section. These forms are listed below, together with their names and key facts.

Special Form	Name of Form; Relevant Facts
$ax + by = c$	standard form; at least one of a and b is not zero.
$y = mx + b$	slope-intercept form; slope is m, y-intercept is b.
$y - y_1 = m(x - x_1)$	point-slope form; slope is m, passes through (x_1, y_1).

EXERCISES 5-3

Find an equation of the straight line which satisfies the given conditions.

1. slope -1, through $(-1, 1)$
2. through $(-1, 3)$, slope -2
3. slope 2, y-intercept 4
4. slope 3, x-intercept -5
5. y-intercept 2, x-intercept -3
6. through $(2, 2)$ and $(-1, 3)$
7. through $(-7, 5)$, slope 0
8. through $(3, 4)$, slope does not exist

Find an equation, in standard form, of the straight line which satisfies the given conditions.

9. y-intercept 3, slope $\frac{1}{2}$
10. slope -7, through $(3, -1)$
11. through $(3, 4)$ and $(4, 3)$
12. through $(-4, 2)$ and $(3, 0)$
13. x-intercept 7, slope 4
14. x-intercept -3, slope -2
15. x-intercept 2, y-intercept -1
16. y-intercept 3, x-intercept -3
17. y-intercept 3, through $(-1, -1)$
18. y-intercept -2, through $(-2, -3)$

19. y-intercept 3, no x-intercept

20. y-intercept -7, slope 0

21. x-intercept 5, slope does not exist

22. x-intercept -3, no y-intercept

23. through $(-2, 3)$, parallel to $3x - y = 6$

24. through $(-7, 2)$, parallel to $-2x = -3y + 7$

25. through $(-5, -3)$, perpendicular to $5x = 6y - 1$

26. through $(-1, 3)$, perpendicular to $5x - 2y = 11$

27. through $(0, 0)$, perpendicular to $y = x$

28. x-intercept 3, perpendicular to $3x - y = 2$

29. y-intercept -1, parallel to $5x - 7y = 5$

30. x-intercept 5, parallel to $9x - 7 = 3y$

*31. Do the points $(-1, 3)$, $(4, 7)$, and $(6, 10)$ lie on the same straight line? Justify your answer. (*Hint:* Check the slopes between the points.)

*32. Do the points $(3, 7)$, $(-1, 2)$, and $(7, 12)$ lie on the same straight line? Justify your answer.

5-4
Quadratic
Relations
(Parabolas)

Although they are not quite as simple as the linear functions, the quadratic functions are fully as useful and important as the linear functions.

DEFINITION 5-11   A function f is a *quadratic function* if there are real numbers a, b, and c, with $a \neq 0$, such that the function value for f is given by

$$f(x) = ax^2 + bx + c.$$

That is, a quadratic function is a function which has a defining equation of the form

$$y = ax^2 + bx + c,$$

with $a \neq 0$.

EXAMPLE 1   Which of the following equations define a quadratic function?

(a) $y = 2x^2$ (b) $2x + x^2 = y$ (c) $y^2 = 4x$
(d) $x^3 + 4x = y$ (e) $y = 2x + 7$

SOLUTION   The equations in parts (a) and (b) can be written in the form $y = ax^2 + bx + c$, with $a \neq 0$, so each of them defines a quadratic function. The relation defined by $y^2 = 4x$ in part (c) is not a function, and so is not a quadratic function. The equations in parts (d) and (e) cannot be written in the required form, so they do not define quadratic functions.

The graph of a quadratic function is called a *parabola*. We have seen the graph of the quadratic function $y = x^2$ in Figure 5.2, Example 6, of Section 5-1. Another example is provided below.

EXAMPLE 2 Sketch the graph of

$$y = 2x^2 - 12x + 17.$$

SOLUTION One technique that can be used to great advantage in graphing a quadratic function is to complete the square on the x-terms, using the same sort of procedure as was used in solving quadratic equations in Section 4-6. Instead of dividing both sides by the coefficient of x^2, we factor the coefficient away from the x-terms:

$$y = 2x^2 - 12x + 17$$
$$= 2(x^2 - 6x \quad) + 17.$$

To complete the square on $x^2 - 6x$, we add $[\frac{1}{2}(-6)]^2 = 9$ inside the parentheses. In doing this, we are actually adding $2(9) = 18$, so we must subtract 18 from 17 to have an equivalent equation:

$$y = 2(x^2 - 6x + 9) + 17 - 18$$
$$= 2(x - 3)^2 - 1.$$

This form of the equation gives some important information about the graph. Since $2(x - 3)^2 \geq 0$ for all x, the smallest possible value for y is -1, and this occurs when $2(x - 3)^2 = 0$, that is, when $x = 3$. This means that *the lowest point on the graph is at* $(3, -1)$.

We observe now that $(x - 3)^2 = |x - 3|^2$ is the square of the distance from x to 3, and this means that we get the same value for y when we move a certain distance to the left from $x = 3$ as we get when we move the same distance to the right from $x = 3$. In other words, *the graph is symmetrical, or balanced, with respect to the line $x = 3$.*

We now assign some values to x and compute the corresponding y-values. These are recorded in the table in Figure 5.8, and a sketch of the graph is shown there. The extreme point on the graph is called the *vertex*.

As mentioned just before Example 2, the graph of a quadratic function is called a *parabola*. The parabolas in Figures 5.2 and 5.8 are typical as to shape. Parabolas always have a "bullet" shape, with an extreme point (the *vertex*), and a *line of symmetry* through the vertex. However, the vertex may be the highest point on the graph instead of the lowest point. The graph opens downward and has a highest point when $a < 0$ in $y = ax^2 + bx + c$. A parabola of this type is shown in Example 3 below.

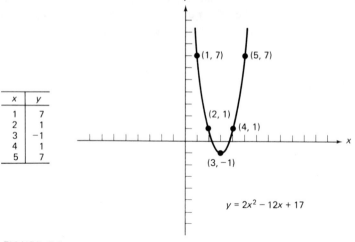

x	y
1	7
2	1
3	-1
4	1
5	7

$y = 2x^2 - 12x + 17$

FIGURE 5.8

The discussion given in Example 2 can be extended to the general case. If the equation $y = ax^2 + bx + c$ is rewritten as

$$y = a(x - h)^2 + k,$$

then

1. *the vertex is at (h, k);*
2. *the line $x = h$ is an axis of symmetry;* and
3. *the graph opens upward if $a > 0$, downward if $a < 0$.*

Convenient values to assign for x when plotting points are $x = h$, $x = h \pm 1$, $x = h \pm 2$, and so on.

EXAMPLE 3   Sketch the graph of

$$y = -2x^2 + 8x - 7.$$

SOLUTION   We first rewrite the equation, completing the square on the x-terms to obtain the form $y = a(x - h)^2 + k$.

$$y = -2(x^2 - 4x\ \ ) - 7$$
$$= -2(x^2 - 4x + 4) - 7 + 8$$
$$= -2(x - 2)^2 + 1$$

We observe that $a = -2$, $h = 2$, and $k = 1$. The vertex is at $(2, 1)$, the line $x = 2$ is an axis of symmetry, and the parabola opens downward since $a < 0$. With this much knowledge of the graph, it is sufficient to plot five points and then sketch the graph. (See Figure 5.9.)

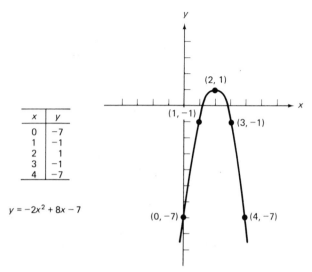

x	y
0	−7
1	−1
2	1
3	−1
4	−7

$y = -2x^2 + 8x - 7$

FIGURE 5.9

It is possible to obtain formulas which give the vertex and the axis of symmetry for

$$y = ax^2 + bx + c, \qquad a \neq 0,$$

in terms of a, b, and c. All we need do is complete the square in the general case. We have

$$y = a\left(x^2 + \frac{b}{a}x\right) + c$$

$$= a\left(x^2 + \frac{b}{a}x + \frac{b^2}{4a^2}\right) + c - \frac{b^2}{4a}$$

$$= a\left(x + \frac{b}{2a}\right)^2 + \frac{4ac - b^2}{4a}.$$

Careful examination of this equation leads to the statements made in Theorem 5-12.

THEOREM 5-12 The graph of $y = ax^2 + bx + c$, $a \neq 0$, is a parabola which

(1) has vertex at $\left(-\dfrac{b}{2a}, \dfrac{4ac - b^2}{4a}\right)$,

(2) has the line $x = -\dfrac{b}{2a}$ as an axis of symmetry, and

(3) opens upward if $a > 0$, downward if $a < 0$.

It is actually not necessary to memorize the y-coordinate of the vertex, because it can be readily obtained by substituting the value $x = -b/2a$ in the equation $y = ax^2 + bx + c$.

In many cases, it is useful to find the intercepts for a parabola $y = ax^2 + bx + c$. The y-intercept is easy: when $x = 0$, $y = c$. In looking for the x-intercepts, we are at once confronted with the problem of solving

$$ax^2 + bx + c = 0.$$

We recall from Section 4-6 that this equation may have two equal real roots, two unequal real roots, or no real roots, depending on the discriminant $b^2 - 4ac$. The graphical interpretation of this situation is that the parabola may be tangent to the x-axis at the vertex, or it may cross the x-axis in two places, or it may not cross the x-axis at all. We return to this discussion in Section 5-5.

In many instances, it has been found that a quadratic function can be used to satisfactorily represent a quantity in real life. Examples of this can be found in reflecting telescopes with parabolic mirrors, parabolic arches, and parabolic shapes of suspension cables on bridges. An application of a different nature is given in the next example.

EXAMPLE 4   A business has found that their profit $P(x)$ is a quadratic function of the number x of units it sells, and that the profit function, in dollars, is given by

$$P(x) = 240x - x^2.$$

Find the number of units that yields a maximum profit, and find the maximum profit. Sketch a graph of the profit function.

SOLUTION   We note that P is a quadratic function with $a = -1$, $b = 240$, $c = 0$. Since $a < 0$, P has a maximum value at the vertex. To locate the vertex, we set

$$x = -\frac{b}{2a} = -\frac{240}{2(-1)} = 120,$$

and compute

$$P(120) = 240(120) - (120)^2 = 14{,}400.$$

Thus the maximum profit occurs when 120 units are sold, and the maximum profit is \$14,400. The graph is sketched in Figure 5.10. Only the vertex and the intercepts are needed to provide an adequate graph. In order to give a better picture, we have used a unit of length on the y-axis which is different from that on the x-axis.

One can interchange the roles of x and y and obtain a complete dual to the results of this section. That is, one can study relations defined

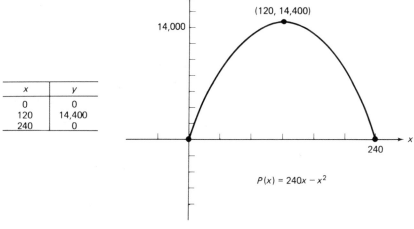

x	y
0	0
120	14,400
240	0

FIGURE 5.10

by an equation of the form

$$x = ay^2 + by + c,$$

where $a \neq 0$. Such a relation is not a function, since there would usually be two distinct y-values for the same x-value. To go through the details for relations of this form would be very boring, to say the least. We simply state a theorem corresponding to Theorem 5-12.

THEOREM 5-13   The graph of $x = ay^2 + by + c$, $a \neq 0$, is a parabola which

(1) has vertex at $\left(\dfrac{4ac - b^2}{4a}, \ -\dfrac{b}{2a}\right)$,

(2) has the line $y = -\dfrac{b}{2a}$ as an axis of symmetry, and

(3) opens to the right if $a > 0$, to the left if $a < 0$.

When relations of the form $x = ay^2 + by + c$ are encountered in the calculus, they usually appear in a form similar to that in the following example.

EXAMPLE 5   Sketch the graph of the function

$$y = 1 + \sqrt{x - 2}.$$

SOLUTION   The graph is not familiar in the given form, but it can be recognized after the radical is eliminated by squaring both sides. We have

$$y - 1 = \sqrt{x - 2},$$

and

$$(y - 1)^2 = x - 2.$$

Solving for x yields

$$x = (y - 1)^2 + 2.$$

We recognize this as the equation of a parabola which has vertex at $(2, 1)$ and opens to the right. However, this parabola is not the graph of the original equation, because $y = 1 + \sqrt{x - 2}$ requires that $y \geq 1$. (Recall that $\sqrt{x - 2}$ indicates the nonnegative square root of $x - 2$.) The graph of the original equation consists only of the top half of the parabola, and is shown in Figure 5.11.

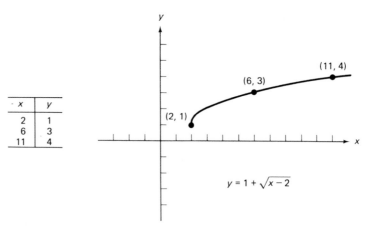

x	y
2	1
6	3
11	4

FIGURE 5.11

EXERCISES 5-4

Graph each of the following parabolas. Locate the vertex and the axis of symmetry, and plot at least two other points on each graph.

1. $y = x^2 + 1$
2. $y = x^2 - 2x + 3$
3. $y = x^2 + 2x + 4$
4. $y = -x^2 + 4x - 3$
5. $y = -x^2 - 2x - 2$
6. $y = 2x^2 + 8x + 9$
7. $y = 3x^2 + 12x + 17$
8. $y = -4x^2 - 4x + 1$
9. $y = 4x - x^2$
10. $y = x^2 - 9x$
11. $y = x^2 - 4$
12. $y = -2x^2 + 6$
13. $y = \frac{2}{3}(x - 2)^2 - 1$
14. $y = -\frac{4}{3}(x + 1)^2 + 2$
15. $y = x^2 + 8x + 13$
16. $y = -x^2 + 6x - 6$
17. $y = 5 + 6x + x^2$
18. $y = 5 - 4x + 2x^2$
19. $x = y^2 - 4y + 6$
20. $x = 3y^2 + 6y - 20$
21. $x = -2y^2 + 8y - 11$
22. $x = -2y^2 + 4y - 3$

23. $x = y^2 - 2y - 4$ 24. $x = y^2 + 6y + 10$

25. $x = 12 - 3(y + 2)^2$ 26. $x = 4 - 2(y - 1)^2$

27. $x = y^2 - 10y + 25$ 28. $x = y^2 - 4y + 4$

29. $x = -2y^2 - 4$ 30. $x = -3y^2 + 9$

31. $x = 6y + y^2$ 32. $x = 2y^2 - 8y$

33. $x = -9 - 6y - y^2$ 34. $x = 2y - y^2 - 1$

Sketch the graph of each of the following relations.

35. $y = \sqrt{x - 3}$ 36. $y = -\sqrt{x + 3}$

37. $y = 1 - \sqrt{x - 2}$ 38. $y = 2 + \sqrt{x + 1}$

39. $y = -1 + 2\sqrt{x - 1}$ 40. $y = -1 - 2\sqrt{x - 1}$

41. $y = 2 - \sqrt{x}$ 42. $y = 2 + \sqrt{x}$

43. $x = \sqrt{y - 2}$ 44. $x = -\sqrt{y + 2}$

45. $x = 2 - 2\sqrt{y - 1}$ 46. $x = 1 + 2\sqrt{y - 2}$

47. A rectangular pen is fenced off along the side of a barn, and no fence is required along the barn. If 100 feet of fence is available, find the dimensions of the pen which encloses the largest possible area.

48. A business finds that if it sells x items, its revenue R is given by $R(x) = 100x - 2x^2 - 1050$. Find x so that revenue is a maximum. What is the maximum revenue?

49. If a rock is thrown upward from the ground at 128 ft/sec, the distance s of the rock from the ground after t seconds is given by $s(t) = -16t^2 + 128t$. Find t when the rock is highest above the ground, and find the maximum height.

50. The number, P, of frogs, in millions, produced in Louisiana in the month of June depends on the number, x, of inches of rainfall in June, and is given by $P(x) = 20x - x^2$. Find the number of inches of rainfall in June that produces the most frogs.

5-5
*Nonlinear
Inequalities
in One Variable*

We noted in the last section that finding the x-intercepts of a parabola

$$y = ax^2 + bx + c$$

is the same problem as that of solving the quadratic equation

$$ax^2 + bx + c = 0.$$

The solutions of a quadratic inequality are related to parabolas in a similar way.

A *quadratic inequality* in x is an inequality of the form

$$ax^2 + bx + c > 0,$$

with $a \neq 0$, or a similar statement with one of the symbols "$\geq$," "$<$," or "$\leq$," instead of "$>$." The *solution set* of a quadratic inequality is the set of all real numbers x for which the statement of inequality is true.

The solution set of the quadratic inequality

$$ax^2 + bx + c > 0$$

can be obtained immediately from the graph of $y = ax^2 + bx + c$, provided that the x-intercepts are known. These solutions are precisely those which give a positive value for y in the equation $y = ax^2 + bx + c$. In other words, the solutions are the values of x for which the graph of $y = ax^2 + bx + c$ is *above* the x-axis.

Similarly, the solutions to

$$ax^2 + bx + c < 0$$

are the values of x for which the graph of $y = ax^2 + bx + c$ is *below* the x-axis.

The procedure of solving a quadratic inequality by using the graph of the related parabola is called the *graphical method* of solution. This method is illustrated in the following examples.

EXAMPLE 1   Use the graphical method to solve

$$x^2 + 6x + 5 > 0.$$

SOLUTION   As explained above, the solution set is the set of all x for which the graph of

$$y = x^2 + 6x + 5$$

is *above* the x-axis. To find these values of x, we must first find the values of x for which $y = 0$. That is, we must find where the parabola crosses the x-axis. Setting $y = 0$, we have

$$x^2 + 6x + 5 = 0,$$

and

$$(x + 5)(x + 1) = 0.$$

Thus the x-intercepts are -5 and -1. The graph opens upward, and is sketched roughly in Figure 5.12. (Note that we did not bother to find the vertex.)

The graph is above the x-axis when $x < -5$, or when $x > -1$. Therefore, $x^2 + 6x + 5 > 0$ when $x < -5$, or $x > -1$. That is, the solution set to

$$x^2 + 6x + 5 > 0$$

is

$$\{x \mid x < -5 \quad \text{or} \quad x > -1\}.$$

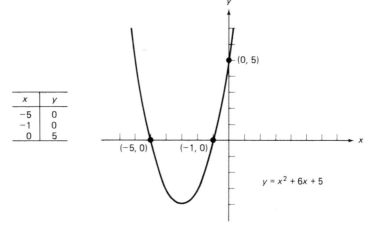

FIGURE 5.12

Note that the solution set for $x^2 + 6x + 5 < 0$ is $\{x \mid -5 < x < -1\}$.

EXAMPLE 2 Use the graphical method to solve

$$x^2 - 5x + 4 \leq 0.$$

SOLUTION The solution set is the set of all x for which the graph of

$$y = x^2 - 5x + 4$$

is *at or below* the x-axis. To find the values of x where the graph crosses the x-axis, we set $y = 0$ and solve for x. This gives

$$x^2 - 5x + 4 = 0,$$

or

$$(x - 1)(x - 4) = 0.$$

Thus the graph crosses the x-axis at $x = 1$ and $x = 4$. The parabola opens upward, and is sketched in Figure 5.13. The graph is below the x-axis for $1 < x < 4$, and crosses the x-axis at $x = 1$ and $x = 4$. This means that the solution set to

$$x^2 - 5x + 4 \leq 0$$

is given by

$$\{x \mid 1 \leq x \leq 4\}.$$

At this point, it is easy to see that the solution of a quadratic inequality in x amounts to determining the values of x which make a quadratic function positive, or negative, or zero, depending on which symbol of inequality is involved. We shall see that these values of x can be deter-

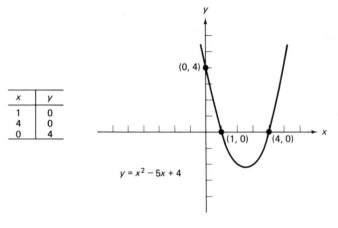

x	y
1	0
4	0
0	4

$y = x^2 - 5x + 4$

FIGURE 5.13

mined without having to draw a graph such as was done in Examples 1 and 2.

In order to solve a quadratic inequality involving the quadratic function

$$y = f(x) = ax^2 + bx + c,$$

we must first find the zeros of the function (i.e., the x-intercepts of the parabola $y = ax^2 + bx + c$). These zeros divide the real number line into regions over which $f(x)$ has the same sign (positive, or negative). This is true because the values of $f(x)$ do not change from positive to negative, or from negative to positive, without going through the value zero. We can determine the sign of $f(x)$ in each region by testing a value of x from that region. Systematically testing all the regions leads to the solution set of the given inequality. This method is called the *algebraic method*, and is illustrated in the remaining examples of this section.

EXAMPLE 3   Use the algebraic method to solve

$$-x^2 + 2x + 3 \leq 0.$$

SOLUTION   We first find the zeros of the quadratic function defined by $y = f(x) = -x^2 + 2x + 3$. We have

$$-x^2 + 2x + 3 = 0,$$

and this is equivalent to

$$x^2 - 2x - 3 = 0,$$

or

$$(x - 3)(x + 1) = 0.$$

The zeros of f are $x = -1$ and $x = 3$, and they separate the real numbers into the regions

$$x < -1, \quad -1 < x < 3, \quad \text{and} \quad x > 3.$$

We choose test points $x = -2$, $x = 0$, and $x = 4$ for each of the regions and compute

$$f(-2) = -(-2)^2 + 2(-2) + 3 = -5,$$
$$f(0) = 3,$$
$$f(4) = -(4)^2 + 2(4) + 3 = -5.$$

The value $x = -2$ is a solution to $f(x) \leq 0$, since $f(-2) = -5 < 0$. This indicates that all points in the region $x < -1$ are solutions. The value $x = 0$ is not a solution to $f(x) \leq 0$, since $f(0) = 3 > 0$. This indicates that all points in the region $-1 < x < 3$ are not solutions. The value $x = 4$ is a solution, since $f(4) = -5 < 0$. This indicates that all points in the region $x > 3$ are solutions. The solution set is given by

$$\{x \mid x \leq -1 \quad \text{or} \quad x \geq 3\}.$$

EXAMPLE 4 Use the algebraic method to solve

$$x^2 - 4x + 5 < 0.$$

SOLUTION As we did before, we look for the zeros of the function $y = f(x) = x^2 - 4x + 5$. But the left member of the equation

$$x^2 - 4x + 5 = 0$$

does not factor, and we resort to the quadratic formula. This gives

$$x = \frac{4 \pm \sqrt{-4}}{2},$$

and the negative number under the radical indicates that there are no real solutions to $x^2 - 4x + 5 = 0$. That is, there are no real zeros of $f(x) = x^2 - 4x + 5$, and this means that $f(x)$ never changes sign. Testing $x = 0$, we have $f(0) = 5$, and this indicates that $f(x)$ is always positive. Thus the solution set to

$$x^2 - 4x + 5 < 0$$

is the empty set, $\varnothing$.

Completing the square on the x-terms sheds some light on the situation. The inequality

$$x^2 - 4x + 5 < 0$$

is equivalent to

$$(x^2 - 4x + 4) + (5 - 4) < 0,$$

or

$$(x - 2)^2 + 1 < 0.$$

This inequality clearly has no solution, and we have solved the problem more quickly than with the algebraic method. The main purpose here was to demonstrate that the algebraic method works, even though it is not always more efficient.

It is sometimes desirable to change a given inequality to an equivalent form, using the properties of inequalities which are stated in Theorem 4-2. An illustration of this sort of change is given in the next example.

EXAMPLE 5   Solve the inequality

$$\frac{x+2}{x-1} > 2.$$

SOLUTION   In the form given, this inequality is not quadratic. However, it can be transformed to an equivalent quadratic inequality. We first subtract 2 from both sides and obtain

$$\frac{x+2}{x-1} - 2 > 0.$$

Getting a common denominator on the left, we have

$$\frac{x+2-2(x-1)}{x-1} > 0,$$

or

$$\frac{-x+4}{x-1} > 0.$$

Now we make the simple observation that $a/b > 0$ if and only if $a \cdot b > 0$. (This is true because either of these statements requires that a and b either both be positive, or both be negative.) Thus we have the equivalent inequality

$$(x-1)(-x+4) > 0,$$

and this is a quadratic inequality. The zeros of $f(x) = (x-1)(-x+4)$ are $x = 1$ and $x = 4$, and they separate the real numbers into the regions

$$x < 1, \quad 1 < x < 4, \quad \text{and} \quad x > 4.$$

Testing a point in each region, we have

$$f(0) = (-1)(4) = -4 < 0,$$
$$f(2) = (1)(2) = 2 > 0,$$
$$f(5) = (4)(-1) = -4 < 0.$$

Thus the solutions to $f(x) > 0$ are given by

$$1 < x < 4,$$

and these are also the solutions to the original inequality

$$\frac{x+2}{x-1} > 2.$$

It is important to notice in Example 5 that *we did not multiply both sides by* $x - 1$. This is not a valid step, because $x - 1$ may be positive, or it may be negative. It is possible to take cases as to whether $x - 1 > 0$, or $x - 1 < 0$, and solve the inequality by considering these cases separately. However, this method is somewhat tedious and inefficient, and we do not go into it here.

The algebraic method can be used to solve inequalities which are not quadratic or linear. This is illustrated in Example 6.

EXAMPLE 6   Solve the inequality

$$\frac{2}{t-1} > \frac{1}{t+2}.$$

SOLUTION   We first obtain an inequality with one member zero by subtracting $\frac{1}{t+2}$ from both sides:

$$\frac{2}{t-1} - \frac{1}{t+2} > 0,$$

or

$$\frac{t+5}{(t-1)(t+2)} > 0.$$

Using the same reasoning as in Example 5, this is equivalent to

$$(t+5)(t-1)(t+2) > 0.$$

The zeros of $f(t) = (t + 5)(t - 1)(t + 2)$ are -5, 1, and -2, and they separate the real numbers into the following regions:

$$t < -5, \quad -5 < t < -2, \quad -2 < t < 1, \quad \text{and} \quad t > 1.$$

Testing one value from each region, we have

$$f(-6) = (-1)(-7)(-4) = -28 < 0,$$
$$f(-3) = (2)(-4)(-1) = 8 > 0,$$
$$f(0) = (5)(-1)(2) = -10 < 0,$$
$$f(2) = (7)(1)(4) = 28 > 0.$$

Thus $f(t)$ is positive for $-5 < t < -2$ or for $t > 1$, and the solution set is

$$\{t \mid -5 < t < -2, \quad \text{or} \quad t > 1\}.$$

Another way to determine the solution set for a given inequality, which is written as a product (or quotient) of linear factors, is to construct a diagram picturing the signs of each factor. Such a diagram is called a *sign graph*. The sign graph for the inequality

$$(t + 5)(t - 1)(t + 2) > 0$$

is given in Figure 5.14. Each of the first three lines in the diagram indicates

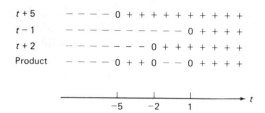

FIGURE 5.14

the sign of a factor in each interval of the number line at the bottom of the diagram. The last line indicates the sign of the product in each interval. Thus we see that

$$(t + 5)(t - 1)(t + 2) > 0 \qquad \text{for } -5 < t < -2 \text{ or } t > 1,$$
$$(t + 5)(t - 1)(t + 2) < 0 \qquad \text{for } t < -5 \text{ or } -2 < t < 1,$$

and

$$(t + 5)(t - 1)(t + 2) = 0 \qquad \text{for } t = -5 \text{ or } t = -2 \text{ or } t = 1.$$

There is another type of inequality that can be solved using the methods of this section. This is illustrated in Example 7 below. The technique used there depends on the facts that $|a|^2 = a^2$ and that if $a > b > 0$, then $a^2 > b^2 > 0$. This allows us to obtain an equivalent inequality by squaring both sides, but *this is valid only when both members of an inequality are nonnegative.*

EXAMPLE 7   Solve the inequality

$$\left|\frac{z + 1}{z - 2}\right| > 2.$$

SOLUTION   Since both members are nonnegative, we may square both sides and obtain the equivalent inequality

$$\frac{(z + 1)^2}{(z - 2)^2} > 4.$$

If $z \neq 2$, then $(z - 2)^2$ is positive. Thus we may multiply both sides by $(z - 2)^2$, under the restriction that $z \neq 2$. This gives

$$(z + 1)^2 > 4(z - 2)^2,$$

or

$$z^2 + 2z + 1 > 4z^2 - 16z + 16.$$

Collecting all nonzero terms in the left member, we have

$$-3z^2 + 18z - 15 > 0.$$

Dividing by -3, we reverse the sense of the inequality and obtain

$$z^2 - 6z + 5 < 0,$$

or
$$(z - 5)(z - 1) < 0.$$

The zeros of $f(z) = (z - 5)(z - 1)$ are 1 and 5, and they separate the real numbers into

$$z < 1, \quad 1 < z < 5, \quad \text{and} \quad z > 5.$$

Testing a point in each region, we have

$$f(0) = (-5)(-1) = 5 > 0,$$
$$f(2) = (-3)(1) = -3 < 0,$$
$$f(6) = (1)(5) = 5 > 0.$$

Thus the solutions to $(z - 5)(z - 1) < 0$ are given by

$$1 < z < 5.$$

But $z = 2$ is included in this range, and we had the restriction $z \neq 2$ imposed when we cleared the inequality of fractions. Looking at the original inequality, we see that $z = 2$ gives a zero denominator, and therefore is not a solution. Thus the solutions to

$$\left| \frac{z + 1}{z - 2} \right| > 2$$

are given by

$$\{z \mid 1 < z < 2 \quad \text{or} \quad 2 < z < 5\}.$$

EXERCISES 5-5

Use the graphical method to solve the following quadratic inequalities.

1. $x^2 - x - 12 > 0$ 2. $x^2 + 7x + 10 > 0$
3. $x^2 - 2x - 3 \leq 0$ 4. $x^2 - 8x + 12 \leq 0$
5. $-x^2 + 4x - 3 \leq 0$ 6. $9 - 8x - x^2 \leq 0$
7. $-2x^2 + 6 < 0$ 8. $x^2 - 4 < 0$
9. $x^2 - 9x > 0$ 10. $6x + x^2 > 0$
11. $8x - 2x^2 \geq 0$ 12. $4x - x^2 \geq 0$
13. $-2x^2 + 4 < 0$ 14. $-3x^2 + 9 < 0$
15. $2x - x^2 \leq 0$ 16. $6x - 2x^2 \leq 0$
17. $x^2 + 6x + 16 < 8$ 18. $x^2 + 4x + 6 \geq 3$

Use the algebraic method to solve the following inequalities.

19. $z^2 + 2z \leq 15$ 20. $w^2 + 2w > 99$
21. $3t^2 + 13t - 10 > 0$ 22. $4s^2 + 3s - 1 \geq 0$
23. $-15x^2 + 28x - 12 > 0$ 24. $-2z^2 + 11z - 15 > 0$
25. $3t^2 > 4t$ 26. $6w^2 < w$

27. $r - 6r^2 > -35$

28. $4x^2 > 2 - 7x$

29. $21 - 4x^2 > -5x$

30. $2x^2 < 12 - 5x$

31. $-4t^2 - 4t + 1 \geq 0$

32. $w^2 + 8w + 13 \geq 0$

33. $(x - 1)(x - 2)(x - 3) > 0$

34. $(x - 1)(x - 2)(x - 3)(x - 4) \leq 0$

Solve the following inequalities by using the methods employed in Examples 5-7.

35. $\dfrac{x + 1}{2x - 1} > 3$

36. $\dfrac{2x - 1}{x} < 0$

37. $\dfrac{3x - 1}{x + 2} < 1$

38. $\dfrac{3t + 2}{t} \leq 2$

39. $\dfrac{t + 1}{t + 3} \geq 2$

40. $\dfrac{1}{w - 2} > \dfrac{1}{3}$

41. $\dfrac{7 - z}{(z - 2)(z - 3)} < 0$

42. $\dfrac{x - 5}{(3x - 1)(2x - 3)} < 0$

43. $\dfrac{2}{w + 2} \geq \dfrac{1}{2w + 1}$

44. $\dfrac{1}{r - 3} \leq \dfrac{3}{r + 1}$

45. $\dfrac{5}{2t + 3} \geq \dfrac{-5}{t}$

46. $\dfrac{3}{x + 1} < \dfrac{2}{x + 2}$

47. $\left| \dfrac{2t - 1}{t} \right| > 1$

48. $\left| \dfrac{2}{w + 2} \right| \leq 4$

49. $\left| \dfrac{z + 2}{z - 3} \right| < 2$

50. $\left| \dfrac{3x - 3}{x + 4} \right| \geq 5$

51. $|z + 2| > |3z + 2|$

52. $|4x + 2| < 5|3 - x|$

*53. $\left| \dfrac{s}{2s + 1} \right| > -5$

*54. $-|3t + 2| > |8t - 1|$

PRACTICE TEST for Chapter 5

1. Find the domain D and range R of the function defined by
$$y = \sqrt{x^2 - 9} - 4.$$

2. If $f(x) = x^2 - 2x$ and $g(x) = \sqrt{x} + 3$, find each of the following.
   (a) $f(4)$ (b) $f(g(9))$ (c) $(g \circ f)(4)$
   (d) $f(g(a))$ (e) $f(x + h) - f(x)$

3. Sketch the graph of the equation $y = |3x + 2|$.

4. Graph each linear relation below, and find the x- and y-intercepts, if they exist.
   (a) $3x - 2y = 12$ (b) $y = 5$

5. Find an equation, in standard form, of the straight line passing through $(3, -1)$ and $(2, 7)$.

6. Locate the vertex and the axis of symmetry, and sketch the graph of the parabola
$$x = \tfrac{2}{3}(y - 1)^2 + 4.$$

7. Sketch the graph of the relation defined by
$$x = -4 + \sqrt{y - 4}.$$

8. Use the graphical method to solve
$$x^2 + 3x + 2 > 6.$$

In problems 9 and 10, use the algebraic method to solve the given inequality.

9. $w^2 + 6w + 8 < 0$

10. $\left| \dfrac{3x - 1}{x + 1} \right| > 2$

Conic Sections and Systems of Equations and Inequalities

6-1
Introduction We have seen in Section 5-4 that the graph of a quadratic relation is a parabola. The parabola is one of three types of curves known as the *conic sections*. The term "conic sections" comes from the fact that each type of curve can be obtained by intersecting a plane with a right circular cone.[1]

If the intersecting plane has a suitable inclination to the axis of the cone, the curve of intersection is a *parabola*, as shown in Figure 6.1(a). With a greater inclination to the axis of the cone, the curve of intersection may be an *ellipse*, which is an oval-shaped curve, as shown in Figure 6.1(b). The *circle* is a special case of the ellipse. With a lesser inclination, the curve of intersection may be a *hyperbola*, which is a curve having two branches as shown in Figure 6.1(c).

Our treatment of the conic sections in this chapter is a limited one that considers only special cases. More extensive developments are done in the study of analytic geometry and calculus.

[1]By a *cone*, we mean a complete cone, with two nappes, as shown in Figure 6.1.

Conic Sections
and Systems
of Equations
and Inequalities

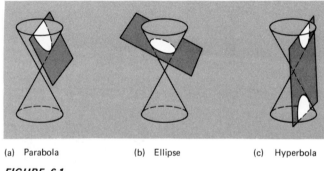

(a) Parabola   (b) Ellipse   (c) Hyperbola

FIGURE 6.1

6-2
The Circle
In this section, we obtain a standard equation for a circle. The derivation of this standard equation requires the use of a formula for the distance between two points in the plane.

Let a, b, and c represent the lengths of the sides of a right triangle, with c the length of the hypotenuse. (See Figure 6.2.) The *Pythagorean Theorem* states the following relationship between the lengths of the sides:

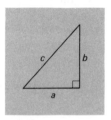

FIGURE 6.2

$$c^2 = a^2 + b^2,$$

or

$$c = \sqrt{a^2 + b^2}.$$

Suppose now that we have a right triangle drawn as indicated in Figure 6.3. The coordinates of the vertices of the right triangle are (x_1, y_1), (x_2, y_1), and (x_2, y_2), as indicated in the figure. For the figure as drawn, the length of the side parallel to the x-axis is $x_2 - x_1$, and the length of the side parallel to the y-axis is $y_2 - y_1$. The Pythagorean Theorem can be used to determine the length, d, of the hypotenuse:

$$d = \sqrt{(x_2 - x_1)^2 + (y_2 - y_1)^2}.$$

The length, d, of the hypotenuse represents the distance between the points with coordinates (x_1, y_1) and (x_2, y_2). This equation holds inde-

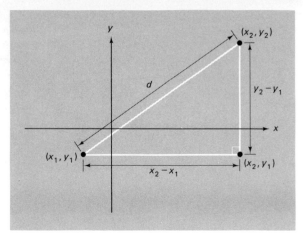

FIGURE 6.3

pendently of the quadrants in which the points lie, and independently
of the points' orientation to each other. This result is known as the
distance formula.

THEOREM 6-1 (THE DISTANCE FORMULA) The distance d be-
tween the points with coordinates (x_1, y_1) and (x_2, y_2) is given by

$$d = \sqrt{(x_2 - x_1)^2 + (y_2 - y_1)^2}.$$

EXAMPLE 1 The distance between the points $(-1, 3)$ and $(7, -3)$ is given by
$$d = \sqrt{[7 - (-1)]^2 + (-3 - 3)^2}$$
$$= \sqrt{64 + 36}$$
$$= 10.$$

In using the distance formula, we chose $(x_1, y_1) = (-1, 3)$, and (x_2, y_2)
$= (7, -3)$. The same result would have been obtained if we had chosen
$(x_1, y_1) = (7, -3)$, and $(x_2, y_2) = (-1, 3)$.

EXAMPLE 2 The distance between the point (x, y) and the origin $(0, 0)$ is given by
$$d = \sqrt{(x - 0)^2 + (y - 0)^2}$$
$$= \sqrt{x^2 + y^2}.$$

Squaring both sides yields
$$x^2 + y^2 = d^2.$$

DEFINITION 6-2 A *circle* is the set of all points equally distant
from a given point, called the *center* of the circle. The distance from the
center of the circle to any point on the circle is called the *radius* of the
circle.

Suppose that we wish to obtain an equation for a circle with radius r and center $(0, 0)$. Referring to Example 2, we see that the set of all points (x, y) whose distance from $(0, 0)$ is r is given by

$$\{(x, y) \mid x^2 + y^2 = r^2\}.$$

and hence an equation for a circle with center at $(0, 0)$ and radius r is

$$x^2 + y^2 = r^2.$$

Similarly, a point (x, y) is at a distance r from the fixed point (h, k) if and only if its coordinates satisfy the equation

$$\sqrt{(x - h)^2 + (y - k)^2} = r,$$

or

$$(x - h)^2 + (y - k)^2 = r^2. \tag{1}$$

That is, a point (x, y) is on the circle with center (h, k) and radius r if and only if its coordinates (x, y) satisfy the equation (1). This important result is stated in the following theorem, and is represented graphically in Figure 6.4.

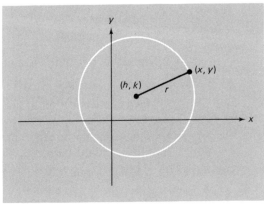

FIGURE 6.4

THEOREM 6-3   The *standard equation* for a circle with center (h, k) and radius r is

$$(x - h)^2 + (y - k)^2 = r^2. \tag{1}$$

EXAMPLE 3   If we put $r = 0$ in equation (1), we have

$$(x - h)^2 + (y - k)^2 = 0. \tag{2}$$

Since the square of a real number cannot be negative, this equation requires both $x - h = 0$ and $y - k = 0$. Thus equation (2) is satisfied only by the coordinates of the single point (h, k), and the graph of equation (2) consists of this one point. For this reason, the graph of equation (2) is sometimes referred to as a *point-circle*, or a *degenerate circle*.

EXAMPLE 4 Determine whether or not the graph of each equation below is a circle. If it is, find the center and radius.

(a) $(x - 5)^2 + (y + 1)^2 = 9$
(b) $x^2 + 2x + y^2 - 6y = 0$
(c) $x^2 + y^2 = 10x - 8y - 41$
(d) $2x^2 + 2y^2 - 4x + 10 = 0$

SOLUTION (a) Expressing $(x - 5)^2 + (y + 1)^2 = 9$ in the form of equation (1), we have

$$(x - 5)^2 + [y - (-1)]^2 = (3)^2.$$

Thus the graph is a circle with radius 3 and center $(5, -1)$.

(b) We must complete the squares on the variables x and y in order to change the equation into the standard form. The given equation

$$x^2 + 2x + y^2 - 6y = 0$$

is equivalent to

$$x^2 + 2x + 1 + y^2 - 6y + 9 = 1 + 9.$$

Expressing this in the form of equation (1), we have

$$[x - (-1)]^2 + (y - 3)^2 = (\sqrt{10})^2,$$

which is the equation of a circle with center at $(-1, 3)$ and radius $\sqrt{10}$.

(c) Completing the squares on the variables x and y, and rewriting in the standard form, we have

$$x^2 - 10x + 25 + y^2 + 8y + 16 = -41 + 25 + 16,$$

and

$$(x - 5)^2 + [y - (-4)]^2 = 0.$$

This is the equation of a circle with center at $(5, -4)$ and radius 0, that is, a point-circle. The values $x = 5$, $y = -4$ are the only ones which satisfy the equation $x^2 + y^2 = 10x - 8y - 41$.

(d) Dividing both sides of the given equation by 2 and rearranging yields

$$x^2 - 2x + y^2 = -5.$$

Completing the square on x gives

$$x^2 - 2x + 1 + y^2 = -5 + 1,$$

or

$$(x - 1)^2 + (y - 0)^2 = -4.$$

Since the sum of the squares of two real numbers can never be negative, there are no values of x and y which satisfy this equation. Hence, the equation

$$2x^2 + 2y^2 - 4x + 10 = 0$$

does not represent a circle. The set of solutions to this equation is the empty set $\varnothing$.

Certain equations have a graph which consists of only a part of a circle. Such an equation is presented in the next example.

EXAMPLE 5   Graph the equation

$$y - 1 = -\sqrt{4 - x^2}.$$

SOLUTION   In the given form, the equation does not much resemble the standard form for the equation of a circle. However, if we square both sides, we obtain

$$(y - 1)^2 = 4 - x^2,$$

or

$$x^2 + (y - 1)^2 = 4.$$

This is the equation of a circle with center $(0, 1)$ and radius 2. But the original equation

$$y - 1 = -\sqrt{4 - x^2}$$

requires that $y - 1 \leq 0$, or that $y \leq 1$. Thus the graph of the original equation consists of only those points on the circle where $y \leq 1$. This graph is shown in Figure 6.5.

FIGURE 6.5

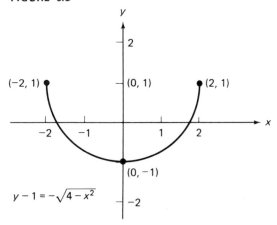

Determine the distance between the given points.

1. $(0, 0)$ and $(4, 3)$ 2. $(-1, 0)$ and $(2, 0)$

3. $(0, -3)$ and $(0, 5)$ 4. $(3, -2)$ and $(1, 0)$

5. $(0, -3)$ and $(3, 1)$ 6. $(-4, -6)$ and $(-1, -2)$

7. $(-9, 22)$ and $(11, 1)$ 8. $(-3, -12)$ and $(-4, 12)$

9. $(-2, -3)$ and $(4, 0)$ 10. $(-1, 7)$ and $(-1, -1)$

11. $(2, 9)$ and $(-2, -9)$ 12. $(4, -8)$ and $(-1, -2)$

13. (a, b) and (b, a) 14. (a, b) and $(-b, a)$

15. Find the shortest distance between $(3, 6)$ and the line $y = 10$.

16. Find the shortest distance between $(-2, 7)$ and the line $x = 4$.

17. If $f(x) \leq 10$, find the vertical distance between $y = f(x)$ and $y = 10$.

18. If $f(x) \leq g(x)$, find the vertical distance between $y = f(x)$ and $y = g(x)$.

Determine whether or not the graph of each equation is a circle. If it is, find the center and radius.

19. $x^2 + y^2 = 16$ 20. $(x - 1)^2 + (y - 2)^2 = 9$

21. $(x + 1)^2 + y^2 = 1$ 22. $(x + 2)^2 + (y - 2)^2 = 2$

23. $x^2 + (y + 4)^2 = \frac{3}{25}$ 24. $x^2 + (y - 7)^2 = \frac{1}{4}$

25. $(x + 3)^2 + (y + 5)^2 = 25$ 26. $(x - a)^2 + y^2 = b^2$

27. $(x - 2a)^2 + (y + 2a)^2 = 4a^2$ 28. $(x + b)^2 + (y - b)^2 = 9$

29. $x^2 + 2x + y^2 + 6y = 0$ 30. $x^2 - 2x + y^2 + 4y = 4$

31. $x^2 + y^2 = 3x$ 32. $x^2 + y^2 = 8y$

33. $x^2 + y^2 = 10x - 6y - 36$ 34. $x^2 + y^2 = 12x + 2y - 40$

35. $x^2 + y^2 - 4x + 16y + 68 = 0$ 36. $x^2 + y^2 - 3x + 3y + 12 = 0$

37. $2x^2 + 2y^2 + 4x + 8y = 8$ 38. $7x^2 + 7y^2 = x$

Write an equation of the circle with the given center and radius.

39. Center $(0, 1)$, radius 2   40. Center $(-1, 3)$, radius 1

41. Center $(2, -2)$, radius $\frac{1}{2}$ 42. Center $(-3, 0)$, radius 10

43. Center $(-3, -4)$, radius 7   44. Center $(10, -3)$, radius $\frac{3}{4}$

45. Center $(-3, 5)$, radius 1   46. Center $(a, -a)$, radius $a/2$

47. Center $(a, 1)$, radius 3   48. Center $(1, -b)$, radius b^2

Graph the following relations.

49. $x^2 + y^2 = 4$ 50. $(x - 2)^2 + (y - 1)^2 = 9$

51. $(x + 1)^2 + (y + 2)^2 = 16$ 52. $x^2 + (y - 4)^2 = 1$

53. $(x - 2)^2 + y^2 = 25$

54. $(x + 3)^2 + (y - 6)^2 = 36$

55. $x^2 + y^2 = 10x$

56. $x^2 + y^2 = -4y$

57. $y = -\sqrt{1 - x^2}$

58. $x = -\sqrt{16 - y^2}$

59. $2x = -\sqrt{1 - 4y^2}$

60. $3y = \sqrt{3 - 9x^2}$

61. $y + 1 = \sqrt{16 - x^2}$

62. $y - 1 = -\sqrt{16 - (x + 2)^2}$

63. $x + 2 = -\sqrt{25 - y^2}$

64. $x - 4 = \sqrt{4 - (y + 5)^2}$

*65. Show that the triangle with vertices $(2, 2)$, $(-2, -2)$, and $(-1, 5)$ is a right triangle.

*66. Show that the triangle with vertices $(-3, 5)$, $(4, 3)$, and $(-3, 1)$ is an isosceles triangle.

*67. Show that $(6, -9)$, $(-1, 4)$, $(9, 1)$, and $(-4, -6)$ are vertices of a square.

*68. Show that $(1, 1)$, $(4, -5)$, and $(-2, 7)$ lie on a straight line.

*69. Show that $(-3, 0)$, $(3, 8)$, $(1, -3)$, and $(7, 5)$ are vertices of a rectangle.

6-3
The Ellipse

There are standard forms for the equation of an ellipse, somewhat analogous to the standard equation of a circle. We obtain these standard equations for an ellipse in this section.

DEFINITION 6-4   Let F_1 and F_2 denote two fixed points in a plane. An *ellipse* is the set of all points P in the plane such that the sum of the distance from P to F_1 and the distance from P to F_2 is a constant. Each of the fixed points is called a *focus* (plural: *foci*) of the ellipse.

To obtain a standard equation for an ellipse, we choose a coordinate system in the plane such that the foci F_1 and F_2 are located at $(c, 0)$ and $(-c, 0)$, where $c > 0$. This is indicated in Figure 6.6. In order to have a form of the equation which is as simple as possible, we let $2a$ denote the constant which is the sum of the distances from P to F_1 and from P

FIGURE 6.6

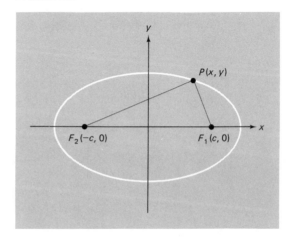

to F_2. This requires that $a > c$, since $2c$ is the distance between F_1 and F_2. A point P with coordinates (x, y) is then on the ellipse if and only if

$$\sqrt{(x - c)^2 + (y - 0)^2} + \sqrt{(x + c)^2 + (y - 0)^2} = 2a.$$

If we isolate the first radical on the left and square both sides, we have

$$[\sqrt{(x - c)^2 + y^2}]^2 = [2a - \sqrt{(x + c)^2 + y^2}]^2,$$

or

$$x^2 - 2cx + c^2 + y^2 = 4a^2 - 4a\sqrt{(x + c)^2 + y^2} + x^2 + 2cx + c^2 + y^2.$$

This simplifies to

$$-2cx = 4a^2 - 4a\sqrt{(x + c)^2 + y^2} + 2cx.$$

In order to remove the remaining radical, we isolate it on the left and obtain

$$4a\sqrt{(x + c)^2 + y^2} = 4a^2 + 4cx.$$

Dividing both sides by 4, we have

$$a\sqrt{(x + c)^2 + y^2} = a^2 + cx.$$

Squaring both sides now yields

$$a^2(x^2 + 2cx + c^2 + y^2) = a^4 + 2a^2cx + c^2x^2,$$

or

$$a^2x^2 + 2a^2cx + a^2c^2 + a^2y^2 = a^4 + 2a^2cx + c^2x^2.$$

This simplifies to

$$(a^2 - c^2)x^2 + a^2y^2 = a^2(a^2 - c^2).$$

We divide both sides by $a^2(a^2 - c^2)$ and obtain

$$\frac{x^2}{a^2} + \frac{y^2}{a^2 - c^2} = 1.$$

As one last simplification, we let b be such that

$$b^2 = a^2 - c^2, \qquad \text{where } b > 0.$$

This is possible, since $a > c$. This substitution yields

$$\frac{x^2}{a^2} + \frac{y^2}{b^2} = 1. \tag{1}$$

Up to this point, our derivation shows that if a point (x, y) is on the ellipse, then the coordinates must satisfy equation (1). It is also true that any point (x, y) with coordinates which satisfy equation (1) must be on the ellipse. We leave this verification, which is somewhat involved, as an exercise, and simply state the following theorem as a result that appears plausible at this point.

THEOREM 6-5 An ellipse with foci on the x-axis symmetrically placed with respect to the origin has a *standard equation* of the form

$$\frac{x^2}{a^2} + \frac{y^2}{b^2} = 1, \tag{1}$$

where $a > b > 0$. Similarly, an ellipse with foci on the y-axis symmetrically placed with respect to the origin has a *standard equation* of the form

$$\frac{y^2}{a^2} + \frac{x^2}{b^2} = 1, \tag{2}$$

where $a > b > 0$.

A typical graph corresponding to equation (1) is shown in Figure 6.7(a), and a typical graph for equation (2) is shown in Figure 6.7(b).

FIGURE 6.7

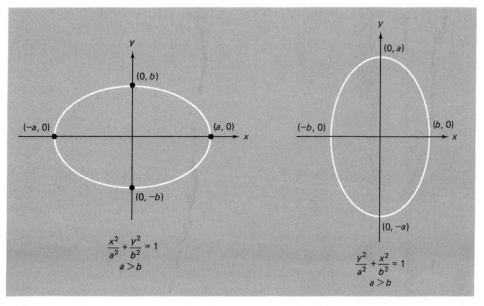

(a)   (b)

The points where these graphs cross the coordinate axes are called the *vertices* of the ellipse.

It is worth noting that the condition $a > b > 0$ is required in both types of equations in Theorem 6-5. The larger square occurs under the variable which corresponds to the location of the foci.

We also note that if one puts $a = b$ in an equation of either type (1) or (2), the resulting equation appears as

$$\frac{x^2}{a^2} + \frac{y^2}{a^2} = 1,$$

where $a > 0$. This is equivalent to

$$x^2 + y^2 = a^2,$$

which we recognize as the equation of a circle. In this sense, then, a circle can be regarded as a special case of an ellipse.

As was the case with the circle, a degenerate conic section is obtained when 1 is replaced by 0 in equation (1) or equation (2). In these cases, the resulting equation is satisfied only by the coordinates of the origin.

EXAMPLE 1   Sketch the graph of the ellipse

$$4x^2 + 9y^2 = 36.$$

SOLUTION   Since the given equation is not in either of the forms given in Theorem 6-5, it may not be immediately clear that this is an equation of an ellipse. In order to obtain one of the standard forms, we divide both members by 36. This gives

$$\frac{4x^2}{36} + \frac{9y^2}{36} = \frac{36}{36},$$

or

$$\frac{x^2}{9} + \frac{y^2}{4} = 1.$$

This fits the form of equation (1) in Theorem 6-5, with $a = 3$ and $b = 2$, so we have an ellipse with foci on the x-axis. It is easy to see that the x-intercepts are ± 3, and the y-intercepts are ± 2. With this information, we can draw the graph as in Figure 6.8.

Example 1 illustrates how easy it is to sketch the graph of an ellipse by using the intercepts. The intercepts can actually be read by inspection from either of the equations in Theorem 6-5. They are always given by $\pm a$ and $\pm b$, *with the square of the x-intercepts under x^2, and the square of the y-intercepts under y^2.*

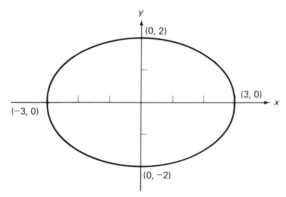

FIGURE 6.8

For any ellipse which has an equation of one of the types given in Theorem 6-5, the graph is symmetric about the x-axis, about the y-axis, and about the origin. Because of this, these are called ellipses with centers at the origin. In analytic geometry, one deals with ellipses which have centers at points other than the origin.

There are variations in the graphs of ellipses which are analogous to the variations in the graphs of parabolas and circles that we have seen earlier. One such variation is demonstrated in the next example.

EXAMPLE 2 Sketch the graph of

$$y = \frac{\sqrt{4 - 9x^2}}{2}.$$

SOLUTION In the given form, the equation does not fit one of our standard forms. In order to recognize the type of graph that we are dealing with, we multiply both sides by 2, and then square both sides. This gives

$$(2y)^2 = (\sqrt{4 - 9x^2})^2,$$

or

$$4y^2 = 4 - 9x^2.$$

Adding $9x^2$ to both sides, we have

$$4y^2 + 9x^2 = 4.$$

Division by 4 gives a 1 in the right member, and the resulting equation reads

$$y^2 + \frac{x^2}{\frac{4}{9}} = 1,$$

or

$$\frac{y^2}{(1)^2} + \frac{x^2}{(\frac{2}{3})^2} = 1.$$

This last equation fits equation (2) in Theorem 6-5 with $a = 1$ and $b = \frac{2}{3}$. The graph of this equation is given in Figure 6.9(a). But this is *not* the graph of the original equation $y = \sqrt{4 - 9x^2}/2$, because the original equation requires that y not be negative. Taking only the top half of the ellipse, we obtain the graph of the original equation, as shown in Figure 6.9(b).

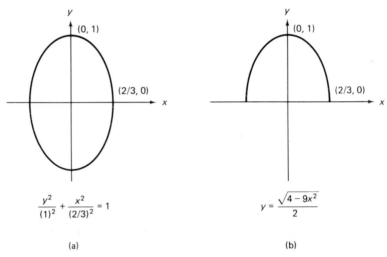

$$\frac{y^2}{(1)^2} + \frac{x^2}{(2/3)^2} = 1$$

(a)

$$y = \frac{\sqrt{4 - 9x^2}}{2}$$

(b)

FIGURE 6.9

EXERCISES 6-3

Sketch the graphs of the following equations.

1. $\dfrac{x^2}{4} + \dfrac{y^2}{9} = 1$

2. $\dfrac{x^2}{25} + \dfrac{y^2}{16} = 1$

3. $25x^2 + 4y^2 = 100$

4. $9x^2 + 16y^2 = 144$

5. $x^2 + 4y^2 = 16$

6. $y^2 = 36 - 9x^2$

7. $4x^2 + 25y^2 = 25$

8. $9x^2 = 9 - 16y^2$

9. $\dfrac{4x^2}{9} + \dfrac{9y^2}{16} = 1$

10. $\dfrac{9x^2}{4} + \dfrac{16y^2}{25} = 1$

11. $x^2 + y^2 = 25$

12. $9x^2 + 16y^2 = 25$

13. $64x^2 + 36y^2 = 100$

14. $36x^2 + 16y^2 = 25$

15. $2x^2 + y^2 = 8$

16. $3x^2 + y^2 = 12$

17. $2x^2 + 3y^2 = 24$

18. $3x^2 + 5y^2 = 45$

19. $x^2 + 4y^2 = 0$

20. $4x^2 + 9y^2 = 0$

21. $2y = \sqrt{16 - x^2}$

22. $3y = \sqrt{25 - 4x^2}$

23. $5x = -\sqrt{9 - 4y^2}$

24. $3x = -\sqrt{25 - 6y^2}$

25. $y = -\sqrt{16 - 9x^2}$

26. $y = -\sqrt{4 - 9x^2}$

27. $x = \dfrac{\sqrt{25 - 16y^2}}{2}$

28. $x = \dfrac{\sqrt{16 - 9y^2}}{5}$

29. $y = \dfrac{-\sqrt{9 - 4x^2}}{3}$

30. $y = \dfrac{-\sqrt{100 - 9x^2}}{4}$

*31. Complete the derivation of the first statement in Theorem 6-5 by showing that any point with coordinates (x, y) which satisfy equation (1) must be on the ellipse.

*32. Suppose that an ellipse has foci at $(0, \pm c)$ on the y-axis, and that the sum of the distances from a point on the ellipse to the foci is $2a$, where $a > c > 0$. Show that the ellipse has an equation of the form

$$\frac{y^2}{a^2} + \frac{x^2}{b^2} = 1,$$

where $b^2 = a^2 - c^2$.

6-4

The Hyperbola

In this section, we formulate some standard forms for an equation of a hyperbola. As with the ellipse, our development is limited to special situations. A hyperbola can be defined in a manner which is analogous to the way in which we defined an ellipse. The formal definition is as follows.

DEFINITION 6-6 Let F_1 and F_2 denote two fixed points in a plane. A *hyperbola* is the set of all points P in the plane such that the difference of the distances from P to the fixed points is a constant. Each of the fixed points is called a *focus* (plural: *foci*) of the hyperbola.

This definition invites a comparison with the definition of an ellipse. The most striking change is that the word "sum" has been replaced by the word "difference." In contrast to a sum, a difference between two given real numbers can be formed in *two distinct ways*, if the numbers are unequal. The difference referred to in Definition 6-6 is the difference obtained when the smaller number is subtracted from the larger.

A standard equation for a hyperbola can be derived in a manner very similar to that done in Section 6-3 for an ellipse. We once again choose a coordinate system in the plane such that the foci F_1 and F_2 are located at $(c, 0)$ and $(-c, 0)$, where $c > 0$. This is illustrated in Figure 6.10. In order to obtain a simple form of the equation, we let $2a$ denote the constant which equals the difference of the distances from P to the foci. For some points, the larger distance involved is between P and F_1, while for others, the larger distance is between P and F_2. To cover both of these possibilities with one equation, we write

$$\left| \sqrt{(x - c)^2 + (y - 0)^2} - \sqrt{(x + c)^2 + (y - 0)^2} \right| = 2a,$$

which is equivalent to

$$\sqrt{(x - c)^2 + (y - 0)^2} - \sqrt{(x + c)^2 + (y - 0)^2} = \pm 2a.$$

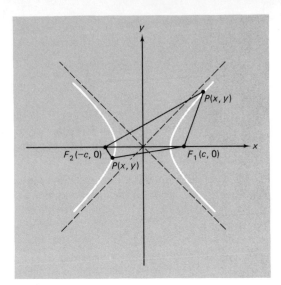

FIGURE 6.10

In writing these equations, we have assumed that $0 < a < c$. This is necessary in order to have points on the hyperbola that are not on the x-axis. (Notice that the distance between F_1 and F_2 is $2c$.)

From this point on, the radicals can be eliminated by the same type of procedure as was employed in the derivation of the equation of an ellipse. We leave as an exercise the details of this procedure, which concludes with the equation

$$\frac{x^2}{a^2} - \frac{y^2}{c^2 - a^2} = 1.$$

Then the substitution

$$b^2 = c^2 - a^2, \qquad \text{where } b > 0,$$

yields

$$\frac{x^2}{a^2} - \frac{y^2}{b^2} = 1. \tag{1}$$

This is the standard form of the equation of a hyperbola with foci on the x-axis. The derivation that we have outlined shows that any point on the hyperbola has coordinates which satisfy the standard equation (1). It can also be demonstrated that any point with coordinates which satisfy equation (1) is on the hyperbola. This is left as one of the more difficult exercises.

The numbers a and b which appear in the standard equation are closely related to the graph of the hyperbola. This is exhibited in Figure 6.11. The points on the branches of the curve which are nearest the other branch are called *vertices*, and are labeled $V_1(a, 0)$ and $V_2(-a, 0)$ in the

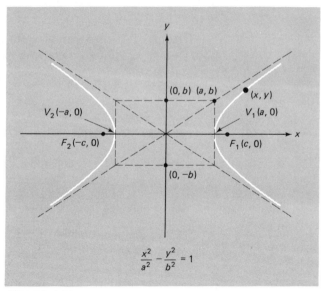

FIGURE 6.11

figure. The dashed lines and line segments are *not* part of the graph, of course, but serve as guides in drawing the hyperbola. The dashed lines passing through the origin and the vertices of the rectangle have equations $y = \pm(b/a)x$. As the distance between x and 0 increases from a, the points (x, y) on the hyperbola are nearer these dashed lines. These dashed lines are called the *asymptotes* of the hyperbola. The rectangle is drawn to serve as a guide in sketching the asymptotes, and it emphasizes that the slopes of the asymptotes are $\pm b/a$.

It is an interesting fact that if the number 1 in the right member of equation (1) is replaced by 0, one obtains an equation of the asymptotes. This is easy to confirm, for the equation

$$\frac{x^2}{a^2} - \frac{y^2}{b^2} = 0$$

is equivalent to

$$b^2x^2 - a^2y^2 = 0,$$

or

$$(bx - ay)(bx + ay) = 0.$$

This product is zero if and only if one of the factors is zero, that is, if and only if

$$y = \pm\frac{b}{a}x.$$

The main results of the foregoing discussion are summarized in the following theorem.

THEOREM 6-7 A hyperbola with foci on the x-axis symmetrically placed with respect to the origin has a *standard equation* of the form

$$\frac{x^2}{a^2} - \frac{y^2}{b^2} = 1.$$

The equations of the asymptotes are given by

$$y = \pm \frac{b}{a} x,$$

which can be obtained from equation (1) by replacing the 1 by 0.

We note that the condition $a > b$, which was required in the case of the ellipse, is *not present* in the case of the hyperbola.

EXAMPLE 1 Sketch the graph of the following hyperbola, locate the vertices and asymptotes, and write the equations of the asymptotes:

$$9x^2 - 4y^2 = 36.$$

SOLUTION We first divide both members by 36 so as to obtain the standard form of the equation. This gives

$$\frac{x^2}{4} - \frac{y^2}{9} = 1,$$

or

$$\frac{x^2}{(2)^2} - \frac{y^2}{(3)^2} = 1.$$

Using $a = 2$ and $b = 3$ in the same manner as in Figure 6.11, we obtain the graph in Figure 6.12. The vertices are located at $(\pm 2, 0)$, and the asymptotes are given by $y = \pm \frac{3}{2}x$.

As one would expect, Theorem 6-7 has a dual for the situation where the foci are on the y-axis. This is stated in Theorem 6-8, and illustrated in Figure 6.13.

THEOREM 6-8 A hyperbola with foci on the y-axis symmetrically placed with respect to the origin has a *standard equation* of the form

$$\frac{y^2}{a^2} - \frac{x^2}{b^2} = 1. \qquad (2)$$

The equations of the asymptotes are given by

$$y = \pm \frac{a}{b} x,$$

which can be obtained from equation (2) by replacing the 1 by 0.

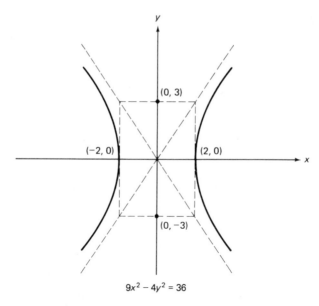

$$9x^2 - 4y^2 = 36$$

FIGURE 6.12

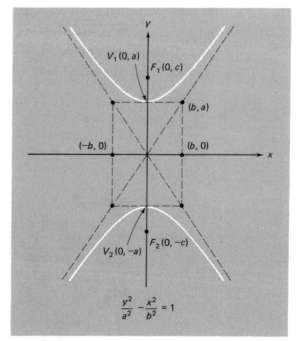

$$\frac{y^2}{a^2} - \frac{x^2}{b^2} = 1$$

FIGURE 6.13

It is easy to see that Theorem 6-8 can be obtained from Theorem 6-7 by simply interchanging x and y.

Once again, there are variations in the equations which correspond to only portions of a hyperbola.

EXAMPLE 2   Sketch the graph of

$$y = -\frac{5\sqrt{x^2 + 16}}{4}.$$

SOLUTION   If we multiply both sides of the given equation by 4 and then square both sides, we have

$$(4y)^2 = 25(x^2 + 16),$$

or

$$16y^2 = 25x^2 + 400.$$

In order to obtain a standard equation, we subtract $25x^2$ from both sides, and then divide both sides by 400. This gives

$$\frac{y^2}{25} - \frac{x^2}{16} = 1,$$

or

$$\frac{y^2}{(5)^2} - \frac{x^2}{(4)^2} = 1.$$

We recognize this as the standard equation of a hyperbola with $a = 5$, $b = 4$, and foci on the y-axis. Using Figure 6.13 as a guide, we obtain the graph of this last equation as given in Figure 6.14(a). But the original equation $y = \frac{-5\sqrt{x^2 + 16}}{4}$ requires that y be nonpositive. Thus we must take only the bottom half of the hyperbola to have the graph of the original equation. This is shown in Figure 6.14(b).

FIGURE 6.14

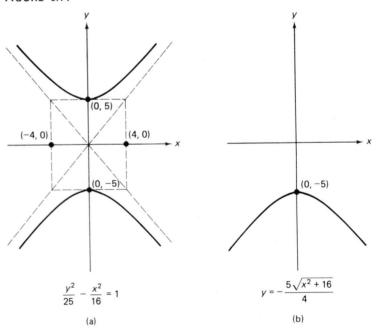

$$\frac{y^2}{25} - \frac{x^2}{16} = 1$$

(a)

$$y = -\frac{5\sqrt{x^2 + 16}}{4}$$

(b)

160

Conic Sections
and Systems
of Equations
and Inequalities

Sketch the graphs of the following equations. Locate the vertices and asymptotes of all hyperbolas, and write the equations of the asymptotes.

1. $\dfrac{x^2}{9} - \dfrac{y^2}{16} = 1$ 2. $\dfrac{x^2}{9} - \dfrac{y^2}{4} = 1$

3. $\dfrac{y^2}{4} - \dfrac{x^2}{25} = 1$ 4. $\dfrac{y^2}{36} - \dfrac{x^2}{25} = 1$

5. $y^2 - x^2 = 9$ 6. $y^2 - x^2 = 16$

7. $4x^2 - 9y^2 = 36$ 8. $36x^2 - 4y^2 = 144$

9. $9x^2 + 4y^2 = 0$ 10. $x^2 + y^2 = 0$

11. $4x^2 - 25y^2 = 0$ 12. $16x^2 - 9y^2 = 0$

13. $x^2 - 4y^2 = 4$ 14. $9x^2 - y^2 = 9$

15. $4x^2 + 25y^2 = 100$ 16. $16x^2 + 9y^2 = 144$

17. $\dfrac{16y^2}{25} - \dfrac{9x^2}{4} = 1$ 18. $\dfrac{25y^2}{16} - \dfrac{4x^2}{9} = 1$

19. $x^2 + y^2 = 16$ 20. $x^2 + y^2 = 9$

21. $2x^2 - y^2 = 8$ 22. $y^2 - 3x^2 = 75$

23. $5y = \sqrt{9 + 4x^2}$ 24. $3y = \sqrt{25 + 16x^2}$

25. $2x = \sqrt{16 + y^2}$ 26. $3x = \sqrt{25 + 4y^2}$

27. $x = -\dfrac{\sqrt{25 + 16y^2}}{2}$ 28. $x = -\dfrac{\sqrt{16 + 9y^2}}{5}$

*29. Show that
$$\sqrt{(x - c)^2 + (y - 0)^2} - \sqrt{(x + c)^2 + (y - 0)^2} = \pm 2a$$
is equivalent to
$$\frac{x^2}{a^2} - \frac{y^2}{b^2} = 1,$$
where a, b, and c are positive, $c > a$, and $b^2 = c^2 - a^2$. This proves the first statement in Theorem 6-7.

*30. Consider a hyperbola with foci at $(0, \pm c)$. Let $2a$ denote the constant difference of the distances from a point on the hyperbola to the foci. Assuming that $0 < a < c$, show that the hyperbola has an equation of the form
$$\frac{y^2}{a^2} - \frac{x^2}{b^2} = 1,$$
with $b^2 = c^2 - a^2$. This proves the first statement in Theorem 6-8.

*31. Show that equation (1) in Theorem 6-7 is equivalent to
$$y = \pm \frac{b}{a} x \sqrt{1 - \frac{a^2}{x^2}}.$$
From this equation, it can be seen that, as $|x|$ increases from a, the points (x, y) on the hyperbola in problem 29 move nearer the lines $y = \pm(b/a)x$.

*32. Show that equation (2) in Theorem 6-8 is equivalent to

$$y = \pm \frac{a}{b} x \sqrt{1 + \frac{b^2}{x^2}}.$$

From this equation, it can be seen that, as $|x|$ increases from a, the points (x, y) on the hyperbola in problem 30 move nearer the lines $y = \pm(a/b)x$.

6-5
Systems
of Equations

It is appropriate at this point to consider a very important problem which frequently arises in connection with the graphs of equations. The problem is to find the coordinates (x, y) of the points which are common to the graphs of two equations in x and y, that is, the *points of intersection* of the two graphs. When one is working with this type of problem, the pair of equations involved is referred to as a *system of equations*. A set of values for x and y which satisfies both equations is called a *simultaneous solution* of the system, or simply a *solution* of the system.

In this section, we consider two methods of solution to the problem described above. Each method is based on the idea that a point which is on both graphs must have coordinates which satisfy both equations. The simpler of these two methods is known as the *substitution method*. The idea here is to use one of the given equations and solve it for one of the variables in terms of the other, then to *substitute* this value in the other equation, thereby obtaining an equation which involves only one variable. This procedure, which sounds more complicated than it really is, is illustrated in the following example.

EXAMPLE 1

Find the point of intersection of the lines $3y + x = 7$ and $3x - 2y + 12 = 0$. That is, solve the system of equations

$$3y + x = 7$$
$$3x - 2y + 12 = 0.$$

SOLUTION

As indicated in the discussion above, we shall employ the substitution method. The first step, then, is to solve for one of the variables in one of the given equations. The simplest possibility here is to solve for x in the first equation, obtaining

$$x = 7 - 3y.$$

When this value for x is substituted in the other equation, we have

$$3(7 - 3y) - 2y + 12 = 0.$$

This simplifies to

$$33 - 11y = 0,$$
$$33 = 11y,$$

and

$$y = 3.$$

Substituting this value for y in $x = 7 - 3y$, we get $x = -2$. Thus the point of intersection is $(-2, 3)$, and it is easy to check that the coordinates of this point do indeed satisfy both of the original equations. The situation is shown geometrically in Figure 6.15.

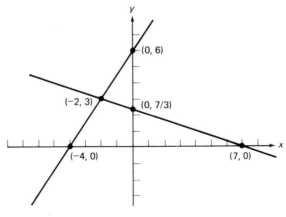

FIGURE 6.15

Another illustration of solution by the substitution method is provided in Example 2.

EXAMPLE 2   Solve the system below. Graph the two equations on the same coordinate system, and label the points of intersection.

$$y^2 - 5y - 4x - 28 = 0$$
$$y - 4x = 1$$

SOLUTION   It is convenient here to solve for y in the second equation, obtaining

$$y = 4x + 1.$$

Substituting in the first equation, we have

$$(4x + 1)^2 - 5(4x + 1) - 4x - 28 = 0,$$

which simplifies to

$$16x^2 - 16x - 32 = 0,$$

and

$$x^2 - x - 2 = 0.$$

This factors as

$$(x - 2)(x + 1) = 0.$$

Thus the x-coordinates of the points of intersection are $x = 2$ and $x = -1$. When these values are substituted in $y = 4x + 1$, we find the points of intersection are given by $(2, 9)$ and $(-1, -3)$.

To draw the graph of the first equation, we solve for x, obtaining

$$x = \tfrac{1}{4}y^2 - \tfrac{5}{4}y - 7.$$

We recognize this as the equation of a parabola which opens to the right with vertex at $y = -b/2a = \tfrac{5}{2}$, $x = -\tfrac{137}{16}$. The graph of $y - 4x = 1$ is a straight line with slope 4 and y-intercept 1. The graphs and the points of intersection are shown in Figure 6.16.

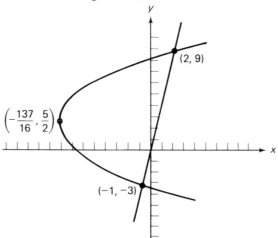

FIGURE 6.16

There is one more point that should be made before leaving this example. It is important to notice that when the values $x = 2$ and $x = -1$ were obtained, they were substituted in the equation of the line, and not in the equation for the parabola. If they had been used in the equation for the parabola, *two* values of y would have been obtained for each x, and only one of these values would satisfy the linear equation. More specifically, when $x = -1$ is substituted in $y^2 - 5y - 4x - 28 = 0$, one obtains $y = -3$ and $y = 8$. The value $y = 8$ is an extraneous solution, since $(-1, 8)$ is not on the line. Similarly, $x = 2$ yields $y = 9$ and $y = -4$, with $y = -4$ an extraneous solution.

The other method of solution referred to at the beginning of this section is the *elimination method*. This name derives from the fact that one of the variables is eliminated by an appropriate combination of the two equations in a system. Suppose that the system of equations is represented symbolically by

$$R(x, y) = c_1$$
$$S(x, y) = c_2,$$

where $R(x, y)$ and $S(x, y)$ denote algebraic expressions in the variables x and y, and c_1 and c_2 denote constants (real numbers). We first observe that if m_1 and m_2 are nonzero real numbers, then any solution of the given system is also a solution to the system

$$m_1 R(x, y) = m_1 c_1$$
$$m_2 S(x, y) = m_2 c_2,$$

and therefore a solution to the sum of these equations

$$m_1 R(x, y) + m_2 S(x, y) = m_1 c_1 + m_2 c_2.$$

The key to success in the elimination method is to choose the multipliers m_1 and m_2 in such a way as to have one of the variables eliminated when the sum is formed.[2] This method is used in the next example.

EXAMPLE 3 Solve the system of equations below by the elimination method. Sketch the graphs of the two equations on the same coordinate system, and label the points of intersection.

$$9x^2 + y^2 = 36$$
$$x^2 + 3y^2 = 56$$

SOLUTION If we multiply the first equation by 3, the second equation by -1, and then add, we have

$$27x^2 + 3y^2 = 108$$
$$-x^2 - 3y^2 = -56$$
$$\overline{ 26x^2 = 52.}$$

Thus y is eliminated, and this equation simplifies to

$$x^2 = 2,$$

and

$$x = \pm\sqrt{2}.$$

In order to complete the solution, we substitute these values in one of the equations in the original system. Substitution of $x = \sqrt{2}$ in the first equation yields

$$9(2) + y^2 = 36,$$
$$y^2 = 18,$$

and

$$y = \pm 3\sqrt{2}.$$

Substitution of $x = -\sqrt{2}$ yields the same values for y. This gives the four pairs

$$(\sqrt{2}, 3\sqrt{2}), (\sqrt{2}, -3\sqrt{2}), (-\sqrt{2}, 3\sqrt{2}), (-\sqrt{2}, -3\sqrt{2}),$$

and each of these checks in the original system. It is good practice to *always check solutions in each of the original equations*. This is especially

---

[2]In some cases, it may be more convenient to form a difference of the equations, rather than a sum. This is illustrated in Example 4 of this section.

true when extraction of roots has been employed in obtaining the solutions. The graphs and points of intersection are shown in Figure 6.17.

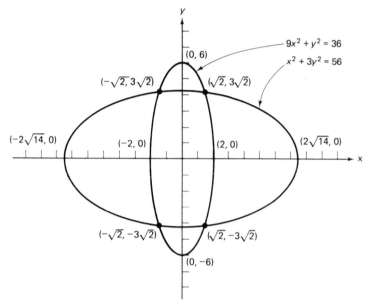

FIGURE 6.17

Graphs that picture the solutions of the system have been provided in each of the preceding examples of this section. These are valuable for understanding, and they provide a rough check on the solutions. They are not indispensible, however, and we omit them in the remainder of the section.

In a system which involves two linear equations, the method of substitution can always be used, as was done in Example 1 of this section. But, if the substitution method requires the use of fractions, it is usually easier to use the elimination method. An example of this situation is given below.

EXAMPLE 4   Solve the following system.

$$4x + 3y = \quad 1$$
$$2x - 5y = -19$$

SOLUTION   If we multiply the second equation by 2 and subtract the result from the first equation, x is eliminated. This appears as

$$4x + \quad 3y = \quad 1$$
$$4x - 10y = -38$$
$$\overline{\quad\quad 13y = \quad 39,}$$

and
$$y = \quad 3.$$
Substitution in the first equation gives
$$4x + 9 = 1,$$
$$4x = -8,$$
$$x = -2.$$
Thus $(-2, 3)$ is the solution of the system.

The two methods of solution that have been presented in this section are not adequate for the solution of all types of systems which might be encountered. The examples which have been presented here were chosen so as to lend themselves to solution by these methods. Some problems of a more difficult nature can be solved by a combination of the two methods. This is illustrated in the next example.

EXAMPLE 5 Solve the following system.
$$x^2 + 2xy + 2y^2 = 5$$
$$x^2 + xy + 2y^2 = 4$$

SOLUTION We see that it is impossible to employ the elimination method directly. Also, the substitution method is not an attractive prospect, since the quadratic formula would have to be used to obtain one variable in terms of the other. However, all terms involving a square can be eliminated by subtracting the second equation from the first. When this is done, we have
$$\begin{array}{r} x^2 + 2xy + 2y^2 = 5 \\ x^2 + xy + 2y^2 = 4 \\ \hline xy = 1, \end{array}$$
and
$$y = \frac{1}{x}.$$
This gives us a value for y which can be substituted in either of the original equations. Substitution in the first equation yields
$$x^2 + 2x\left(\frac{1}{x}\right) + 2\left(\frac{1}{x}\right)^2 = 5,$$
which simplifies to
$$x^2 + \frac{2}{x^2} = 3.$$
Clearing the equation of fractions, we have
$$x^4 + 2 = 3x^2,$$

and
$$x^4 - 3x^2 + 2 = 0.$$

This factors as

$$(x^2 - 2)(x^2 - 1) = 0.$$

Setting $x^2 - 2 = 0$, we obtain $x = \pm\sqrt{2}$, and $x^2 - 1 = 0$ yields $x = \pm 1$. Using each of these values for x in the equation $y = 1/x$, we obtain the solution set

$$\{(\sqrt{2}, \sqrt{2}/2), (-\sqrt{2}, -\sqrt{2}/2), (1, 1), (-1, -1)\}.$$

It is easy to confirm that all these solutions check in the original system. The interested reader may wish to verify that substitution of the values for x in either of the original equations yields extraneous solutions for y.

It may well happen that a given system of equations does not have any solutions. This is illustrated below.

EXAMPLE 6   Solve the following system.

$$x^2 + y^2 = 1$$
$$x^2 - y = 4$$

SOLUTION   Subtracting the second equation from the first, we have

$$\begin{aligned} x^2 + y^2 &= 1 \\ x^2 - y &= 4 \\ \hline y^2 + y &= -3 \end{aligned}$$

or

$$y^2 + y + 3 = 0.$$

The value of the discriminant for this quadratic equation is $b^2 - 4ac = (1)^2 - 4(1)(3) = -11$, so there is no real solution for y. This, in turn, indicates that the original system has no solution.

As a final example, we consider a system in which an absolute value is involved.

EXAMPLE 7   Solve the following system.

$$9x^2 + 4y^2 = 36$$
$$y = |x - 2|$$

SOLUTION   As usual when dealing with absolute values, we consider cases. In the second equation, we have

$$y = x - 2 \quad \text{when} \quad x - 2 \geq 0 \quad \text{or} \quad x \geq 2,$$

and
$$y = -x + 2 \quad \text{when} \quad x - 2 < 0 \quad \text{or} \quad x < 2.$$

We first consider the case where $y = x - 2$ (i.e., when $x \geq 2$). Substituting this value for y in the first equation yields

$$9x^2 + 4(x - 2)^2 = 36,$$

and this simplifies to

$$13x^2 - 16x - 20 = 0,$$

or

$$(13x + 10)(x - 2) = 0.$$

The solutions here are $x = -\frac{10}{13}$ and $x = 2$. But we are restricted in this case to have $x \geq 2$, so the value $x = -\frac{10}{13}$ is discarded. Corresponding to $x = 2$, we obtain $y = 0$, and $(2, 0)$ is one solution.

We consider now the case where $y = -x + 2$ (i.e., where $x < 2$). Substituting this value in the equation of the ellipse, we have

$$9x^2 + 4(2 - x)^2 = 36,$$

which simplifies to

$$13x^2 - 16x - 20 = 0.$$

This is the same equation as in the first case, and we obtain the same values as before: $x = -\frac{10}{13}$, and $x = 2$. But this time, we are restricted to $x < 2$, and we are using $y = -x + 2$. We discard $x = 2$, and obtain

$$y = -(-\tfrac{10}{13}) + 2$$
$$= \tfrac{36}{13}$$

corresponding to $x = -\frac{10}{13}$. This gives the solution $(-\frac{10}{13}, \frac{36}{13})$ to the system. The complete solution set to the system is given by

$$\{(2, 0), (-\tfrac{10}{13}, \tfrac{36}{13})\}.$$

EXERCISES 6-5

Solve the following systems of equations. It is not necessary to graph the equations.

1. $x + 2y = 4$
   $3x - 2y = -12$

2. $2y = x + 4$
   $3x + 2y + 12 = 0$

3. $2x - 5y = 11$
   $5x + y = 14$

4. $7x + 3y = 5$
   $5x - y = 13$

5. $3y = 2x + 6$
   $6y + 5x = 39$

6. $3y = 20 - 5x$
   $5x = 10 - 2y$

7. $2x + 9y = 3$
   $5x + 7y = -8$

8. $3x + 5y = 9$
   $5x + 7y = 13$

9. $x - 2y = 7$
   $2x = 4y - 14$

10. $2x = 3y - 5$
    $9y = 6x - 15$

11. $2x + 7y = 15$
    $xy = 1$

12. $6x + y + 7 = 0$
    $xy = 2$

13. $9y^2 - 4x^2 = 7$
    $xy = -2$

14. $2x^2 + y^2 = 19$
    $xy = 3$

15. $2x^2 + 3xy + y^2 = 12$
    $2x^2 - xy + y^2 = 4$

16. $2x^2 - 3xy - 3y^2 = 11$
    $2x^2 - xy - 3y^2 = 7$

17. $5x^2 - 4xy - 3y^2 = -8$
    $5x^2 + 5xy - 3y^2 = 28$

18. $8x^2 + 6xy - 9y^2 = 8$
    $8x^2 - 3xy - 9y^2 = 17$

*19. $x^2 - 3y^2 = -2$
    $xy + 2y^2 = 3$

*20. $3x^2 - 4xy = 25$
    $2x^2 - 4y^2 = 9$

Solve each of the following systems. Sketch the graphs of the two equations on the same coordinate system, and label the points of intersection.

21. $y = 2x + 6$
    $x^2 = 2y$

22. $x^2 - y = 0$
    $2x - y + 3 = 0$

23. $y = 3x^2 + 12x$
    $2x - y = 16$

24. $y = 3 - 2x - x^2$
    $4x + y = 5$

25. $x^2 + y^2 = 4$
    $2x - y = 2$

26. $x^2 + y^2 = 25$
    $y - x = 7$

27. $x^2 + y^2 = 10x$
    $4y = 3x - 8$

28. $x^2 + y^2 = 25$
    $x + 3y = 5$

29. $x^2 - 2y^2 = -1$
    $2x - y = -1$

30. $3x^2 + 2y^2 = 5$
    $x - y = -2$

31. $4x^2 + 25y^2 = 100$
    $x + 2y = 8$

32. $36x^2 + 16y^2 = 25$
    $x - 2y = 12$

33. $9x^2 + 16y^2 = 144$
    $3x^2 + 4y^2 = 36$

34. $4x^2 + 25y^2 = 83$
    $9x^2 + 16y^2 = 66$

35. $y = x^2 - 2x - 3$
    $y = |x + 1|$

36. $y = 10 + 8x - 2x^2$
    $y = |16x - 32|$

37. $x^2 + y^2 = 25$
    $(3x - 4y)(3x + 4y) = 0$

38. $4x^2 + 36y^2 = 100$
    $(x - 4y)(x + 4y) = 0$

*39. Find a value of the real number m so that the line $y = mx - 5$ is tangent to the circle $x^2 + y^2 = 9$.

*40. Find a value of the real number a so that the line $ax - 12y + 65 = 0$ is tangent to the circle $x^2 + y^2 = 25$.

6-6
Systems
of Inequalities

We have seen that a solution to a given equation in x and y is a set of values for x and y which satisfies the equation, and that the graph of the equation consists of the points with coordinates (x, y) which correspond to solutions of the equation. The situation is much the same for inequalities involving x and y. A *solution* to a given inequality in x and y is a set of values for x and y which makes the given inequality a true statement.

The *graph* of the inequality consists of the points with coordinates which correspond to solutions of the inequality.

The simplest type of inequality in x and y is one which is obtained from a linear equality $ax + by = c$ by replacing the equality sign by one of the inequality symbols ">," "<," "$\geq$," or "$\leq$." Such inequalities are called *linear inequalities*.

Suppose that a certain line in the plane has an equation given by $ax + by = c$. The line separates the points of the plane into three distinct subsets:

(1) the points (x, y), where $ax + by = c$;

(2) the points (x, y), where $ax + by > c$;

(3) the points (x, y), where $ax + by < c$.

Geometrically, the points described in (1) are the points on the line, and the points described in (2) are the points on one side of the line, and the points described in (3) are the points on the other side of the line. Each of the regions described in (2) and (3) is called a *half-plane*, and the line is the *boundary* of the half-planes.

EXAMPLE 1   Sketch the graph of
$$3x + 4y > 24.$$

SOLUTION   We first locate the points on the line $3x + 4y = 24$. It is easy to see that the x-intercept of the line is 8, and the y-intercept is 6. The line is drawn in Figure 6.18 as a dashed line to indicate that the points on the boundary

FIGURE 6.18

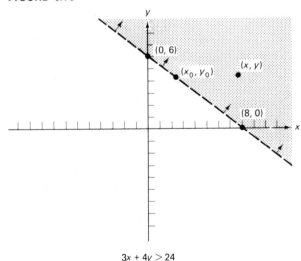

$3x + 4y > 24$

are not part of the solution set. To see how the solutions to $3x + 4y > 24$ consist of all points on one side of the line, let us start with a particular point (x_0, y_0) on the line. If the x-coordinate increases to a new value $x > x_0$, the value of $3x$ increases, and

$$3x + 4y_0 > 3x_0 + 4y_0 = 24.$$

That is, $3x + 4y_0 > 24$ for a point (x, y_0) to the right of (x_0, y_0) on the line. Any point on the right-hand side of the line is located to the right of some point on the line, so we see that all points on the right-hand side of the line satisfy $3x + 4y > 24$. Similar reasoning shows that the inequality $3x + 4y < 24$ holds for all points on the left side of the line. The solution set to $3x + 4y > 24$ is shaded in Figure 6.18, and the side of the line which yields solutions is indicated by arrows based on the line.

The *solution set* to a system of inequalities is the intersection of the solution sets to the individual inequalities in the system. An example of this is furnished below.

EXAMPLE 2   Graph the solution set of the following system.

$$3x + 4y > 24$$
$$x - 2y \geq -2$$

SOLUTION   We use the solution of the first inequality from Example 1. This solution set is indicated in Figure 6.19 by shading with lines that slope downward to the right, and by arrows based on the boundary.

FIGURE 6.19

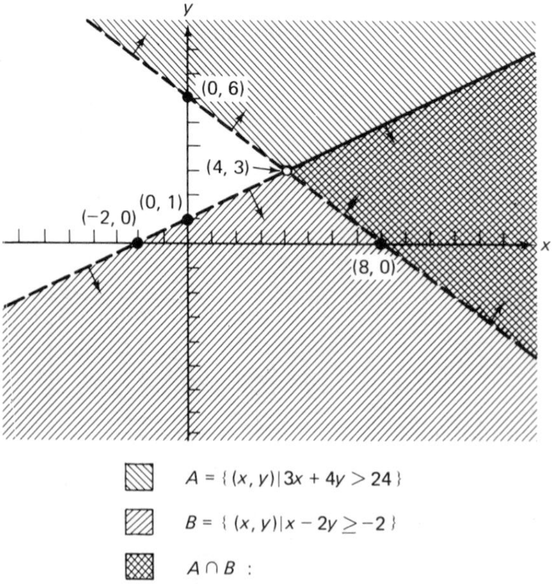

$\boxed{}$ $A = \{(x, y) \mid 3x + 4y > 24\}$

$\boxed{}$ $B = \{(x, y) \mid x - 2y \geq -2\}$

$\boxed{}$ $A \cap B$:

To graph the solution set of the second inequality, we first sketch the line $x - 2y = -2$. The x-intercept is -2, and the y-intercept is 1. To determine the half-plane which is the solution to $x - 2y > -2$, we simply test a point on one side of the line. The origin $(0, 0)$ is the simplest choice to use, and $0 > -2$ indicates that the solutions to $x - 2y > -2$ are on the same side of the line as the origin. This region is indicated in Figure 6.19 by shading with lines which slope upward to the right, and by arrows based on the boundary. The boundary is included in the solution set since equality is included in $x - 2y \geq -2$.

The shading done leads to a crosshatching of the intersection of the individual solution sets, and the region shaded by crosshatching is the solution set of the system. The portion of the boundary of $x - 2y \geq -2$ which is in the solution set of the system is drawn with a solid line to indicate its inclusion in the graph.

The description that was given for the graphs of linear inequalities generalizes to other types of inequalities in x and y. For a given inequality, a corresponding equation may be obtained by replacing the inequality symbol by an equality symbol. The graph of the equation separates the plane into regions, and it forms the boundary of these regions. Each region either consists entirely of solutions, or it contains no solutions at all. The solution set can be determined by simply testing one point from each region. The boundary is included or omitted, according to whether or not equality is permitted in the original statement of inequality.

EXAMPLE 3   Graph the solution set for the following system.

$$x^2 + y^2 \leq \tfrac{25}{9}$$
$$16x - 9y^2 > 0$$

SOLUTION   We begin by graphing the two equations which correspond to the given inequalities.

The equation

$$x^2 + y^2 = \tfrac{25}{9}$$

is an equation of a circle with center $(0, 0)$ and radius $\tfrac{5}{3}$. If we solve for x in $16x - 9y^2 = 0$, we have

$$x = \tfrac{9}{16}y^2,$$

which is an equation of a parabola which has vertex at the origin and opens to the right. To find the points of intersection, we can substitute $y^2 = \tfrac{16}{9}x$ in the equation of the circle. This gives

$$x^2 + \tfrac{16}{9}x = \tfrac{25}{9},$$

or

$$9x^2 + 16x - 25 = 0.$$

This can be factored as

$$(9x + 25)(x - 1) = 0,$$

so we obtain $x = -\frac{25}{9}$ and $x = 1$ as the solutions to this equation. Substituting $x = -\frac{25}{9}$ in $y^2 = \frac{16}{9}x$ yields

$$y^2 = \frac{16}{9}\left(-\frac{25}{9}\right)$$

$$= -\frac{400}{81},$$

which is impossible for a real number y. This indicates that there is no point of intersection at $x = -\frac{25}{9}$. Substituting $x = 1$ in $y^2 = \frac{16}{9}x$ yields

$$y^2 = \frac{16}{9},$$

and

$$y = \pm\frac{4}{3}.$$

The coordinates $(1, \frac{4}{3})$ and $(1, -\frac{4}{3})$ check in both original equations, and give the points of intersection.

It is clear that the solutions to $x^2 + y^2 \leq \frac{25}{9}$ are the points interior to, or on, the circle $x^2 + y^2 = \frac{25}{9}$. To determine which points satisfy $16x - 9y^2 > 0$, we choose $(3, 0)$ as a test point. Since

$$16(3) - 9(0)^2 > 0,$$

$(3, 0)$ is a solution. This indicates that the solution set to $16x - 9y^2 > 0$ is the region to the right of the parabola $16x - 9y^2 = 0$. The solution set of the system, then, is the set of points which lie to the right of the parabola and on, or interior to, the circle. This is shown in Figure 6.20.

FIGURE 6.20

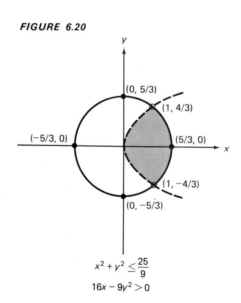

$$x^2 + y^2 \leq \frac{25}{9}$$

$$16x - 9y^2 > 0$$

EXAMPLE 4 Graph the solution set for the following system.

$$x^2 - y^2 \geq 16$$
$$x^2 + 16y^2 \leq 169$$

SOLUTION   The graph of $x^2 - y^2 = 16$ is a hyperbola, and the graph of $x^2 + 16y^2 = 169$ is an ellipse. The points of intersection are easily found to be $(5, \pm 3)$ and $(-5, \pm 3)$, as shown in Figure 6.21. Using $(0, 0)$ as a test point, we find that the origin is not a solution to the first inequality, but it is a solution to the second. This indicates that the solutions are on the opposite side of the hyperbola from the origin, and on the same side of the ellipse as the origin. The solution set is shaded in Figure 6.21.

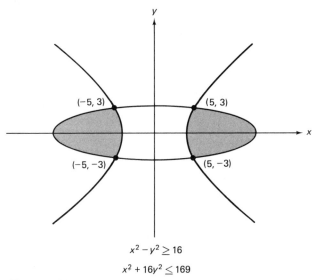

$$x^2 - y^2 \geq 16$$
$$x^2 + 16y^2 \leq 169$$

FIGURE 6.21

EXERCISES 6-6

In problems 1–18, graph the solution set of the given system of inequalities, showing all points of intersection of the boundaries. These problems correspond to the problems numbered 21 through 38 in Exercises 6-5.

1. $y \leq 2x + 6$
    $x^2 \leq 2y$

2. $x^2 - y \leq 0$
    $2x - y + 3 \geq 0$

3. $y > 3x^2 + 12x$
    $2x - y \geq 16$

4. $y < 3 - 2x - x^2$
    $4x + y \geq 5$

5. $x^2 + y^2 \leq 4$
   $2x - y < 2$

6. $x^2 + y^2 \leq 25$
   $y - x < 7$

7. $x^2 + y^2 < 10x$
   $4y > 3x - 8$

8. $x^2 + y^2 < 25$
   $x + 3y > 5$

9. $x^2 - 2y^2 < -1$
   $2x - y < -1$

10. $3x^2 + 2y^2 \leq 5$
    $x - y < -2$

11. $4x^2 + 25y^2 \leq 100$
    $x + 2y > 8$

12. $36x^2 + 16y^2 \leq 25$
    $x - 2y < 12$

13. $9x^2 + 16y^2 < 144$
    $3x^2 + 4y^2 \geq 36$

14. $4x^2 + 25y^2 \geq 83$
    $9x^2 + 16y^2 < 66$

15. $y \geq x^2 - 2x - 3$
    $y < |x + 1|$

16. $y > 10 + 8x - 2x^2$
    $y < |16x - 32|$

17. $\qquad x^2 + y^2 \leq 25$
    $(3x - 4y)(3x + 4y) > 0$

18. $\qquad 4x^2 + 36y^2 \leq 100$
    $(x - 4y)(x + 4y) > 0$

In problems 19–34, graph the solution set of the given system. Label all points of intersection of the boundaries.

19. $x + y = 3$
    $3x - y > 6$

20. $x + y = 1$
    $2x + 3y < 4$

21. $x - y > 3$
    $x + 2y > 4$

22. $3x + 2y < 6$
    $x - y > 2$

23. $y \leq 2x + 2$
    $y + x + 1 \geq 0$
    $2y + 5x \leq 13$

24. $y + 6 > 3x$
    $x + y < 6$
    $x \geq 1$

25. $x > y^2 + 1$
    $x - y = 3$

26. $x > y^2 - 4y + 3$
    $x - y = 3$

27. $x^2 + y^2 > 4$
    $4x^2 + 9y^2 \leq 36$

28. $x^2 + y^2 \geq 9$
    $9x^2 + 16y^2 < 144$

29. $x^2 + 2y^2 \geq 18$
    $2x^2 + y^2 \leq 33$

30. $9x^2 + 16y^2 \leq 25$
    $16x^2 + 9y^2 \geq 25$

31. $x^2 - y^2 > 1$
    $4x^2 + 9y^2 < 36$

32. $x^2 + 4y^2 < 100$
    $4x^2 - 9y^2 > 36$

33. $x^2 + 2y^2 \geq 64$
    $x^2 - y^2 \leq 16$

34. $x^2 - y^2 \geq -1$
    $4x^2 + 9y^2 \geq 36$

6-7
Inverse Functions

In Section 5-1, we considered the inverse g^{-1} of a given relation g. For example, if g is given by

$$g = \{(x, y) \,|\, y = 4x - 2\},$$

then g^{-1} is found by interchanging x and y in the defining equation for g:

$$g^{-1} = \{(x, y) \,|\, x = 4y - 2\}$$
$$= \{(x, y) \,|\, y = \tfrac{1}{4}x + \tfrac{1}{2}\}.$$

The graphs of g and g^{-1} are straight lines, as shown in Figure 6.22. Notice that the graph of the inverse relation, $y = \frac{1}{4}x + \frac{1}{2}$, is the mirror image of the graph of the original relation, $y = 4x - 2$, reflected through the line $y = x$.

In general, the graph of the inverse of a relation can be sketched by simply reflecting the graph of the original relation through the line $y = x$. This is true because the points (a, b) and (b, a) are always symmetric to each other with respect to the line $y = x$. (See problem 41 in Exercises 6-7.)

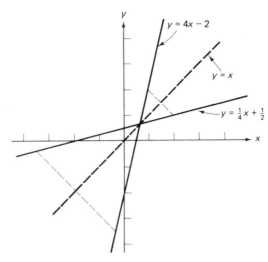

FIGURE 6.22

EXAMPLE 1 Sketch the graph of the inverse of each of the following relations by reflecting the graph of the original relation through the line $y = x$. Use the vertical line test to determine if the relation and/or its inverse is a function.

(a) $g(x) = 2x + 5$ (b) $f(x) = x^2$ (c) $h(x) = |x + 2|$

SOLUTION   The graph of each relation and its inverse are shown in Figure 6.23.

(a) Using the vertical line test, we see that the relations g and g^{-1} are both functions.

(b) Although the relation f is a function, its inverse f^{-1} is not a function.

(c) The relation h is a function, and the inverse relation h^{-1} is not a function.

When both a relation and its inverse are functions, a special result holds for the composition of the two functions. To see how this goes,

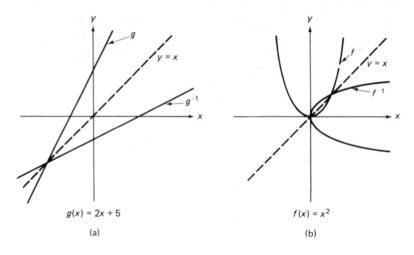

$$g(x) = 2x + 5$$

(a)

$$f(x) = x^2$$

(b)

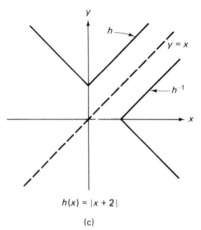

$$h(x) = |x + 2|$$

(c)

FIGURE 6.23

consider the function

$$g(x) = 2x + 5$$

from part (a) of Example 1, and its inverse, which is given by

$$g^{-1}(x) = \tfrac{1}{2}x - \tfrac{5}{2}.$$

The two compositions of the functions g and g^{-1} are given by

$$g(g^{-1}(x)) = g\left(\frac{x}{2} - \frac{5}{2}\right) = 2\left(\frac{x}{2} - \frac{5}{2}\right) + 5 = x$$

for any x, and

$$g^{-1}(g(x)) = g^{-1}(2x + 5) = \frac{2x + 5}{2} - \frac{5}{2} = x$$

for any x. The special result of both evaluations is x.

The result illustrated in the preceding paragraph occurs for any pair of inverse functions. Suppose that f and f^{-1} are functions. If x is any element in the domain of f, then we have

$$(x, f(x)) \in f.$$

From the definition of f^{-1}, it follows that

$$(f(x), x) \in f^{-1},$$

and this means that $f(x)$ is an element in the domain of f^{-1} such that

$$f^{-1}(f(x)) = x.$$

Similarly, it can be shown (see problem 42 of Exercises 6-7) that

$$f(f^{-1}(x)) = x$$

for any x in the domain of f^{-1}, and we have the following theorem.

THEOREM 6-9   Suppose that f and f^{-1} are both functions. Then

$$f^{-1}(f(x)) = x$$

for all x in the domain of f, and

$$f(f^{-1}(x)) = x$$

for all x in the domain of f^{-1}.

EXAMPLE 2   For the function f defined by

$$y = \sqrt{3 - x},$$

where $y = f(x)$, (a) state the domain and range of f and f^{-1}, and (b) write an equation for the inverse function in the form $y = f^{-1}(x)$.

SOLUTION   The function f is given by

$$f = \{(x, y) \mid y = \sqrt{3 - x}\}.$$

In order for $f(x)$ to be defined, we must have $3 - x \geq 0$, or $x \leq 3$. Thus the domain of f is

$$\{x \mid x \leq 3\}.$$

Since the principal square root is never negative, the range of f is

$$\{y \mid y \geq 0\}.$$

Interchanging x and y in the defining equation for f yields the inverse

function:
$$f^{-1} = \{(x, y) \mid x = \sqrt{3 - y}\}.$$

Since $\sqrt{3 - y} \geq 0$ for all y, the domain of f^{-1} is
$$\{x \mid x \geq 0\},$$

and since $3 - y \geq 0$ must hold in order for $\sqrt{3 - y}$ to be defined, the range of f^{-1} is
$$\{y \mid y \leq 3\}.$$

To obtain an equation of the form $y = f^{-1}(x)$ for the inverse function, we must solve the equation
$$x = \sqrt{3 - y}$$

for y. Squaring both sides, we easily obtain
$$y = f^{-1}(x) = 3 - x^2, \qquad \text{where } x \geq 0.$$

The condition $x \geq 0$ *must be stated explicitly*, because its implicit presence was lost when both sides were squared. The graphs of f and f^{-1} are shown in Figure 6.24.

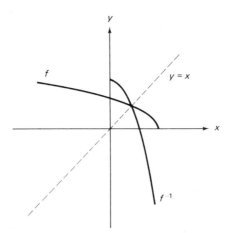

FIGURE 6.24

The composition properties stated in Theorem 6-9 are sometimes used to define the inverse function f^{-1}. The formulation from this point of view is given in the following theorem.

THEOREM 6-10   Let f be a given function with domain D and range R. If g is a function with domain R and range D such that
$$f(g(x)) = x, \qquad \text{for all } x \in R$$
and
$$g(f(x)) = x, \qquad \text{for all } x \in D,$$
then $g = f^{-1}$ and $f = g^{-1}$.

This formulation emphasizes that if two functions are inverses of each other, the domain of one is the range of the other. It is particularly helpful in determining whether or not two given functions are inverses of each other.

EXAMPLE 3 Which of the following pairs of functions are inverses of each other?

(a) $f(x) = x^3 - 2,\quad g(x) = \sqrt[3]{x + 2}$
(b) $f(x) = x^2,\quad g(x) = \sqrt{x}$

SOLUTION In order for f and g to be inverses of each other, they must satisfy the conditions of Theorem 6-10.

(a) We have

$$f(g(x)) = f(\sqrt[3]{x + 2}) = (\sqrt[3]{x + 2})^3 - 2 = x,$$

and

$$g(f(x)) = g(x^3 - 2) = \sqrt[3]{x^3 - 2 + 2} = x.$$

The function f has both domain and range the set of all real numbers, and g does also. Thus f and g are inverses of each other.

(b) We have

$$f(g(x)) = f(\sqrt{x}) = (\sqrt{x})^2 = x,$$

so equation (1) of Theorem 6-10 holds for all x in the domain of g. However,

$$g(f(x)) = g(x^2) = \sqrt{x^2} = |x| \neq x$$

for any negative x. Thus equation (2) fails, and the functions f and g are not inverses of each other.

EXERCISES 6-7

Find the inverse of each of the following relations. Also determine if each relation and/or its inverse is a function. (See Example 8, Section 5-1.)

1. $f = \{(x, y)\,|\,y = x^2 + 2x\}$
2. $f = \{(x, y)\,|\,y = 2x^2 - 8x\}$
3. $g = \{(x, y)\,|\,x^2 + y^2 = 4\}$
4. $g = \{(x, y)\,|\,4x^2 + 9y^2 = 36\}$
5. $h = \{(x, y)\,|\,y = -\sqrt{4 - x^2}\}$
6. $h = \{(x, y)\,|\,y = 2\sqrt{9 - x^2}/3\}$
7. $f = \{(x, y)\,|\,y = \sqrt{x^2 + 2x}\}$
8. $f = \{(x, y)\,|\,y = \sqrt{2x^2 - 8x}\}$

In problems 9–18, (a) find the inverse f^{-1} of each of the functions f defined by the equations $y = f(x)$, (b) sketch the graph of each inverse relation f^{-1} by reflecting the graph of f through the line $y = x$, and (c) use the vertical line test to determine if f^{-1} is a function. (See Example 1.)

9. $y = 2x - 4$
10. $y = 3x + 9$

11. $4x + 3y = 12$

12. $3x + 2y = 6$

13. $y = |2x + 6|$

14. $y = |x + 4|$

15. $y = \sqrt{x - 1}$

16. $y = \sqrt{x + 2}$

17. $y = \sqrt{16 - x^2}$

18. $y = \dfrac{3\sqrt{4 - x^2}}{2}$

Which of the following pairs of functions f, g are inverse functions of each other? (See Example 3.)

19. $f(x) = 2x - 4$, $\quad g(x) = \frac{1}{2}x + 2$

20. $f(x) = \frac{2}{3}x + 6$, $\quad g(x) = \frac{3}{2}x - 9$

21. $f(x) = \dfrac{6x - 2}{3}$, $\quad g(x) = \dfrac{3x + 2}{6}$

22. $f(x) = \dfrac{3x + 4}{2}$, $\quad g(x) = \dfrac{2x - 4}{3}$

23. $f(x) = \dfrac{x}{x + 1}$, $\quad g(x) = \dfrac{x + 1}{x}$

24. $f(x) = \dfrac{2x}{x - 1}$, $\quad g(x) = \dfrac{x - 1}{2x}$

25. $f(x) = \sqrt{x - 1}$; $g(x) = x^2 + 1$ $\quad$ where $x \geq 0$.

26. $f(x) = \sqrt{x^2 - 4}$ $\quad$ where $x \geq 2$; $g(x) = \sqrt{x^2 + 4}$ $\quad$ where $x \geq 0$.

27. $f(x) = \sqrt{x + 2}$, $\quad g(x) = x^2 - 2$

28. $f(x) = \sqrt{2x - 6}$, $\quad g(x) = \frac{1}{2}x^2 + 3$

Each of the following equations defines a function f by the rule $y = f(x)$. (a) In each problem, state the domain and range of f, and the domain and range of f^{-1}. (b) For each function f, write an equation for the inverse function in the form $y = f^{-1}(x)$. (See Example 2.)

29. $y = 2x - 6$

30. $y = 3x + 9$

31. $y = 12 - \frac{3}{4}x$

32. $y = 6 - \frac{4}{3}x$

33. $y = \dfrac{1}{x + 1}$

34. $y = \dfrac{x - 1}{x + 1}$

35. $xy = 1$

36. $xy = y + 1$

37. $y = \sqrt{x - 2}$

38. $y = \sqrt{x + 1}$

39. $y = x^3$

40. $y = \sqrt[3]{x}$

*41. Prove that the points (a, b) and (b, a) are symmetric to each other with respect to the line $y = x$. That is, prove that the line $y = x$ is the perpendicular bisector of the line segment joining (a, b) and (b, a).

*42. Prove that if both f and f^{-1} are functions, then $f(f^{-1}(x)) = x$ for all x in the domain of f^{-1}.

*43. Prove that $(f^{-1})^{-1} = f$, for any relation f.

1. Find the distance between the points $(7, 2)$ and $(3, -1)$.

2. Determine whether or not the graph of the equation below is a circle. If it is, find the center and the radius.

$$x^2 - 2x + y^2 = 12y - 38$$

3. Write an equation of the circle with center $(2, -3)$ and radius 4.

4. Sketch the graph of the relation defined by

$$2y = \sqrt{1 - 4x^2}.$$

In problems 5 and 6, sketch the graph of the given equation.

5. $25y^2 - 4x^2 = 100$
6. $2x = -\sqrt{16 - y^2}$

In problems 7 and 8, solve the given system of equations. It is not necessary to graph the equations.

7. $x^2 - 3y = 1$
   $2x - y = 3$

8. $x^2 + 3xy - 6y^2 = 8$
   $x^2 - xy - 6y^2 = 4$

9. Graph the solution set of the following system of inequalities, showing all points of intersection of the boundaries.

$$x^2 + y^2 > 9$$
$$16x^2 + 9y^2 \leq 144$$

10. (a) Find the inverse f^{-1} of the function f defined by $y = \sqrt{x - 2}$.
    (b) Sketch the graph of f^{-1} by reflecting the graph of f through the line $y = x$.
    (c) State the domain and range of f^{-1}.

7-1
Definitions
and Basic
Properties

While watching a football game on television, a football fan often sees flashed on the screen a tabulation similar to the following:

	Cowboys	Steelers
Number of offensive plays	8	14
Yards rushing	63	84
Yards passing	102	83
Total yards	165	167
Time of possession	5 : 20	9 : 40
Number of turnovers	2	0

If the fan is interested in the number of "yards passing" by the Steelers, he looks at the entry in the third row and second column, and finds 83.

A traveler who needs to know the distance from Los Angeles to Chicago looks at a mileage chart similar to the following:

184

	Chicago	New York	Los Angeles	Miami
Chicago	0	840	2090	1375
New York	840	0	2805	1332
Los Angeles	2090	2805	0	2743
Miami	1375	1332	2743	0

The entry in the third row and first column gives the desired mileage, 2090.

Suppose that we consider only the array of numbers in each of the above examples, omitting the row and column headings, and placing square brackets around the arrays. We have

$$\begin{bmatrix} 8 & 14 \\ 63 & 84 \\ 102 & 83 \\ 165 & 167 \\ 5\!:\!20 & 9\!:\!40 \\ 2 & 0 \end{bmatrix}, \qquad \begin{bmatrix} 0 & 840 & 2090 & 1375 \\ 840 & 0 & 2805 & 1332 \\ 2090 & 2805 & 0 & 2743 \\ 1375 & 1332 & 2743 & 0 \end{bmatrix}.$$

These arrays of entries are called *matrices*, and each number in the array is called an *element* of the *matrix*. The formal definition is as follows.

DEFINITION 7-1 An *m* by *n matrix* is a rectangular array of elements arranged in *m* rows and *n* columns. Such a matrix can be written in the form

$$A = \begin{bmatrix} a_{11} & a_{12} & \cdots & a_{1n} \\ a_{21} & a_{22} & \cdots & a_{2n} \\ a_{31} & a_{32} & \cdots & a_{3n} \\ \cdot & \cdot & & \cdot \\ \cdot & \cdot & & \cdot \\ \cdot & \cdot & & \cdot \\ a_{m1} & a_{m2} & \cdots & a_{mn} \end{bmatrix},$$

where a_{ij} denotes the element in row i and column j of the matrix A. The matrix A is referred to as a matrix of *dimension* $m \times n$ (read: m by n).

It is customary to denote matrices by capital letters. An $m \times n$ matrix can be denoted compactly by

$$A = [a_{ij}]_{(m, n)} \quad \text{or} \quad A_{(m, n)},$$

or A if the dimension is known from the context.

In the second example above, the mileage from Los Angeles to Chicago is represented by the element $a_{31} = 2090$, and the matrix representing the mileage has dimension 4×4. The matrix of football statistics is of dimension 6×2.

DEFINITION 7-2 A matrix which has n rows and n columns is called a *square matrix of order n*. A matrix which has only one column is called a *column matrix*, and a matrix which has only one row is called a *row matrix*. A square matrix which has zeros for all its elements, except possibly those elements with the same row and column number, is called a *diagonal matrix*.

EXAMPLE 1 State the dimensions of each matrix below. If a matrix is a column matrix, a row matrix, a square matrix, or a diagonal matrix, identify it as such.

(a) $A = \begin{bmatrix} 2 & 1 \\ -3 & 4 \\ -2 & 0 \end{bmatrix}$ (b) $B = [-1 \ 0 \ 4 \ 7]$ (c) $C = \begin{bmatrix} -\sqrt{3} \\ \pi \end{bmatrix}$

(d) $D = [0]$ (e) $E = \begin{bmatrix} 1 & 0 \\ 0 & -1 \end{bmatrix}$

SOLUTION

(a) The matrix A has dimension 3×2. None of the special terms apply to A.
(b) B is a row matrix of dimension 1×4.
(c) C is a 2×1 column matrix.
(d) D is a 1×1 matrix. It is a column matrix, a row matrix, a square matrix, and a diagonal matrix.
(e) E is a diagonal matrix (and hence a square matrix) of order 2.

DEFINITION 7-3 Two matrices are *equal* if they have equal numbers of rows, equal numbers of columns, and have elements placed in corresponding positions which are equal.

EXAMPLE 2 The meaning of the definition of equality is illustrated by the following pairs of matrices.

(a) $\begin{bmatrix} 1 \\ 2 \\ -5 \end{bmatrix}$ and $[1 \ \ 2 \ \ -5]$ are not equal since they have different dimensions.

(b) $[0]$ and $\begin{bmatrix} 0 & 0 \\ 0 & 0 \end{bmatrix}$ are not equal since they have different dimensions.

(c) $\begin{bmatrix} x & 2y \\ 1 & z \\ 0 & 2 \end{bmatrix}$ and $\begin{bmatrix} -3 & 4 \\ 1 & z \\ 0 & \frac{4}{2} \end{bmatrix}$ are equal only if $x = -3$ and $y = 2$.

(d) $\begin{bmatrix} 1 & 3 \end{bmatrix}$ and $\begin{bmatrix} 3 & 1 \end{bmatrix}$ are not equal since elements in corresponding positions are not equal.

Suppose that we had two matrices, each representing the statistics from a quarter in a football game, as in the example at the beginning of the section.

First
Quarter

$\begin{bmatrix} 8 & 14 \\ 63 & 84 \\ 102 & 83 \\ 165 & 167 \\ 5:20 & 9:40 \\ 2 & 0 \end{bmatrix}$

Second
Quarter

$\begin{bmatrix} 10 & 13 \\ 33 & 95 \\ 80 & 47 \\ 113 & 142 \\ 6:50 & 8:10 \\ 1 & 1 \end{bmatrix}$

If corresponding elements in these two matrices were added, the result would be a matrix whose elements would represent the statistics for the two quarters together. Addition of matrices is defined in this manner.

DEFINITION 7-4   The *sum* of two $m \times n$ matrices A and B is the $m \times n$ matrix $A + B$ formed by adding the elements which are placed in corresponding positions of A and B.

Notice that the sum of two matrices of different dimensions is *not* defined.

EXAMPLE 3   Addition of matrices is illustrated below.

(a) $\begin{bmatrix} 2 & -2 & 1 \\ 1 & 0 & -3 \end{bmatrix} + \begin{bmatrix} 1 & -5 & 7 \\ 2 & 1 & 0 \end{bmatrix} = \begin{bmatrix} 3 & -7 & 8 \\ 3 & 1 & -3 \end{bmatrix}$.

(b) $\begin{bmatrix} 1 & 4 & 0 \\ -1 & 2 & 1 \end{bmatrix} + \begin{bmatrix} 0 & 1 \\ 2 & -2 \end{bmatrix}$ is not defined since the the dimensions of the two matrices are not equal.

Recall that addition of real numbers is associative and commutative (see Section 1-2). These two properties also hold for addition in the set of all $m \times n$ matrices, as stated in the next theorem. Proofs are requested in problems 53 and 54 at the end of this section.

THEOREM 7-5   Let A, B, and C be $m \times n$ matrices. Then

(a) $A + (B + C) = (A + B) + C$ (Associative property)

and

(b) $A + B = B + A$ (Commutative property).

In Section 1-2, the additive identity and the additive inverse of an element in the real numbers were identified as follows.

1. *Additive Identity.* There is a real number 0 such that $a + 0 = 0 + a = a$ for all $a \in \mathcal{R}$.

2. *Additive Inverse.* For each $a \in \mathcal{R}$, there is an element $-a \in \mathcal{R}$ such that $a + (-a) = (-a) + a = 0$.

For the set of all $m \times n$ matrices, we make similar definitions.

DEFINITION 7-6   The *additive identity* for the set of all $m \times n$ matrices is an $m \times n$ matrix with all elements equal to 0. This additive identity matrix is also called the *zero matrix* of dimension $m \times n$.

Thus $\begin{bmatrix} 0 & 0 \\ 0 & 0 \end{bmatrix}$ is the zero matrix of order 2, and $\begin{bmatrix} 0 & 0 \\ 0 & 0 \\ 0 & 0 \end{bmatrix}$ is the 3×2 zero matrix.

DEFINITION 7-7   To multiply a matrix A by a real number c, multiply each element in A by c. This *product* is denoted by cA.

If
$$A = \begin{bmatrix} 2 & -1 & -2 \\ 1 & 4 & 0 \end{bmatrix},$$
then
$$3A = \begin{bmatrix} 6 & -3 & -6 \\ 3 & 12 & 0 \end{bmatrix}$$
and
$$(-1)A = \begin{bmatrix} -2 & 1 & 2 \\ -1 & -4 & 0 \end{bmatrix}.$$

DEFINITION 7-8   The *additive inverse* of the $m \times n$ matrix A is the $m \times n$ matrix $-A$ such that $A + (-A) = (-A) + A$ is the $m \times n$ zero matrix.

It is easy to see that $-A$ is the same as $(-1)A$, the matrix obtained by multiplying each element of A by -1. For example, the additive inverse of
$$A = \begin{bmatrix} -2 & 4 & 1 & -1 \\ -1 & 0 & 3 & 2 \\ 0 & -2 & 0 & 1 \end{bmatrix}.$$

is

$$-A = \begin{bmatrix} 2 & -4 & -1 & 1 \\ 1 & 0 & -3 & -2 \\ 0 & 2 & 0 & -1 \end{bmatrix}.$$

Subtraction of matrices is defined using additive inverses: $A - B = A + (-B)$. For example, if

$$A = \begin{bmatrix} -2 & 1 \\ 3 & -2 \end{bmatrix} \quad \text{and} \quad B = \begin{bmatrix} 1 & 7 \\ 2 & -4 \end{bmatrix},$$

then

$$A - B = A + (-B) = \begin{bmatrix} -2 & 1 \\ 3 & -2 \end{bmatrix} + \begin{bmatrix} -1 & -7 \\ -2 & 4 \end{bmatrix} = \begin{bmatrix} -3 & -6 \\ 1 & 2 \end{bmatrix}.$$

This is an exact parallel to the definition of subtraction which we stated for real numbers in Section 1-2.

EXERCISES 7-1

State the dimension of each matrix. If a matrix is a column matrix, a row matrix, a square matrix, or a diagonal matrix, identify it as such.

1. $\begin{bmatrix} 1 & 0 & -3 \\ 2 & 1 & 4 \end{bmatrix}$

2. $\begin{bmatrix} 5 & -3 & -10 & -1 \\ \frac{1}{2} & 0 & -2 & 1 \\ 4 & 0 & 0 & 5 \end{bmatrix}$

3. $\begin{bmatrix} 1 \\ 3 \\ 7 \end{bmatrix}$

4. $[0 \quad -1 \quad -2]$

5. $\begin{bmatrix} -1 & 0.3 & 2.7 & 1.2 \\ 0.1 & 0 & -1 & 0 \end{bmatrix}$

6. $\begin{bmatrix} \sqrt{3} & \sqrt{2} \\ \pi & -\sqrt{5} \end{bmatrix}$

7. $[0]$

8. $\begin{bmatrix} -1 & 0 \\ 0 & -1 \end{bmatrix}$

9. $[1 \quad 0]$

10. $\begin{bmatrix} 1 \\ 1 \end{bmatrix}$

11. $\begin{bmatrix} 1 & 0 & 0 \\ 0 & 1 & 0 \\ 0 & 0 & 1 \end{bmatrix}$

12. $\begin{bmatrix} 3 & -2 \\ 1 & 0 \\ 0 & -1 \end{bmatrix}$

13. $\begin{bmatrix} x & y & z \\ r & s & t \\ a & b & c \end{bmatrix}$

14. $\begin{bmatrix} a & -a \\ b & -b \\ c & -c \end{bmatrix}$

15. $\begin{bmatrix} x^2 - 1 \\ y^3 \end{bmatrix}$

16. $\begin{bmatrix} x^2 & -1 \\ y^3 & 0 \end{bmatrix}$

Write the matrix whose elements are defined by each equation.

17. $a_{ij} = 2i + j; i = 1, 2; j = 1, 2, 3, 4.$

18. $a_{ij} = i \cdot j; i = 1, 2, 3; j = 1, 2, 3, 4.$

19. $a_{ij} = (-1)^i \cdot j; i = 1, 2, 3, 4; j = 1, 2.$

20. $a_{ij} = (-1)^{i+j}; i = 1, 2, 3; j = 1, 2, 3.$

21. Write a 2×4 matrix such that $a_{ij} = 1$ if $i < j$, 0 otherwise.

22. Write a 3×4 matrix such that $a_{ij} = i + j$ if $i \geq j$, 0 otherwise.

Determine whether or not the following pairs of matrices are equal.

23. $\begin{bmatrix} 1 & 2 \\ -1 & 3 \\ 1 & 7 \end{bmatrix}, \begin{bmatrix} 2 & 1 \\ 3 & -1 \\ 7 & 1 \end{bmatrix}$

24. $\begin{bmatrix} 0 & 0 \\ 0 & 0 \end{bmatrix}, \begin{bmatrix} 0 & 0 & 0 \\ 0 & 0 & 0 \\ 0 & 0 & 0 \end{bmatrix}$

25. $\begin{bmatrix} 1 & -7 \\ 5 & 3.2 \end{bmatrix}, \begin{bmatrix} 3-2 & -7 \\ 5 & \frac{32}{10} \end{bmatrix}$

26. $[1 \quad 3 \quad -1], \begin{bmatrix} 1 \\ 3 \\ -1 \end{bmatrix}$

27. $\begin{bmatrix} -2 & -1 \\ 3 & 4 \\ -2 & 0 \end{bmatrix}, \begin{bmatrix} -2 & -1 & 0 \\ 3 & 4 & 0 \\ -2 & 0 & 0 \end{bmatrix}$

28. $\begin{bmatrix} 10 & -2 & 4 \\ 1 & 3 & -3 \\ 0 & 0 & 0 \end{bmatrix}, \begin{bmatrix} 10 & -2 & 4 \\ 1 & 3 & -4 \end{bmatrix}$

29. $\begin{bmatrix} -1 & 4 & 3 \\ 2 & -1 & 1 \end{bmatrix}, \begin{bmatrix} -1 & 2 \\ 4 & -1 \\ 3 & 1 \end{bmatrix}$

30. $\begin{bmatrix} 1 & 0 \\ 0 & 1 \end{bmatrix}, \begin{bmatrix} 0 & 1 \\ 1 & 0 \end{bmatrix}$

31. $\begin{bmatrix} x & 1 \\ 0 & -y \end{bmatrix}, \begin{bmatrix} 2 & 1 \\ 0 & -3 \end{bmatrix}$

32. $[2x \quad x], [6 \quad 3]$

33. $[2x - 5], [-1]$

34. $\begin{bmatrix} x^2 \\ -y \end{bmatrix}, \begin{bmatrix} 4 \\ 2 \end{bmatrix}$

35. $\begin{bmatrix} 1 & -x \\ 4 & 2 \end{bmatrix}, [1 \quad -x]$

36. $\begin{bmatrix} x & x \\ x & x \end{bmatrix}, \begin{bmatrix} x \\ x \end{bmatrix}$

Perform the indicated operations, if possible.

37. $\begin{bmatrix} 3 & -1 & 1 \\ 2 & 7 & -4 \end{bmatrix} + \begin{bmatrix} 2 & 1 & 0 \\ 1 & 3 & -1 \end{bmatrix}$

38. $\begin{bmatrix} 1 & 0 \\ 0 & 1 \end{bmatrix} + \begin{bmatrix} 2 & 1 \\ -3 & 2 \end{bmatrix}$

39. $\begin{bmatrix} -2 \\ 3 \\ -6 \end{bmatrix} + [1 \quad 1 \quad -2]$

40. $\begin{bmatrix} -3 & 1 \\ 0 & -2 \\ 1 & -1 \end{bmatrix} + \begin{bmatrix} -3 & 1 & 0 \\ -2 & 1 & -1 \end{bmatrix}$

41. $\begin{bmatrix} 4 & -2 \\ 1 & 7 \end{bmatrix} - \begin{bmatrix} 2 & 6 \\ -3 & 0 \end{bmatrix}$

42. $\begin{bmatrix} 3 & 7 & 11 \\ -2 & 4 & -1 \end{bmatrix} - \begin{bmatrix} 1 & 1 & -2 \\ 1 & -3 & -1 \end{bmatrix}$

43. $2 \begin{bmatrix} -11 & 2 \\ 0 & -3 \end{bmatrix} + 3 \begin{bmatrix} -11 & 2 & 1 \\ 0 & -3 & 0 \end{bmatrix}$

44. $2 \begin{bmatrix} -11 & 2 \\ 0 & -3 \end{bmatrix} - 3 \begin{bmatrix} -11 & 2 \\ 0 & -3 \end{bmatrix}$

45. $-3 \begin{bmatrix} 4 \\ -1 \\ 6 \end{bmatrix} + 2 \begin{bmatrix} -2 \\ 5 \end{bmatrix}$

46. $5 \begin{bmatrix} -1 & 3 & 6 \end{bmatrix} - 2 \begin{bmatrix} 0 & -3 & -1 \end{bmatrix}$

47. $0 \begin{bmatrix} 1 & 3 \\ -2 & 1 \\ 0 & 4 \\ -1 & 0 \end{bmatrix} + 3 \begin{bmatrix} 0 & 0 \\ 0 & 0 \\ 0 & 0 \\ 0 & 1 \end{bmatrix}$

48. $3 \begin{bmatrix} 0 \end{bmatrix} - 5 \begin{bmatrix} 2 \end{bmatrix}$

49. $\begin{bmatrix} 2x & -2 \\ -x & 2 \\ 3x & 3 \end{bmatrix} - \begin{bmatrix} 4 & -x \\ 0 & -2 \\ x & x \end{bmatrix}$

50. $\begin{bmatrix} a & b & c \\ d & e & f \\ g & h & i \end{bmatrix} + \begin{bmatrix} 1 & 0 & 0 \\ 0 & 1 & 0 \\ 0 & 0 & 1 \end{bmatrix}$

51. $\begin{bmatrix} a & 0 & 0 \\ 0 & b & 0 \\ 0 & 0 & c \end{bmatrix} + 2 \begin{bmatrix} a & 0 & 0 \\ 0 & b & 0 \\ 0 & 0 & c \end{bmatrix}$

52. $\begin{bmatrix} x^2 - 2 \\ y \end{bmatrix} + \begin{bmatrix} x^2 & -2 \\ 0 & y \end{bmatrix}$

*53. Prove part (a) of Theorem 7-5.

*54. Prove part (b) of Theorem 7-5.

7-2
Matrix
Multiplication

The definition of matrix multiplication is much more involved than the definition of addition. We begin with a formal statement of the definition, and then illustrate the definition with some examples.

DEFINITION 7-9   The *product* of an $m \times n$ matrix A and an $n \times p$ matrix B is an $m \times p$ matrix $C = AB$, where the element c_{ij} in row i and column j of AB is found by using the elements in row i of A and the elements in column j of B in the manner described below. The elements in

$$\text{row } i \text{ of } A: \quad a_{i1} \quad a_{i2} \quad a_{i3} \quad \cdots \quad a_{in}$$

and the elements in

$$\text{column } j \text{ of } B: \quad \begin{array}{c} b_{1j} \\ b_{2j} \\ b_{3j} \\ \cdot \\ \cdot \\ \cdot \\ b_{nj} \end{array}$$

are combined to yield the

$$\left.\begin{array}{l}\text{element in}\\ \text{row } i \text{ and}\\ \text{column } j\\ \text{of } C = AB\end{array}\right\}: \quad c_{ij} = a_{i1}b_{1j} + a_{i2}b_{2j} + a_{i3}b_{3j} + \cdots + a_{in}b_{nj}.$$

That is, the element

$$c_{ij} = a_{i1}b_{1j} + a_{i2}b_{2j} + a_{i3}b_{3j} + \cdots + a_{in}b_{nj}$$

in row i and column j of AB is found by adding the products formed from corresponding elements of row i in A and column j in B (first times first, second times second, etc.).

Notice that the number of columns in A *must* equal the number of rows in B in order to form the product AB. If this is the case, then A and B are said to be *conformable for multiplication*. A simple diagram illustrates this fact.

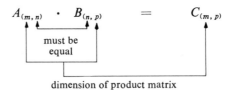

Some examples are helpful in understanding the definition of matrix multiplication.

EXAMPLE 1   Perform all possible multiplications using two of the following three matrices.

$$A = \begin{bmatrix} 3 & 1 \\ -2 & -1 \\ 4 & -2 \end{bmatrix}, \quad B = \begin{bmatrix} 2 & 1 & 0 \\ -1 & 3 & 5 \end{bmatrix}, \quad C = \begin{bmatrix} 2 & 7 \\ 0 & -4 \end{bmatrix}$$

SOLUTION   (a) The product AB exists since A has 2 columns and B has 2 rows:

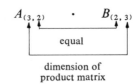

Performing the multiplication yields

$$\begin{bmatrix} 3 & 1 \\ -2 & -1 \\ 4 & -2 \end{bmatrix} \begin{bmatrix} 2 & 1 & 0 \\ -1 & 3 & 5 \end{bmatrix}$$

$$= \begin{bmatrix} 3(2) + 1(-1) & 3(1) + 1(3) & 3(0) + 1(5) \\ -2(2) + (-1)(-1) & -2(1) + (-1)(3) & -2(0) + (-1)(5) \\ 4(2) + (-2)(-1) & 4(1) + (-2)(3) & 4(0) + (-2)(5) \end{bmatrix},$$

and

$$AB = \begin{bmatrix} 5 & 6 & 5 \\ -3 & -5 & -5 \\ 10 & -2 & -10 \end{bmatrix}.$$

(b) The product BA exists since B has 3 columns and A has 3 rows:

$$B_{(2,\,3)} \quad \cdot \quad A_{(3,\,2)}$$

equal

dimension of
product matrix

Performing the multiplication, we have

$$\begin{bmatrix} 2 & 1 & 0 \\ -1 & 3 & 5 \end{bmatrix} \begin{bmatrix} 3 & 1 \\ -2 & -1 \\ 4 & -2 \end{bmatrix}$$

$$= \begin{bmatrix} 2(3) + 1(-2) + 0(4) & 2(1) + 1(-1) + 0(-2) \\ -1(3) + 3(-2) + 5(4) & -1(1) + 3(-1) + 5(-2) \end{bmatrix},$$

and

$$BA = \begin{bmatrix} 4 & 1 \\ 11 & -14 \end{bmatrix}.$$

Notice that the dimensions of the products AB and BA are not the same. Hence we have $AB \neq BA$. (The commutative law of multiplication does *not* hold for matrices.)

(c) The product AC exists since A has 2 columns and C has 2 rows. The multiplication is given by

$$AC = \begin{bmatrix} 3 & 1 \\ -2 & -1 \\ 4 & -2 \end{bmatrix} \begin{bmatrix} 2 & 7 \\ 0 & -4 \end{bmatrix} = \begin{bmatrix} 6 & 17 \\ -4 & -10 \\ 8 & 36 \end{bmatrix}.$$

(d) The product CA does not exist since C has 2 columns and A has 3 rows:

$$C_{(2,\,2)} \quad \cdot \quad A_{(3,\,2)} \text{ does not exist.}$$

not equal

(e) The product BC is not defined since B has 3 columns and C has 2 rows.

(f) The product CB is defined since C has 2 columns and B has 2 rows. We have

$$CB = \begin{bmatrix} 2 & 7 \\ 0 & -4 \end{bmatrix} \begin{bmatrix} 2 & 1 & 0 \\ -1 & 3 & 5 \end{bmatrix} = \begin{bmatrix} -3 & 23 & 35 \\ 4 & -12 & -20 \end{bmatrix}.$$

Even when the two factor matrices have the same dimension, AB and BA are usually different. This is illustrated in the next example. (See also problems 43 to 47 in Exercises 7-2.)

EXAMPLE 2   Compare the products AB and BA, where

$$A = \begin{bmatrix} 1 & 0 \\ -3 & -1 \end{bmatrix} \quad \text{and} \quad B = \begin{bmatrix} -2 & -1 \\ 1 & -3 \end{bmatrix}.$$

SOLUTION   Multiplying, we have

$$AB = \begin{bmatrix} 1 & 0 \\ -3 & -1 \end{bmatrix} \begin{bmatrix} -2 & -1 \\ 1 & -3 \end{bmatrix} = \begin{bmatrix} -2 & -1 \\ 5 & 6 \end{bmatrix},$$

and

$$BA = \begin{bmatrix} -2 & -1 \\ 1 & -3 \end{bmatrix} \begin{bmatrix} 1 & 0 \\ -3 & -1 \end{bmatrix} = \begin{bmatrix} 1 & 1 \\ 10 & 3 \end{bmatrix}.$$

Thus $AB \neq BA$.

Although multiplication of matrices does not have the commutative property, it can be shown that matrix multiplication is associative, provided that the products involved are defined. This is stated in Theorem 7-10.

THEOREM 7-10   Let A be an $m \times n$ matrix, B be an $n \times p$ matrix, and C be a $p \times q$ matrix. Then

$$A(BC) = (AB)C. \qquad \text{(Associative property)}$$

A proof of this theorem for the case where A, B, and C are 2×2 matrices is requested in problem 55 at the end of this section. It can also be shown that the following distributive properties hold (see problems 56 to 59).

THEOREM 7-11   Let A be an $m \times n$ matrix, B and C be $n \times p$ matrices, and let D be a $p \times q$ matrix. Let a and b be real numbers. Then
(a) $A(B + C) = AB + AC$;
(b) $(B + C)D = BD + CD$;
(c) $a(B + C) = aB + aC$;
(d) $(a + b)A = aA + bA$.

In Section 1-2, we saw that the number 1 is the multiplicative identity element for the set of real numbers: $1 \cdot a = a \cdot 1 = a$, for all real numbers a. Similarly, for every square matrix A of order n, there is a square matrix I_n of order n such that $AI_n = I_nA = A$. The matrix I_n is called the *identity matrix* of order n. It is a diagonal matrix with 1's on the diagonal from upper left to lower right, and 0's at all other positions:

$$I_n = \begin{bmatrix} 1 & 0 & 0 & \cdots & 0 \\ 0 & 1 & 0 & \cdots & 0 \\ 0 & 0 & 1 & \cdots & 0 \\ \cdot & \cdot & \cdot & & \cdot \\ \cdot & \cdot & \cdot & & \cdot \\ \cdot & \cdot & \cdot & & \cdot \\ 0 & 0 & 0 & \cdots & 1 \end{bmatrix}.$$

If A has dimension $m \times n$, then I_m is called a *left identity* for A since $I_mA = A$, and I_n is called a *right identity* for A since $AI_n = A$.

Matrix multiplication is important primarily because many "real-world" problems can be formulated in terms of matrices so that matrix multiplication is instrumental in the solution of the problems.

EXAMPLE 3 Consider the following problem of determining the cost of 10 pounds of four chicken feed mixtures. The number of pounds of oats, barley, and corn contained in 10 pounds of each of the four mixtures A, B, C, and D are given by the following matrix.

		Ingredients		
		Oats	Barley	Corn
	A	2	4	4
	B	3	2	5
Mixture	C	2	2	6
	D	5	1	4

For example, there are 2 pounds of oats, 2 pounds of barley, and 6 pounds of corn in 10 pounds of mixture C.

There are two suppliers of ingredients, and the cost per pound of each ingredient from each supplier is given in the following matrix:

		Supplier	
		1	2
	Oats	$0.07	$0.08
Ingredient	Barley	$0.04	$0.06
	Corn	$0.05	$0.03

The product of these two matrices,

$$\begin{bmatrix} 2 & 4 & 4 \\ 3 & 2 & 5 \\ 2 & 2 & 6 \\ 5 & 1 & 4 \end{bmatrix} \begin{bmatrix} \$0.07 & \$0.08 \\ \$0.04 & \$0.06 \\ \$0.05 & \$0.03 \end{bmatrix}$$

will yield the matrix whose elements represent the cost of 10 lbs. of each mixture with ingredients supplied by each supplier:

		Supplier	
		1	2
	A	$0.50	$0.52
	B	$0.54	$0.51
Mixture	C	$0.52	$0.46
	D	$0.59	$0.58

For example, 10 lbs. of mixture *B* costs $0.54 when ingredients are furnished by supplier 1, and $0.51 when furnished by supplier 2.

EXERCISES 7-2

State whether the following matrices are conformable for multiplication in the order given. If the multiplication is possible, also give the dimension of the product matrix.

1. $A_{(2, 3)} \cdot B_{(3, 7)}$

2. $C_{(4, 1)} \cdot D_{(1, 4)}$

3. $B_{(3, 2)} \cdot C_{(3, 3)}$

4. $D_{(1, 1)} \cdot A_{(2, 1)}$

5. $A_{(1, 3)} \cdot C_{(3, 1)}$

6. $C_{(3, 1)} \cdot A_{(1, 3)}$

7. $I_2 \cdot A_{(2, 7)}$

8. $X_{(2, 7)} \cdot I_7$

9. $X_{(8, 2)} \cdot Y_{(2, 1)}$

10. $Z_{(10, 2)} \cdot Z_{(10, 2)}$

11. $A_{(3, 3)} \cdot A_{(3, 3)}$

12. $B_{(4, 1)} \cdot A_{(2, 2)}$

13. $X_{(2, 2)} \cdot Z_{(3, 3)}$

14. $Z_{(3, 3)} \cdot X_{(2, 2)}$

15. $D_{(1, 1)} \cdot C_{(1, 12)}$

16. $B_{(2, 12)} \cdot E_{(2, 2)}$

Perform the following matrix multiplications, if possible.

17. $\begin{bmatrix} 1 & 3 & 0 \\ -2 & 4 & 1 \end{bmatrix} \begin{bmatrix} 1 & 2 \\ 2 & -1 \\ 1 & 3 \end{bmatrix}$

18. $\begin{bmatrix} 1 & 2 \\ 2 & -1 \\ 1 & 3 \end{bmatrix} \begin{bmatrix} 1 & 3 & 0 \\ -2 & 4 & 1 \end{bmatrix}$

19. $\begin{bmatrix} -2 & 1 & 3 \\ 4 & 7 & 0 \\ -1 & 4 & -2 \end{bmatrix} \begin{bmatrix} 3 & 5 \\ -1 & 0 \\ 2 & 4 \end{bmatrix}$

20. $\begin{bmatrix} 3 & 5 \\ -1 & 0 \\ 2 & 4 \end{bmatrix} \begin{bmatrix} -2 & 1 & 3 \\ 4 & 7 & 0 \\ -1 & 4 & -2 \end{bmatrix}$

21. $\begin{bmatrix} 1 & 0 \\ 5 & -1 \end{bmatrix} I_3$

22. $I_2 \begin{bmatrix} 1 & 7 & 0 \\ 0 & 0 & 1 \end{bmatrix}$

23. $[-3 \quad -2 \quad 0 \quad 1][1 \quad 0 \quad 3 \quad -1]$

24. $\begin{bmatrix} 1 \\ 5 \\ 9 \end{bmatrix} \begin{bmatrix} -9 \\ -5 \\ -1 \end{bmatrix}$

25. $[0 \quad -3 \quad 1] \begin{bmatrix} 10 \\ -4 \\ 10 \end{bmatrix}$

26. $[5 \quad 3 \quad -4 \quad 0] \begin{bmatrix} 8 \\ 4 \\ 3 \end{bmatrix}$

27. $\begin{bmatrix} -2 \\ 1 \\ 1 \end{bmatrix} [4 \quad 11 \quad -2]$

28. $[4 \quad 11 \quad -2] \begin{bmatrix} -2 \\ 1 \\ 1 \end{bmatrix}$

29. $\begin{bmatrix} 1 & 0 \\ 0 & 1 \\ 0 & 0 \end{bmatrix} \begin{bmatrix} 1 & 3 \\ 2 & -2 \\ 1 & 0 \end{bmatrix}$

30. $\begin{bmatrix} 1 & 3 \\ 2 & -2 \\ 1 & 0 \end{bmatrix} \begin{bmatrix} 1 & 0 \\ 0 & 1 \\ 0 & 0 \end{bmatrix}$

31. $\begin{bmatrix} 0 & 0 & 0 & 0 \\ 0 & 0 & 0 & 0 \\ 0 & 0 & 0 & 0 \end{bmatrix} \begin{bmatrix} 1 & 0 \\ 5 & 1 \\ -3 & 4 \\ 2 & -1 \end{bmatrix}$

32. $[0][9 \quad 2]$

33. $\begin{bmatrix} 3 & -1 \\ 8 & 1 \end{bmatrix} [0 \quad 0]$

34. $\begin{bmatrix} 1 & 0 \\ 0 & -1 \end{bmatrix} \begin{bmatrix} 1 & 1 & 1 \\ 1 & 1 & 1 \end{bmatrix}$

35. $\begin{bmatrix} 11 & 3 \\ 4 & 1 \end{bmatrix} \begin{bmatrix} -1 & 3 \\ 4 & -11 \end{bmatrix}$

36. $\begin{bmatrix} 1 & 1 & 2 \\ 1 & 1 & 0 \\ 0 & -1 & 1 \end{bmatrix} \begin{bmatrix} -\frac{1}{2} & \frac{3}{2} & 1 \\ \frac{1}{2} & -\frac{1}{2} & -1 \\ \frac{1}{2} & -\frac{1}{2} & 0 \end{bmatrix}$

37. $\begin{bmatrix} 10 & 1 & -4 \\ -8 & 1 & 5 \\ -1 & -1 & 4 \end{bmatrix} \begin{bmatrix} \frac{1}{9} & 0 & \frac{1}{9} \\ \frac{3}{9} & \frac{4}{9} & -\frac{2}{9} \\ \frac{1}{9} & \frac{1}{9} & \frac{2}{9} \end{bmatrix}$

38. $\begin{bmatrix} -1 & 0 \\ 0 & -1 \end{bmatrix} \begin{bmatrix} -1 & 0 \\ 0 & -1 \end{bmatrix}$

Form all possible products using two of the three given matrices.

39. $A = \begin{bmatrix} -1 & 3 \\ -2 & 1 \end{bmatrix}$, $B = \begin{bmatrix} -5 & -2 \\ 1 & 0 \\ 4 & -2 \end{bmatrix}$, $C = \begin{bmatrix} 1 & 1 & 0 \\ 0 & 1 & 1 \\ -1 & 0 & -1 \end{bmatrix}$

40. $A = [1 \quad -3 \quad 3]$, $B = \begin{bmatrix} 2 & -2 \\ 3 & -4 \\ 1 & 1 \end{bmatrix}$, $C = \begin{bmatrix} 2 & 4 & 0 \\ 1 & -1 & 1 \end{bmatrix}$

41. $A = \begin{bmatrix} 1 & -1 & 0 & 2 \\ 1 & 1 & 0 & 4 \\ 0 & 0 & 1 & 5 \\ 0 & 1 & -1 & 0 \end{bmatrix}$, $B = \begin{bmatrix} 1 & -1 & 2 & 0 \\ 0 & 1 & 5 & 3 \end{bmatrix}$, $C = \begin{bmatrix} 1 & 4 \\ 4 & 0 \\ 1 & 2 \\ 0 & -1 \end{bmatrix}$

42. $A = \begin{bmatrix} -2 & 3 \\ 1 & 1 \\ -1 & 0 \end{bmatrix}$, $B = \begin{bmatrix} 3 & 1 & 0 \\ -3 & 5 & -5 \end{bmatrix}$, $C = \begin{bmatrix} 1 & 1 \\ -1 & 2 \end{bmatrix}$

43. Find two matrices A and B such that $AB \neq BA$.

44. Find two square matrices A and B, of order 2, such that $AB \neq BA$.

45. Find two square matrices A and B, of order 3, such that $AB \neq BA$.

46. Find two square matrices A and B, of order 4, such that $AB \neq BA$.

*47. Find two nonzero square matrices A and B, of order 2, such that $AB = BA$.

*48. Find two nonzero matrices A and B such that $AB = O$, where O is the zero matrix.

49. Evaluate $AB + AC$ and $A(B + C)$, and compare the results for

$$A = \begin{bmatrix} -1 & 3 \\ -2 & 0 \\ -1 & 1 \end{bmatrix}, \quad B = \begin{bmatrix} 2 & 1 \\ -3 & 0 \end{bmatrix}, \quad C = \begin{bmatrix} 5 & 3 \\ 0 & -1 \end{bmatrix}.$$

50. Evaluate $A(BC)$ and $(AB)C$, and compare the results for

$$A = \begin{bmatrix} 4 & 1 \\ 0 & -1 \\ -1 & 1 \end{bmatrix}, \quad B = \begin{bmatrix} 1 & 6 & 0 & 1 \\ 5 & -3 & 4 & 0 \end{bmatrix}, \quad C = \begin{bmatrix} 2 \\ 3 \\ -1 \\ 1 \end{bmatrix}.$$

51. Evaluate $(A - B)(A + B)$ and $A^2 - B^2$, and compare the results for

$$A = \begin{bmatrix} -6 & 4 \\ 1 & 3 \end{bmatrix}. \quad B = \begin{bmatrix} 0 & 1 \\ 1 & 2 \end{bmatrix}.$$

(*Note:* $A^2 = A \cdot A$.)

52. For the same matrices as in problem 51, evaluate $(A + B)^2$ and $A^2 + 2AB + B^2$, and compare the results.

53. A housewife found that she needed to purchase 2 cans of peaches, 1 sack of flour, $\frac{1}{2}$ dozen eggs, and 1 sack of sugar in order to have the ingredients for a new recipe. From advertisements in the newspaper, she found prices of each item at three supermarkets, as follows:

		Item			
		1 Can Peaches	1 Sack Flour	1 Dozen Eggs	1 Sack Sugar
	A	\$0.65	\$0.69	\$0.84	\$1.03
Store	*B*	\$0.63	\$0.89	\$0.62	\$0.78
	C	\$0.72	\$0.90	\$0.78	\$0.82

Use matrix multiplication to determine the grocery bill at each of the three stores.

54. Suppose that a nut vendor wants to determine the cost per pound of three mixtures, each containing peanuts, cashews, and pecans, as given below:

		Nut		
		Peanuts	Cashews	Pecans
	A	$\frac{1}{2}$	$\frac{1}{6}$	$\frac{1}{3}$
Mixture	*B*	$\frac{1}{3}$	$\frac{1}{3}$	$\frac{1}{3}$
	C	$\frac{3}{10}$	$\frac{3}{10}$	$\frac{2}{5}$

If peanuts cost $0.90 per pound, cashews cost $1.20 per pound, and pecans cost $1.80 per pound, use matrix multiplication to determine the cost per pound of each mixture.

*55. Prove Theorem 7-10 for square matrices A, B, and C of order 2.

*56. Prove part (a) of Theorem 7-11 for square matrices A, B, and C of order 2.

*57. Prove part (b) of Theorem 7-11 for square matrices B, C, and D of order 2.

*58. Prove part (c) of Theorem 7-11 for 3×2 matrices B and C.

*59. Prove part (d) of Theorem 7-11.

7-3
Solution
of Linear Systems
of n Equations
in n Unknowns
by Gaussian
Elimination

A system of n real linear equations in n unknowns $x_1, x_2, \ldots, x_n$ is a system of the form

$$a_{11}x_1 + a_{12}x_2 + a_{13}x_3 + \cdots + a_{1n}x_n = b_1$$
$$a_{21}x_1 + a_{22}x_2 + a_{23}x_3 + \cdots + a_{2n}x_n = b_2$$
$$\vdots \qquad\qquad \vdots$$
$$a_{n1}x_1 + a_{n2}x_2 + a_{n3}x_3 + \cdots + a_{nn}x_n = b_n,$$

where each a_{ij} and each b_i is a real number. In Section 6-5, two methods were presented for solving systems of 2 linear equations in 2 unknowns:

$$a_{11}x_1 + a_{12}x_2 = b_1$$
$$a_{21}x_1 + a_{22}x_2 = b_2.$$

The "elimination method" can also be used to solve any system of n linear equations with n unknowns. Example 1 illustrates this method for the case of 3 equations and 3 unknowns.

EXAMPLE 1

Solve the system below by the elimination method.

$$x - y + 2z = 9$$
$$-2x + 3y - z = -11$$
$$3x + y + z = 4$$

SOLUTION

Adding 2 times the first equation to the second equation yields an equation which involves only the variables y and z:

$$y + 3z = 7.$$

Adding -3 times the first equation to the third equation yields another equation in the variables y and z:

$$4y - 5z = -23.$$

If we replace the second and third equations in the original system by our two new equations, we have the following system:

$$x - y + 2z = 9$$
$$y + 3z = 7$$
$$4y - 5z = -23.$$

By the same reasoning as was used in Section 6-5, this system is equivalent to the original system (i.e., they have the same solution set). The advantage that we have in the new system is that x has been eliminated from the last two equations, and we can solve for y and z in

$$y + 3z = 7$$
$$4y - 5z = -23$$

as we did in Section 6-5. The value for x can then be determined from the first equation in our new system.

In solving for y and z, we can eliminate y from the last equation by adding -4 times the top equation to the bottom equation. This yields

$$y + 3z = 7$$
$$-17z = -51.$$

We now easily find $z = 3$, $y = -2$, as the solution values for y and z. Substituting in $x - y + 2z = 9$, we have

$$x + 2 + 6 = 9,$$

and

$$x = 1.$$

Thus the solution to the original system is given by

$$x = 1, \quad y = -2, \quad z = 3.$$

Matrices prove to be useful in solving systems of linear equations. One method which uses matrices is called the *Gaussian elimination method*. Before describing this method, we need the following definition.

DEFINITION 7-12 The *augmented matrix* $[A \mid B]$ for the system

$$a_{11}x_1 + a_{12}x_2 + a_{13}x_3 + \cdots + a_{1n}x_n = b_1$$
$$a_{21}x_1 + a_{22}x_2 + a_{23}x_3 + \cdots + a_{2n}x_n = b_2$$
$$\vdots \qquad\qquad\qquad \vdots$$
$$a_{n1}x_1 + a_{n2}x_2 + a_{n3}x_3 + \cdots + a_{nn}x_n = b_n$$

is the matrix

$$\begin{bmatrix} a_{11} & a_{12} & a_{13} & \cdots & a_{1n} & b_1 \\ a_{21} & a_{22} & a_{23} & \cdots & a_{2n} & b_2 \\ \cdot & \cdot & \cdot & & \cdot & \cdot \\ \cdot & \cdot & \cdot & & \cdot & \cdot \\ \cdot & \cdot & \cdot & & \cdot & \cdot \\ a_{n1} & a_{n2} & a_{n3} & \cdots & a_{nn} & b_n \end{bmatrix}.$$

EXAMPLE 2 The augmented matrix for

$$3x + 2y = 4$$
$$x - y = 7$$

is the matrix

$$\begin{bmatrix} 3 & 2 & | & 4 \\ 1 & -1 & | & 7 \end{bmatrix},$$

and the augmented matrix for

$$x - y + 2z = 9$$
$$-2x + 3y - z = -11$$
$$3x + y + z = 4$$

is the matrix

$$\begin{bmatrix} 1 & -1 & 2 & | & 9 \\ -2 & 3 & -1 & | & -11 \\ 3 & 1 & 1 & | & 4 \end{bmatrix}.$$

Conversely, the system associated with the augmented matrix

$$\begin{bmatrix} 1 & 3 & 0 & 4 & | & 1 \\ 0 & 1 & -1 & 2 & | & -1 \\ 1 & -1 & 5 & 3 & | & 1 \\ 2 & 1 & 1 & 0 & | & -2 \end{bmatrix}$$

is

$$x_1 + 3x_2 \qquad + 4x_4 = 1$$
$$x_2 - x_3 + 2x_4 = -1$$
$$x_1 - x_2 + 5x_3 + 3x_4 = 1$$
$$2x_1 + x_2 + x_3 \qquad = -2.$$

The system associated with the augmented matrix

$$\begin{bmatrix} 1 & 0 & 0 & | & a \\ 0 & 1 & 0 & | & b \\ 0 & 0 & 1 & | & c \end{bmatrix}$$

is

$$x \qquad = a$$
$$y \qquad = b$$
$$z = c,$$

which has the obvious solution $x = a$, $y = b$, $z = c$.

EXAMPLE 3 Suppose that we wish to solve the system

$$2x + 3y = 0$$
$$x + y = 1$$

with augmented matrix $\begin{bmatrix} 2 & 3 & | & 0 \\ 1 & 1 & | & 1 \end{bmatrix}.$

SOLUTION   Since it makes no difference which equation is written first, we could write

$$x + y = 1$$
$$2x + 3y = 0$$

with augmented matrix $\begin{bmatrix} 1 & 1 & | & 1 \\ 2 & 3 & | & 0 \end{bmatrix}$.

Adding -2 times the first equation to the second equation yields

$$x + y = 1$$
$$y = -2$$

with augmented matrix $\begin{bmatrix} 1 & 1 & | & 1 \\ 0 & 1 & | & -2 \end{bmatrix}$.

Adding -1 times the second equation to the first equation yields

$$x \quad\;\; = 3$$
$$y = -2$$

with augmented matrix $\begin{bmatrix} 1 & 0 & | & 3 \\ 0 & 1 & | & -2 \end{bmatrix}$.

The solution is $x = 3$ and $y = -2$.

Focusing our attention on the matrices, the system can be solved by making the following changes in the augmented matrix:

$$\begin{bmatrix} 2 & 3 & | & 0 \\ 1 & 1 & | & 1 \end{bmatrix} \xrightarrow{R_1 \leftrightarrow R_2} \begin{bmatrix} 1 & 1 & | & 1 \\ 2 & 3 & | & 0 \end{bmatrix} \xrightarrow{-2R_1 + R_2} \begin{bmatrix} 1 & 1 & | & 1 \\ 0 & 1 & | & -2 \end{bmatrix} \xrightarrow{-R_2 + R_1} \begin{bmatrix} 1 & 0 & | & 3 \\ 0 & 1 & | & -2 \end{bmatrix}$$

where $R_1 \leftrightarrow R_2$ indicates that rows 1 and 2 are interchanged, $-2R_1 + R_2$ indicates that row 2 is replaced by -2 times row 1 plus row 2, and so on.

In the ordinary elimination method that was illustrated in Example 1, equivalent systems are obtained by using the following operations.

---

1. Interchange two equations.

2. Multiply (or divide) both members of an equation by the same nonzero number.

3. Add (or subtract) a multiple of one equation to another equation.

Example 3 shows how these operations on systems correspond to performing the following row operations on the augmented matrix.

---

1. Interchange two rows.

2. Multiply (or divide) every element in a row by the same nonzero number.

3. Add (or subtract) a multiple of one row to another row.

The *Gaussian elimination method* uses row operations on the augmented matrix to solve a system of linear equations. The procedure followed is described below.

Gaussian Elimination Method

1. Write the augmented matrix $[A \mid B]$ for the system, where A is the square matrix whose elements are the coefficients of the unknowns in the system, and B is the column matrix whose elements are the constants.

2. Use any of the row operations (1), (2), (3) above to change $[A \mid B]$ to $[I \mid C]$, where I is an identity matrix of the same size as A, and C is a column matrix.

3. Read the solutions from the column matrix C.

This method is illustrated in the following example.

EXAMPLE 4 Use Gaussian elimination to solve the following system.

$$
\begin{aligned}
x + y + 2z &= 3 \\
2x - y + z &= 6 \\
-x + 3y &= -5
\end{aligned}
$$

SOLUTION The augmented matrix is

$$
[A \mid B] = \begin{bmatrix} 1 & 1 & 2 & 3 \\ 2 & -1 & 1 & 6 \\ -1 & 3 & 0 & -5 \end{bmatrix}.
$$

We shall use row operations to transform $[A \mid B]$ into

$$
[I \mid C] = \begin{bmatrix} 1 & 0 & 0 & a \\ 0 & 1 & 0 & b \\ 0 & 0 & 1 & c \end{bmatrix},
$$

which is the augmented matrix for the system whose solution is

$$
x = a, \qquad y = b, \qquad z = c.
$$

The most straightforward approach to use to transform $[A \mid B]$ into $[I \mid C]$ is to change one column of A at a time into a column of the identity matrix I, working from left to right. In the first column, we use appropriate row operations to obtain a 1 in the first row. Then, with that 1, we use the third type of row operation to obtain 0's in the remaining positions of column 1:

$$\begin{bmatrix} 1 & 1 & 2 & \vdots & 3 \\ 2 & -1 & 1 & \vdots & 6 \\ -1 & 3 & 0 & \vdots & -5 \end{bmatrix} \xrightarrow{-2R_1 + R_2} \begin{bmatrix} 1 & 1 & 2 & \vdots & 3 \\ 0 & -3 & -3 & \vdots & 0 \\ -1 & 3 & 0 & \vdots & -5 \end{bmatrix}$$

$$\xrightarrow{R_1 + R_3} \begin{bmatrix} 1 & 1 & 2 & \vdots & 3 \\ 0 & -3 & -3 & \vdots & 0 \\ 0 & 4 & 2 & \vdots & -2 \end{bmatrix}.$$

Next, we use row operations to obtain a 1 in the second row, second column, position. Then, with that 1, we use row operations of the third type to obtain 0's in the remaining positions of column 2:

$$\xrightarrow{-\frac{1}{3}R_2} \begin{bmatrix} 1 & 1 & 2 & \vdots & 3 \\ 0 & 1 & 1 & \vdots & 0 \\ 0 & 4 & 2 & \vdots & -2 \end{bmatrix} \xrightarrow{-R_2 + R_1} \begin{bmatrix} 1 & 0 & 1 & \vdots & 3 \\ 0 & 1 & 1 & \vdots & 0 \\ 0 & 4 & 2 & \vdots & -2 \end{bmatrix}$$

$$\xrightarrow{-4R_2 + R_3} \begin{bmatrix} 1 & 0 & 1 & \vdots & 3 \\ 0 & 1 & 1 & \vdots & 0 \\ 0 & 0 & -2 & \vdots & -2 \end{bmatrix}.$$

Proceeding to column 3, we use row operations to obtain a 1 in the third row. Then, with that 1, we use row operations of type 3 to obtain 0's in the remaining positions of columm 3:

$$\xrightarrow{-\frac{1}{2}R_3} \begin{bmatrix} 1 & 0 & 1 & \vdots & 3 \\ 0 & 1 & 1 & \vdots & 0 \\ 0 & 0 & 1 & \vdots & 1 \end{bmatrix} \xrightarrow{-R_3 + R_2} \begin{bmatrix} 1 & 0 & 1 & \vdots & 3 \\ 0 & 1 & 0 & \vdots & -1 \\ 0 & 0 & 1 & \vdots & 1 \end{bmatrix}$$

$$\xrightarrow{-R_3 + R_1} \begin{bmatrix} 1 & 0 & 0 & \vdots & 2 \\ 0 & 1 & 0 & \vdots & -1 \\ 0 & 0 & 1 & \vdots & 1 \end{bmatrix}.$$

Thus the solution is

$$x = 2, \qquad y = -1, \qquad z = 1.$$

Sometimes it is impossible to transform the augmented matrix $[A \mid B]$ into $[I \mid C]$ by row operations. If any of the row operations on $[A \mid B]$ yield a row with all zero elements except possibly the last element, then either there is no solution to the system, or the solution is not unique. This is illustrated in the next example.

EXAMPLE 5 Solve the following system by Gaussian elimination.

$$\begin{aligned} x - 2y - 2z &= -1 \\ x + y + z &= 2 \\ x + 2y + 2z &= 1 \end{aligned}$$

Following the procedure described in Example 4 yields

$$\begin{bmatrix} 1 & -2 & -2 & \vdots & -1 \\ 1 & 1 & 1 & \vdots & 2 \\ 1 & 2 & 2 & \vdots & 1 \end{bmatrix} \xrightarrow{-R_1+R_2} \begin{bmatrix} 1 & -2 & -2 & \vdots & -1 \\ 0 & 3 & 3 & \vdots & 3 \\ 1 & 2 & 2 & \vdots & 1 \end{bmatrix}$$

$$\xrightarrow{-R_1+R_3} \begin{bmatrix} 1 & -2 & -2 & \vdots & -1 \\ 0 & 3 & 3 & \vdots & 3 \\ 0 & 4 & 4 & \vdots & 2 \end{bmatrix}$$

$$\xrightarrow{\frac{1}{3}R_2} \begin{bmatrix} 1 & -2 & -2 & \vdots & -1 \\ 0 & 1 & 1 & \vdots & 1 \\ 0 & 4 & 4 & \vdots & 2 \end{bmatrix}$$

$$\xrightarrow{-4R_2+R_3} \begin{bmatrix} 1 & -2 & -2 & \vdots & -1 \\ 0 & 1 & 1 & \vdots & 1 \\ 0 & 0 & 0 & \vdots & -2 \end{bmatrix}.$$

Since the third row in the last matrix contains all zero elements except the last element, it is impossible to obtain $[I \mid C]$ by row operations. The last matrix obtained is the augmented matrix for a system with a last equation which reads $0 = -2$. This indicates there is no solution to the original system.

If, in Example 5, the last row in the last matrix had turned out to read $[0 \quad 0 \quad 0 \mid 0]$ instead of $[0 \quad 0 \quad 0 \mid -2]$, the corresponding equation would have read $0 = 0$ instead of $0 = -2$. There then would have been many solutions, instead of no solution.

EXERCISES 7-3

Solve each system by the elimination method that was used in Example 1.

1. $2x + 3y - z = 1$
   $x - 4y + 2z = 2$
   $x - y + z = 3$

2. $2x - y + 3z = 8$
   $-3x + 2y + 4z = -4$
   $5x + 3y - 2z = 5$

3. $3x_1 - x_2 = 0$
   $5x_1 - x_3 = 0$
   $-x_1 - x_2 + x_3 = 1$

4. $2x_1 - 2x_3 = 0$
   $4x_2 - x_3 = 2$
   $3x_1 + 5x_2 = -6$

5. $4x - 3y + 2z = 0$
   $11x + 4y - 7z = 0$
   $-3x - 2y + 2z = 0$

6. $-6x + 5z = 0$
   $3y - 4z = 0$
   $7x + 5y = 0$

Write the augmented matrix for each of the following systems.

7. $3x - y = 0$
   $-x + y = 1$

8. $r + s = 10$
   $r - s = -4$

9. $3x - 2y + 5z = 0$
$\quad 4x + 7y - z = 0$
$\quad x \quad\quad + z = 0$

10. $x_1 + x_3 = 4$
$\quad x_2 + x_3 = 0$
$\quad x_1 + x_2 = 2$

11. $\quad p - q = 0$
$\quad r + s = 0$
$\quad 3p + 2s = 0$
$\quad 5q - r = 0$

12. $\quad a + 3b + c + d = 1$
$\quad a - 3b - c + 2d = 5$
$\quad 2a \quad\quad + c + 3d = 7$
$\quad b + c \quad\quad = 0$

Write a system of linear equations which has the given augmented matrix.

13. $\begin{bmatrix} 1 & 2 & | & 5 \\ 3 & 4 & | & 6 \end{bmatrix}$

14. $\begin{bmatrix} -2 & 1 & | & -1 \\ 4 & 3 & | & 0 \end{bmatrix}$

15. $\begin{bmatrix} 1 & 0 & 0 & | & a \\ 0 & 0 & 1 & | & b \\ 0 & 1 & 0 & | & c \end{bmatrix}$

16. $\begin{bmatrix} 1 & 1 & 0 & | & 0 \\ 3 & 0 & -3 & | & 0 \\ -1 & 1 & -1 & | & -6 \end{bmatrix}$

17. $\begin{bmatrix} 1 & 0 & 1 & 0 & | & 0 \\ 0 & 2 & 1 & 3 & | & 7 \\ 3 & 1 & 1 & 1 & | & 1 \\ -3 & 1 & -1 & 2 & | & 5 \end{bmatrix}$

18. $\begin{bmatrix} 0 & 0 & 0 & 1 & | & 1 \\ 0 & -1 & 0 & 0 & | & 2 \\ 1 & 0 & 0 & 0 & | & 3 \\ 0 & 0 & -1 & 0 & | & 4 \end{bmatrix}$

Use Gaussian elimination to find a solution, if one exists, for each system.

19. $x + y = 1$
$\quad 2x + 3y = -2$

20. $3x + 2y = 1$
$\quad x + 2y = 7$

21. $4a + 5b = -22$
$\quad 3a - 4b = -1$

22. $\quad 7x_1 - x_2 = -2$
$\quad -3x_1 + x_2 = 6$

23. $x - 2y = 0$
$\quad 2x - 4y = 1$

24. $-3g + h = 2$
$\quad 6g - 2h = 0$

25. $\quad 2x - 3y = -1$
$\quad -5x + 8y = 0$

26. $9x - y = 0$
$\quad 3x + 3y = -10$

27. $2x + 7y = 0$
$\quad x - 2y = 0$

28. $3x \quad\quad = 0$
$\quad x + 5y = -15$

29. $x - 2y \quad\quad = 1$
$\quad\quad y + z = 0$
$\quad 2x \quad\quad + 3z = 3$

30. $\quad x - 2y + 3z = 0$
$\quad\quad 3y - 2z = 0$
$\quad -3x + 4y - z = 0$

31. $\quad x - y - 4z = -4$
$\quad -3x + 4y + 2z = 6$
$\quad -x + 3y + 2z = 10$

32. $2a + 2b - c = -2$
$\quad\quad - b + 4c = 11$
$\quad a - b \quad\quad = 6$

33. $\quad -2y - 2z = -2$
$\quad 2x - y + z = -3$
$\quad x + y + 3z = -2$

34. $3x_1 + 2x_2 + 5x_3 = 2$
$\quad 2x_1 \quad\quad + 4x_3 = 2$
$\quad x_1 + 3x_2 - x_3 = -2$

35. $r - t = 1$
$\quad 3r + s = 6$
$\quad 5s + 6t = -12$

36. $-3x_2 + 4x_3 = -3$
$\quad 3x_1 - x_3 = 6$
$\quad -x_1 + 3x_2 = 1$

37. $2x - 2y - 2z = -2$
$\quad x - 4y + z = -2$
$\quad 5x - 8y - 3z = -2$

38. $\quad x + 4y - z = 0$
$\quad 3x - 5y + z = 1$
$\quad 5x - 14y + 3z = 1$

39. $\quad x + y \quad - t = 2$
$\quad\quad y + z + 2t = -3$
$\quad -2x \quad\quad + t = -4$
$\quad\quad x + y + z + t = 0$

40. $2x_1 - 4x_2 + x_3 + 10x_4 = -2$
$\quad 3x_1 - x_2 + 3x_3 \quad\quad = -3$
$\quad x_1 \quad\quad + 2x_3 \quad\quad = 0$
$\quad\quad x_2 - x_3 + x_4 = 4$

41. $-a + b - c \quad\quad = 5$
$\quad\quad b - c + 3d = 10$
$\quad a \quad - c - d = 5$
$\quad a \quad\quad + 3d = 5$

42. $x - y = 0$
$\quad y - z = 0$
$\quad z - t = 0$
$\quad x - t = 0$

Show that the systems with the following augmented matrices have no solutions.

43. $\begin{bmatrix} 1 & 0 & 0 & | & 1 \\ 0 & 1 & 0 & | & 2 \\ 0 & 0 & 0 & | & 3 \end{bmatrix}$

44. $\begin{bmatrix} 1 & 2 & 1 & | & 1 \\ 2 & 4 & 2 & | & 0 \\ 0 & 0 & 0 & | & 0 \end{bmatrix}$

Some of the coefficients and constant terms in the systems below involve an unspecified constant c. For each system, find the values of c for which the system has (a) no solution, (b) exactly one solution, (c) many solutions.

*45. $\quad x + 2y = 1$
$\quad 2x + c^2 y = c$

*46. $\quad x + y = 1$
$\quad x + cy = c^2$

*47. $\quad x + 2y \quad\quad - z = 1$
$\quad x + y \quad\quad = 0$
$\quad x + 2y + (c^2 - 1)z = c + 1$

*48. $\quad x + 2y \quad\quad - z = 1$
$\quad 2x + 4y \quad\quad - 2z = 0$
$\quad x + 2y + (c^2 - 1)z = c + 1$

7-4
*Calculation
of Inverses*

In Sections 7-1 and 7-3, we saw that

(1) a matrix with all elements zero acts as an additive identity for matrix addition;

(2) every matrix has an additive inverse $-A$;

(3) an identity matrix I_n acts as a multiplicative identity for matrix multiplication.

A question that arises in connection with (3) is this: Does every nonzero matrix A have a multiplicative inverse? In other words, for any nonzero matrix A, is there a matrix B such that $AB = BA = I_n$? The answer to this question might be somewhat disappointing, since *not all nonzero matrices have multiplicative inverses.*

We first notice that multiplicative inverses can exist only for square matrices, since the only way that AB and BA can be equal is for A and B to be square and of the same order. But we shall see shortly that some nonzero square matrices do not have multiplicative inverses.

From now on, we shall use the term "inverse" to mean "multiplicative inverse." If the matrix A has an inverse, we shall denote it by A^{-1}.

A given 2×2 matrix $A = \begin{bmatrix} a & b \\ c & d \end{bmatrix}$ has an inverse if the number $ad - bc$ is not zero. In fact,

$$A^{-1} = \frac{1}{ad - bc}\begin{bmatrix} d & -b \\ -c & a \end{bmatrix}.$$

This is readily verified by the following multiplications:

$$\frac{1}{ad - bc}\begin{bmatrix} d & -b \\ -c & a \end{bmatrix}\begin{bmatrix} a & b \\ c & d \end{bmatrix} = \frac{1}{ad - bc}\begin{bmatrix} da - bc & db - bd \\ -ca + ac & -cb + ad \end{bmatrix} = I_2,$$

and

$$\begin{bmatrix} a & b \\ c & d \end{bmatrix}\frac{1}{ad - bc}\begin{bmatrix} d & -b \\ -c & a \end{bmatrix} = \frac{1}{ad - bc}\begin{bmatrix} ad - bc & -ab + ba \\ cd - dc & -cb + da \end{bmatrix} = I_2.$$

It is also true that A does not have an inverse if $ad - bc = 0$. The proof of this statement is left as an exercise (see problem 50). Combining these two statements, we have the following theorem.

THEOREM 7-13   For a given 2×2 matrix $A = \begin{bmatrix} a & b \\ c & d \end{bmatrix}$, let $\delta(A)$ denote the number $\delta(A) = ad - bc$. The inverse of A exists if $\delta(A) \neq 0$, and is given by

$$A^{-1} = \frac{1}{\delta(A)}\begin{bmatrix} d & -b \\ -c & a \end{bmatrix}, \qquad \text{where } \delta(A) = ad - bc \neq 0.$$

If $\delta(A) = 0$, then A does not have an inverse.

Formulas for A^{-1} which are similar to the one in Theorem 7-13 exist for matrices of order greater than 2, but they are much more complicated, and they more properly belong in a course in linear algebra. At any rate, a much easier method for finding A^{-1} is presented later in this section.

EXAMPLE 1   Use Theorem 7-13 to find A^{-1}, if it exists.

(a) $A = \begin{bmatrix} 1 & 4 \\ 3 & -2 \end{bmatrix}$

(b) $A = \begin{bmatrix} 3 & -2 \\ 4 & -3 \end{bmatrix}$

(c) $A = I_2 = \begin{bmatrix} 1 & 0 \\ 0 & 1 \end{bmatrix}$

(d) $A = \begin{bmatrix} 1 & -3 \\ -2 & 6 \end{bmatrix}$

SOLUTION

(a) For $A = \begin{bmatrix} 1 & 4 \\ 3 & -2 \end{bmatrix}$, we have $\delta(A) = (1)(-2) - (3)(4) = -14$.

Thus A^{-1} exists and is given by

$$A^{-1} = \frac{1}{-14} \begin{bmatrix} -2 & -4 \\ -3 & 1 \end{bmatrix} = \begin{bmatrix} \frac{1}{7} & \frac{2}{7} \\ \frac{3}{14} & -\frac{1}{14} \end{bmatrix}.$$

(b) For $A = \begin{bmatrix} 3 & -2 \\ 4 & -3 \end{bmatrix}$, we have $\delta(A) = (3)(-3) - (4)(-2) = -1$

and

$$A^{-1} = \frac{1}{-1} \begin{bmatrix} -3 & 2 \\ -4 & 3 \end{bmatrix} = \begin{bmatrix} 3 & -2 \\ 4 & -3 \end{bmatrix}.$$

Thus we have the unexpected result here that $A^{-1} = A$.

(c) For $I_2 = \begin{bmatrix} 1 & 0 \\ 0 & 1 \end{bmatrix}$, we have $\delta(I_2) = 1$ and

$$I^{-1} = \begin{bmatrix} 1 & 0 \\ 0 & 1 \end{bmatrix},$$

a result that should be expected.

(d) For $A = \begin{bmatrix} 1 & -3 \\ -2 & 6 \end{bmatrix}$, we have $\delta(A) = (1)(6) - (-2)(-3) = 0$.

Thus A^{-1} does not exist for this matrix A.

The easier method for finding A^{-1}, which was referred to before Example 1, is known as the *Gaussian elimination method*. It can be used to calculate the inverse of a square matrix A of any order n, provided that A^{-1} exists. The procedure to be followed is given by the three steps below.

---

1. Augment A with an identity matrix of the same size: $[A \,|\, I_n]$.
2. Use row operations to transform $[A \,|\, I_n]$ into the form $[I_n \,|\, B]$. If this is not possible, then A^{-1} does not exist.
3. If $[A \,|\, I_n]$ is transformed into $[I_n \,|\, B]$ by row operations, the inverse can be read from the last n columns of $[I_n \,|\, B]$. That is, $B = A^{-1}$.

EXAMPLE 2   Use the Gaussian elimination method to find the inverse of each matrix below, if it exists.

(a) $\begin{bmatrix} -2 & 3 \\ 1 & 2 \end{bmatrix}$

(b) $\begin{bmatrix} 1 & 0 & 2 \\ 0 & -1 & 3 \\ 2 & 1 & 3 \end{bmatrix}$

(c) $\begin{bmatrix} 1 & 1 & 1 \\ 1 & -1 & 0 \\ 2 & 0 & 1 \end{bmatrix}$

SOLUTION (a) We augment $A = \begin{bmatrix} -2 & 3 \\ 1 & 2 \end{bmatrix}$ with a 2×2 identity, and use row operations to transform $[A \mid I]$ into $[I \mid A^{-1}]$.

$$[A \mid I] = \begin{bmatrix} -2 & 3 & \vdots & 1 & 0 \\ 1 & 2 & \vdots & 0 & 1 \end{bmatrix} \xrightarrow{R_1 \leftrightarrow R_2} \begin{bmatrix} 1 & 2 & \vdots & 0 & 1 \\ -2 & 3 & \vdots & 1 & 0 \end{bmatrix}$$

$$\xrightarrow{2R_1 + R_2} \begin{bmatrix} 1 & 2 & \vdots & 0 & 1 \\ 0 & 7 & \vdots & 1 & 2 \end{bmatrix} \xrightarrow{\frac{1}{7}R_2} \begin{bmatrix} 1 & 2 & \vdots & 0 & 1 \\ 0 & 1 & \vdots & \frac{1}{7} & \frac{2}{7} \end{bmatrix}$$

$$\xrightarrow{-2R_2 + R_1} \begin{bmatrix} 1 & 0 & \vdots & -\frac{2}{7} & \frac{3}{7} \\ 0 & 1 & \vdots & \frac{1}{7} & \frac{2}{7} \end{bmatrix}.$$

Thus $A^{-1} = \begin{bmatrix} -\frac{2}{7} & \frac{3}{7} \\ \frac{1}{7} & \frac{2}{7} \end{bmatrix}$. This should be checked by computing AA^{-1} and $A^{-1}A$.

(b) For $A = \begin{bmatrix} 1 & 0 & 2 \\ 0 & -1 & 3 \\ 2 & 1 & 3 \end{bmatrix}$, we have

$$[A \mid I] = \begin{bmatrix} 1 & 0 & 2 & \vdots & 1 & 0 & 0 \\ 0 & -1 & 3 & \vdots & 0 & 1 & 0 \\ 2 & 1 & 3 & \vdots & 0 & 0 & 1 \end{bmatrix}$$

$$\xrightarrow{-2R_1 + R_3} \begin{bmatrix} 1 & 0 & 2 & \vdots & 1 & 0 & 0 \\ 0 & -1 & 3 & \vdots & 0 & 1 & 0 \\ 0 & 1 & -1 & \vdots & -2 & 0 & 1 \end{bmatrix}$$

$$\xrightarrow{-R_2} \begin{bmatrix} 1 & 0 & 2 & \vdots & 1 & 0 & 0 \\ 0 & 1 & -3 & \vdots & 0 & -1 & 0 \\ 0 & 1 & -1 & \vdots & -2 & 0 & 1 \end{bmatrix}$$

$$\xrightarrow{-R_2 + R_3} \begin{bmatrix} 1 & 0 & 2 & \vdots & 1 & 0 & 0 \\ 0 & 1 & -3 & \vdots & 0 & -1 & 0 \\ 0 & 0 & 2 & \vdots & -2 & 1 & 1 \end{bmatrix}$$

$$\xrightarrow{\frac{1}{2}R_3} \begin{bmatrix} 1 & 0 & 2 & \vdots & 1 & 0 & 0 \\ 0 & 1 & -3 & \vdots & 0 & -1 & 0 \\ 0 & 0 & 1 & \vdots & -1 & \frac{1}{2} & \frac{1}{2} \end{bmatrix}$$

$$\xrightarrow{3R_3+R_2} \left[\begin{array}{ccc|ccc} 1 & 0 & 2 & 1 & 0 & 0 \\ 0 & 1 & 0 & -3 & \frac{1}{2} & \frac{3}{2} \\ 0 & 0 & 1 & -1 & \frac{1}{2} & \frac{1}{2} \end{array}\right]$$

$$\xrightarrow{-2R_3+R_1} \left[\begin{array}{ccc|ccc} 1 & 0 & 0 & 3 & -1 & -1 \\ 0 & 1 & 0 & -3 & \frac{1}{2} & \frac{3}{2} \\ 0 & 0 & 1 & -1 & \frac{1}{2} & \frac{1}{2} \end{array}\right].$$

This gives $A^{-1} = \begin{bmatrix} 3 & -1 & -1 \\ -3 & \frac{1}{2} & \frac{3}{2} \\ -1 & \frac{1}{2} & \frac{1}{2} \end{bmatrix}$.

(c) Following the same procedure with $A = \begin{bmatrix} 1 & 1 & 1 \\ 1 & -1 & 0 \\ 2 & 0 & 1 \end{bmatrix}$, we have

$$[A\,|\,I] = \left[\begin{array}{ccc|ccc} 1 & 1 & 1 & 1 & 0 & 0 \\ 1 & -1 & 0 & 0 & 1 & 0 \\ 2 & 0 & 1 & 0 & 0 & 1 \end{array}\right]$$

$$\xrightarrow{-R_1+R_2} \left[\begin{array}{ccc|ccc} 1 & 1 & 1 & 1 & 0 & 0 \\ 0 & -2 & -1 & -1 & 1 & 0 \\ 2 & 0 & 1 & 1 & 0 & 1 \end{array}\right]$$

$$\xrightarrow{-2R_1+R_3} \left[\begin{array}{ccc|ccc} 1 & 1 & 1 & 1 & 0 & 0 \\ 0 & -2 & -1 & -1 & 1 & 0 \\ 0 & -2 & -1 & -2 & 0 & 1 \end{array}\right]$$

$$\xrightarrow{-R_2+R_3} \left[\begin{array}{ccc|ccc} 1 & 1 & 1 & 1 & 0 & 0 \\ 0 & -2 & -1 & -1 & 1 & 0 \\ 0 & 0 & 0 & -1 & -1 & 1 \end{array}\right].$$

The zeros in the shaded portion of the last matrix indicate that it is impossible to transform $[A\,|\,I]$ into $[I\,|\,B]$ by row operations, and A^{-1} does not exist.

EXERCISES 7-4

Use Theorem 7-13 to find the inverse of each of the following matrices, if it exists.

1. $\begin{bmatrix} 1 & 1 \\ 0 & 2 \end{bmatrix}$

2. $\begin{bmatrix} 0 & -1 \\ 1 & 0 \end{bmatrix}$

3. $\begin{bmatrix} -1 & -3 \\ -1 & -2 \end{bmatrix}$

4. $\begin{bmatrix} 2 & 3 \\ -1 & 4 \end{bmatrix}$

5. $\begin{bmatrix} 5 & -3 \\ 4 & -2 \end{bmatrix}$

6. $\begin{bmatrix} 2 & 1 \\ -4 & 2 \end{bmatrix}$

7. $\begin{bmatrix} -3 & 5 \\ 12 & -20 \end{bmatrix}$

8. $\begin{bmatrix} 4 & -3 \\ 0 & 0 \end{bmatrix}$

9. $\begin{bmatrix} 0 & 1 \\ 0 & 3 \end{bmatrix}$

10. $\begin{bmatrix} -8 & 2 \\ -4 & 1 \end{bmatrix}$

11. $\begin{bmatrix} 10 & 5 \\ -2 & 1 \end{bmatrix}$

12. $\begin{bmatrix} 1 & 1 \\ 1 & -1 \end{bmatrix}$

Use the Gaussian elimination method to find the inverse of each matrix, if it exists.

13. $\begin{bmatrix} 2 & 0 \\ 0 & -4 \end{bmatrix}$

14. $\begin{bmatrix} 1 & 0 \\ -1 & 2 \end{bmatrix}$

15. $\begin{bmatrix} 0 & 3 \\ -2 & 4 \end{bmatrix}$

16. $\begin{bmatrix} 2 & 1 \\ 1 & 1 \end{bmatrix}$

17. $\begin{bmatrix} -1 & 3 \\ 2 & 2 \end{bmatrix}$

18. $\begin{bmatrix} 4 & -2 \\ 1 & -1 \end{bmatrix}$

19. $\begin{bmatrix} 5 & -15 \\ 2 & 3 \end{bmatrix}$

20. $\begin{bmatrix} 5 & -15 \\ -1 & 3 \end{bmatrix}$

21. $\begin{bmatrix} 1 & -1 \\ 0 & 0 \end{bmatrix}$

22. $\begin{bmatrix} 3 & 0 \\ -1 & 0 \end{bmatrix}$

23. $\begin{bmatrix} -2 & 3 \\ 5 & 7 \end{bmatrix}$

24. $\begin{bmatrix} -4 & 6 \\ 2 & 1 \end{bmatrix}$

25. $\begin{bmatrix} 0 & 1 & 0 \\ 1 & 4 & 1 \\ 0 & 3 & -1 \end{bmatrix}$

26. $\begin{bmatrix} 1 & 3 & -2 \\ 2 & 5 & -7 \\ 1 & 4 & 0 \end{bmatrix}$

27. $\begin{bmatrix} 1 & -4 & 2 \\ 2 & -9 & 5 \\ 1 & -5 & 4 \end{bmatrix}$

28. $\begin{bmatrix} 2 & -3 & 3 \\ -3 & 1 & 0 \\ 1 & -1 & 1 \end{bmatrix}$

29. $\begin{bmatrix} 1 & 0 & 1 \\ -1 & 2 & 1 \\ 0 & 1 & 3 \end{bmatrix}$

30. $\begin{bmatrix} 0 & -1 & -2 \\ 2 & 4 & 8 \\ -1 & 1 & 0 \end{bmatrix}$

31. $\begin{bmatrix} 1 & 1 & 1 \\ -1 & 2 & -1 \\ 0 & 3 & -1 \end{bmatrix}$

32. $\begin{bmatrix} 2 & 0 & -3 \\ -1 & 2 & 1 \\ 2 & 0 & 0 \end{bmatrix}$

33. $\begin{bmatrix} 1 & 1 & -1 \\ -1 & 1 & 1 \\ 1 & 1 & 1 \end{bmatrix}$

34. $\begin{bmatrix} 1 & -4 & 0 \\ -3 & 10 & 1 \\ 0 & 0 & 4 \end{bmatrix}$

35. $\begin{bmatrix} 5 & 0 & -10 \\ 1 & 3 & -2 \\ 0 & 1 & 1 \end{bmatrix}$

36. $\begin{bmatrix} 1 & 1 & 3 \\ 1 & 3 & -2 \\ 1 & 2 & -3 \end{bmatrix}$

37. $\begin{bmatrix} 3 & 2 & 6 \\ -2 & 1 & -3 \\ 1 & 2 & 3 \end{bmatrix}$

38. $\begin{bmatrix} -1 & 2 & 0 \\ 0 & -1 & 2 \\ 2 & 0 & -1 \end{bmatrix}$

39. $\begin{bmatrix} 0 & -1 & 1 \\ 3 & 2 & -1 \\ 0 & 1 & -1 \end{bmatrix}$

40. $\begin{bmatrix} 1 & -1 & 1 \\ -1 & 0 & 1 \\ 3 & -2 & 1 \end{bmatrix}$

41. $\begin{bmatrix} 5 & 3 & -2 \\ -1 & 2 & 5 \\ 11 & 4 & -9 \end{bmatrix}$

42. $\begin{bmatrix} -5 & -3 & 1 \\ 6 & 3 & 0 \\ 1 & 2 & -3 \end{bmatrix}$

43. $\begin{bmatrix} 1 & 0 & 0 & -4 \\ 0 & 2 & 3 & 0 \\ 0 & 2 & -3 & 0 \\ 1 & 0 & 0 & 4 \end{bmatrix}$

44. $\begin{bmatrix} 1 & -1 & -1 & 0 \\ 0 & 1 & -2 & 4 \\ 2 & -2 & -1 & -2 \\ -1 & 2 & 1 & 1 \end{bmatrix}$

45. $\begin{bmatrix} 1 & 0 & -1 & 2 \\ 3 & -1 & -1 & 6 \\ 2 & 0 & -3 & 8 \\ 1 & 2 & -2 & -9 \end{bmatrix}$

46. $\begin{bmatrix} 1 & -2 & 1 & 0 \\ 0 & 1 & 0 & -2 \\ 1 & -1 & 0 & 1 \\ 2 & -4 & 1 & 5 \end{bmatrix}$

*47. Find D^{-1} if $D = \begin{bmatrix} a & 0 & 0 \\ 0 & b & 0 \\ 0 & 0 & c \end{bmatrix}$, where a, b, and c are nonzero real numbers.

*48. Let A and B be $n \times n$ matrices such that A^{-1} and B^{-1} exist. Prove that $(AB)^{-1}$ exists, and that $(AB)^{-1} = B^{-1}A^{-1}$.

*49. Let A, B, and C be $n \times n$ matrices such that A^{-1}, B^{-1}, and C^{-1} exist. Prove that $(ABC)^{-1}$ exists, and that $(ABC)^{-1} = C^{-1}B^{-1}A^{-1}$.

*50. Let $A = \begin{bmatrix} a & b \\ c & d \end{bmatrix}$. Prove that A does not have an inverse if $ad - bc = 0$.

7-5
Solution
of Linear Systems
by Inverses

Inverses of matrices can be used in solving certain types of systems of n linear equations in n unknowns. The use of inverses depends on the fact that any system of linear equations can be represented by a single matrix equation. For example, the system

$$x + 4y = 7$$
$$3x - 2y = -7 \tag{1}$$

is equivalent to the single matrix equation

$$\begin{bmatrix} x + 4y \\ 3x - 2y \end{bmatrix} = \begin{bmatrix} 7 \\ -7 \end{bmatrix},$$

and this equation can be written in factored form as

$$\begin{bmatrix} 1 & 4 \\ 3 & -2 \end{bmatrix} \begin{bmatrix} x \\ y \end{bmatrix} = \begin{bmatrix} 7 \\ -7 \end{bmatrix}.$$

Thus the system of linear equations given in (1) is equivalent to the matrix equation $AX = B$, where

$$A = \begin{bmatrix} 1 & 4 \\ 3 & -2 \end{bmatrix}, \quad X = \begin{bmatrix} x \\ y \end{bmatrix}, \quad B = \begin{bmatrix} 7 \\ -7 \end{bmatrix}.$$

The matrix A is called the *coefficient matrix*, X is called the *unknown matrix*, and B is the *constant matrix*.

The discussion in the preceding paragraph extends easily to the general situation. A given system of n linear equations in n unknowns

$$a_{11}x_1 + a_{12}x_2 + a_{13}x_3 + \cdots + a_{1n}x_n = b_1$$
$$a_{21}x_1 + a_{22}x_2 + a_{23}x_3 + \cdots + a_{2n}x_n = b_2$$
$$\vdots \qquad\qquad\qquad \vdots \qquad\qquad\qquad \tag{2}$$
$$a_{n1}x_1 + a_{n2}x_2 + a_{n3}x_3 + \cdots + a_{nn}x_n = b_n$$

can be expressed in matrix form as

$$AX = B,$$

where

$$A = \begin{bmatrix} a_{11} & a_{12} & \cdots & a_{1n} \\ a_{21} & a_{22} & \cdots & a_{2n} \\ \vdots & \vdots & & \vdots \\ a_{n1} & a_{n2} & \cdots & a_{nn} \end{bmatrix}, \quad X = \begin{bmatrix} x_1 \\ x_2 \\ \vdots \\ x_n \end{bmatrix}, \quad B = \begin{bmatrix} b_1 \\ b_2 \\ \vdots \\ b_n \end{bmatrix}.$$

The square matrix A is called the *coefficient matrix*, the column matrix X is called the *unknown matrix*, and the column matrix B is called the *constant matrix*.

To find the values of $x_1, x_2, \ldots, x_n$ which satisfy the system of equations (2), we need only solve for X in the matrix equation $AX = B$. This can always be done when A^{-1} exists. For we can multiply both sides of the equation

$$AX = B$$

by A^{-1} on the left and obtain

$$A^{-1}(AX) = A^{-1}B.$$

Using the associative property of matrix multiplication, we have

$$(A^{-1}A)X = A^{-1}B.$$

Since $A^{-1}A = I_n$, this yields

$$I_nX = A^{-1}B.$$

But $I_nX = X$, so we have

$$X = A^{-1}B.$$

To check that $X = A^{-1}B$ does indeed satisfy the equation, we can substitute $X = A^{-1}B$ in the left member of $AX = B$:

$$A(A^{-1}B) = (AA^{-1})B, \qquad \text{by the associative property,}$$
$$= I_nB, \qquad \text{since } AA^{-1} = I_n,$$
$$= B, \qquad \text{since } I_n \text{ is the identity for multiplication.}$$

These results are recorded in the following theorem. (The reader should compare this theorem with the first paragraph of Section 4-1.)

THEOREM 7-14   If A^{-1} exists, then $X = A^{-1}B$ satisfies the equation $AX = B$, and this is the only value of X which satisfies the equation.

The use of this theorem to solve systems of linear equations is illustrated in the following examples.

EXAMPLE 1   Use the inverse of the coefficient matrix to solve the system

$$x + 4y = 7$$
$$3x - 2y = -7.$$

SOLUTION   As we saw in the first paragraph of this section, this system can be expressed as a matrix equation $AX = B$, where

$$A = \begin{bmatrix} 1 & 4 \\ 3 & -2 \end{bmatrix}, \qquad X = \begin{bmatrix} x \\ y \end{bmatrix}, \qquad B = \begin{bmatrix} 7 \\ -7 \end{bmatrix}.$$

By Theorem 7-14, the solution to $AX = B$ is $X = A^{-1}B$, if A^{-1} exists. In Example 1(a) of Section 7-4, we found

$$A^{-1} = \frac{1}{14} \begin{bmatrix} 2 & 4 \\ 3 & -1 \end{bmatrix}.$$

Thus

$$X = A^{-1}B = \frac{1}{14} \begin{bmatrix} 2 & 4 \\ 3 & -1 \end{bmatrix} \begin{bmatrix} 7 \\ -7 \end{bmatrix} = \begin{bmatrix} -1 \\ 2 \end{bmatrix}.$$

That is,

$$\begin{bmatrix} x \\ y \end{bmatrix} = \begin{bmatrix} -1 \\ 2 \end{bmatrix},$$

and the solution to the system is

$$x = -1, \qquad y = 2.$$

The solution given in Theorem 7-14 is valid for a system of any size. Consider the next example, involving a system of 3 linear equations in 3 unknowns.

EXAMPLE 2 Use the inverse of the coefficient matrix to solve

$$\begin{aligned} x \qquad + 2z &= \quad 1 \\ -y + 3z &= -3 \\ 2x + y + 3z &= \quad 3. \end{aligned}$$

SOLUTION The matrices involved in the matrix form are given by

$$A = \begin{bmatrix} 1 & 0 & 2 \\ 0 & -1 & 3 \\ 2 & 1 & 3 \end{bmatrix}, \qquad X = \begin{bmatrix} x \\ y \\ z \end{bmatrix}, \qquad B = \begin{bmatrix} 1 \\ -3 \\ 3 \end{bmatrix}.$$

The inverse of the coefficient matrix was found to be

$$A^{-1} = \begin{bmatrix} 3 & -1 & -1 \\ -3 & \frac{1}{2} & \frac{3}{2} \\ -1 & \frac{1}{2} & \frac{1}{2} \end{bmatrix}$$

in Example 2(b) of Section 7-4. Thus

$$X = A^{-1}B = \begin{bmatrix} 3 & -1 & -1 \\ -3 & \frac{1}{2} & \frac{3}{2} \\ -1 & \frac{1}{2} & \frac{1}{2} \end{bmatrix} \begin{bmatrix} 1 \\ -3 \\ 3 \end{bmatrix} = \begin{bmatrix} 3 \\ 0 \\ -1 \end{bmatrix},$$

and the solution to the system is

$$x = 3, \qquad y = 0, \qquad z = -1.$$

EXERCISES 7-5

Solve each system by using the inverse of the coefficient matrix. Notice that the coefficient matrices come from the odd-numbered problems in Exercises 7-4.

1. $\begin{aligned} x + y &= -3 \\ 2y &= -10 \end{aligned}$

2. $\begin{aligned} -x - 3y &= -9 \\ -x - 2y &= -5 \end{aligned}$

3. $5a - 3b = 11$
$\quad 4a - 2b = 8$

4. $10x + 5y = 0$
$\quad -2x + y = -8$

5. $2x \qquad = \frac{2}{3}$
$\quad -4y = -\frac{2}{3}$

6. $\qquad 3y = -6$
$\quad -2x + 4y = -22$

7. $-x_1 + 3x_2 = -2$
$\quad 2x_1 + 2x_2 = 12$

8. $5x_1 - 15x_2 = -2$
$\quad 2x_1 + 3x_2 = 1$

9. $-2x + 3y = -8$
$\quad 5x + 7y = 78$

10. $\qquad y = 2$
$\quad x + 4y + z = 10$
$\quad 3y - z = 6$

11. $x - 4y + 2z = -1$
$\quad 2x - 9y + 5z = 0$
$\quad x - 5y + 4z = 6$

12. $x_1 \qquad + x_3 = -2$
$\quad -x_1 + 2x_2 + x_3 = -4$
$\quad x_2 + 3x_3 = -5$

13. $r + s + t = -4$
$\quad -r + 2s - t = -32$
$\quad 3s - t = -34$

14. $a + b - c = 1$
$\quad -a + b + c = -15$
$\quad a + b + c = 1$

15. $5x \qquad - 10z = -5$
$\quad x + 3y - 2z = -\frac{5}{2}$
$\quad y + z = 0$

16. $3x + 2y + 6z = -14$
$\quad -2x + y - 3z = 13$
$\quad x + 2y + 3z = -5$

17. $x - 4t = -1$
$\quad 2y + 3z = -17$
$\quad 2y - 3z = 1$
$\quad x + 4t = 7$

18. $x \qquad - z + 2w = 7$
$\quad 3x - y - z + 6w = 24$
$\quad 2x \qquad - 3z + 8w = 22$
$\quad x + 2y - 2z - 9w = -31$

Solve each system by using the inverse of the coefficient matrix.

19. $2x + 3y = 1$
$\quad 2x + y = 7$

20. $8a - 5b = 2$
$\quad -3a + 2b = 1$

21. $x_1 - 3x_2 = -6$
$\quad 6x_1 - 3x_2 = 9$

22. $x + y = 1$
$\quad x + 2y = -3$

23. $3x + 2y = 22$
$\quad x + 2y = 10$

24. $2x + 8y = 3$
$\quad 3x - 2y = 1$

25. $x + y = 1$
$\quad 7x + 3y = 0$

26. $10x + 5y = 3$
$\quad 4x - y = 0$

27. $x_1 + 5x_2 = -4$
$\quad 2x_1 - 2x_3 = 0$
$\quad 4x_2 - x_3 = 4$

28. $x + 3y - 2z = 5$
$\quad x + 2y - 5z = -2$
$\quad x + 4y = 5$

29. $x - y + z = -2$
$\quad y - 3z = 10$
$\quad 3x - 3y + 2z = -3$

30. $a + b + 3c = -3$
$\quad 2b + c = -6$
$\quad a - b = -1$

31. $\qquad -y + 3z = 0$
$\quad -x \qquad + 5z = 0$
$\quad x - y - z = 1$

32. $x \qquad - 3z = -2$
$\quad 2y - 3z = -16$
$\quad -x - 2y + 7z = 20$

33. $x + 2y + z = 4$
$\quad y + z = 0$
$\quad x + y + z = 3$

34. $x + 2y + z = 0$
$\quad x + y - 2z = -1$
$\quad 2x + 4y + 3z = \frac{1}{2}$

*35.
$$x + y + 2z - w = 0$$
$$-2x - y - 2z + 2w = -1$$
$$4x - 2y + z = -4$$
$$y + z - w = 1$$

*36.
$$a - 2b + c = -1$$
$$b - 2d = 0$$
$$b - c + d = -2$$
$$2a - 4b + c + 5d = -4$$

*37. Assuming that A^{-1} exists, solve for X in the matrix equation $XA = B$. (X and B are not column matrices here.)

*38. Given that A^{-1} and C^{-1} exist, solve for X in the matrix equation $AXC = B$.

The following systems are taken from problems in Exercises 6-5. Use Theorem 7-14 to solve these systems. $\left(Hint\colon \text{Let } X = \begin{bmatrix} x^2 \\ y^2 \end{bmatrix}. \right)$

39.
$$4x^2 + 9y^2 = 36$$
$$2x^2 - y^2 = 7$$

40.
$$9x^2 + y^2 = 9$$
$$x^2 + 3y^2 = 14$$

41.
$$9x^2 + 16y^2 = 144$$
$$3x^2 + 4y^2 = 36$$

42.
$$4x^2 + 25y^2 = 83$$
$$9x^2 + 16y^2 = 66$$

43.
$$x^2 + y^2 = 0$$
$$(3x - 4y)(3x + 4y) = 0$$

44.
$$4x^2 + 36y^2 = 100$$
$$(x - 4y)(x + 4y) = 0$$

Perform the indicated operations, if possible.

1. $\begin{bmatrix} 2 & -3 \\ -4 & -5 \\ 0 & 2 \end{bmatrix} - 2 \begin{bmatrix} -1 & 0 \\ 3 & -11 \\ 1 & 0 \end{bmatrix}$

2. $[1 \quad 3 \quad 0] + [2 \quad -7]$

Perform the matrix multiplications, if possible.

3. $\begin{bmatrix} 1 & 4 & 2 \\ -2 & 1 & 0 \end{bmatrix} \begin{bmatrix} -3 & 1 & 2 \\ 0 & 5 & -2 \end{bmatrix}$

4. $\begin{bmatrix} 1 & -2 & 0 \\ 0 & 3 & 2 \\ 5 & 0 & 1 \end{bmatrix} \begin{bmatrix} -1 & 0 \\ 3 & 4 \\ 0 & -1 \end{bmatrix}$

Use Gaussian elimination to find a solution, if one exists.

5. $\begin{aligned} x - 2y - 4z &= -1 \\ 3x \quad\quad - z &= 4 \\ x + 4y + 7z &= 2 \end{aligned}$

6. $\begin{aligned} -x + 2y + z &= -1 \\ 3x + y - z &= 7 \\ y + z &= -3 \end{aligned}$

Find the inverse of the given matrix, if it exists.

7. $\begin{bmatrix} 2 & -3 \\ -1 & 2 \end{bmatrix}$

8. $\begin{bmatrix} 1 & 1 & -3 \\ 1 & 0 & 3 \\ -2 & 1 & -12 \end{bmatrix}$

Solve each system by using the inverse of the coefficient matrix.

9. $\begin{aligned} 4x - y &= 2 \\ 7x - 3y &= 1 \end{aligned}$

10. $\begin{aligned} 2x + y - z &= 1 \\ y - 2z &= 9 \\ x + 3y \quad\quad &= 1 \end{aligned}$

Determinants

In Theorem 7-13, $\delta(A)$ denoted the number $ad - bc$ which is associated with the 2×2 matrix

$$A = \begin{bmatrix} a & b \\ c & d \end{bmatrix}.$$

This number, $\delta(A)$, is called the determinant of the 2×2 matrix A. Every square matrix A, of any order, has such a number associated with it. This number is called the *determinant* of A. The *order* of the determinant of A is the same as the order of the matrix A. The determinant of A is denoted by $\delta(A)$, or by $|A|$, or by simply replacing the brackets around the elements of A by straight lines. Thus, if

$$A = \begin{bmatrix} a & b \\ c & d \end{bmatrix},$$

then

$$\delta(A) = |A| = \begin{vmatrix} a & b \\ c & d \end{vmatrix} = ad - bc.$$

The material in this section leads to a method of evaluation for the determinant of any square matrix. It is important to note that the determinant of a matrix is a number.

EXAMPLE 1 Evaluate the determinant of each of the following matrices.

(a) $A = \begin{bmatrix} -1 & 7 \\ 2 & -3 \end{bmatrix}$ (b) $B = \begin{bmatrix} 1 & 0 \\ 0 & 1 \end{bmatrix}$ (c) $C = \begin{bmatrix} 2 & 0 \\ -1 & 0 \end{bmatrix}$

SOLUTION   (a) $\delta(A) = (-1)(-3) - (7)(2) = -11$, or $\begin{vmatrix} -1 & 7 \\ 2 & -3 \end{vmatrix} = -11.$

(b) $\delta(B) = |B| = (1)(1) - (0)(0) = 1$, or $\begin{vmatrix} 1 & 0 \\ 0 & 1 \end{vmatrix} = 1.$

(c) $\delta(C) = |C| = (2)(0) - (0)(-1) = 0$, or $\begin{vmatrix} 2 & 0 \\ -1 & 0 \end{vmatrix} = 0.$

The determinant of a square matrix of order 3 is given in the next definition.

DEFINITION 8-1   For a given 3×3 matrix

$$A = \begin{bmatrix} a_{11} & a_{12} & a_{13} \\ a_{21} & a_{22} & a_{23} \\ a_{31} & a_{32} & a_{33} \end{bmatrix},$$

the determinant of A is the number

$$|A| = a_{11}a_{22}a_{33} + a_{12}a_{23}a_{31} + a_{13}a_{21}a_{32}$$
$$- a_{11}a_{23}a_{32} - a_{12}a_{21}a_{33} - a_{13}a_{22}a_{31}.$$

EXAMPLE 2 Evaluate $|A|$ for

$$A = \begin{bmatrix} 3 & 2 & 1 \\ -4 & -1 & 5 \\ -3 & -4 & -2 \end{bmatrix}.$$

SOLUTION   Using Definition 8-1, we have

$$|A| = (3)(-1)(-2) + (2)(5)(-3) + (1)(-4)(-4)$$
$$- (3)(5)(-4) - (2)(-4)(-2) - (1)(-1)(-3)$$
$$= 6 - 30 + 16 + 60 - 16 - 3$$
$$= 33.$$

Similar and even more unwieldy definitions can be made for determinants of higher-order matrices. Although we could use these definitions to evaluate a determinant, they are very inefficient. A much more efficient way to proceed is to evaluate determinants by a method called the "cofactor expansion." The cofactor expansion applies to determinants of any order. In order to describe the method, some preliminary definitions are needed.

DEFINITION 8-2   Let A be a square matrix of order n, and let a_{ij} be the element in row i and column j of A. Then the *minor*, denoted by M_{ij}, of the element a_{ij} is the determinant of the matrix formed by deleting row i and column j from the matrix A. The *cofactor*, denoted by C_{ij}, of the element a_{ij} is the product of $(-1)^{i+j}$ and the minor M_{ij}. That is,

$$C_{ij} = (-1)^{i+j}M_{ij}.$$

Note that if $i + j$ is even, then $C_{ij} = M_{ij}$, and if $i + j$ is odd, then $C_{ij} = -M_{ij}$.

EXAMPLE 3   Consider the 3×3 matrix

$$A = \begin{bmatrix} 6 & -3 & 2 \\ 0 & 1 & -2 \\ 2 & 4 & 5 \end{bmatrix}.$$

(a) The element in row 2 and column 1 is 0; that is, $a_{21} = 0$.

(b) The minor of a_{21} is $\begin{vmatrix} -3 & 2 \\ 4 & 5 \end{vmatrix}$; hence $M_{21} = -23$.

(c) The cofactor of a_{21} is $(-1)^{2+1}\begin{vmatrix} -3 & 2 \\ 4 & 5 \end{vmatrix}$; hence $C_{21} = 23$.

With the definitions of minor and cofactor in mind, we consider again the definition of a determinant of order 3 (Definition 8-1). In the equation

$$|A| = a_{11}a_{22}a_{33} + a_{12}a_{23}a_{31} + a_{13}a_{21}a_{32}$$
$$- a_{11}a_{23}a_{32} - a_{12}a_{21}a_{33} - a_{13}a_{22}a_{31},$$

there are two terms involving a_{11}, two terms involving a_{12}, and two terms involving a_{13}. Regrouping the terms and factoring, we have

$$|A| = a_{11}(a_{22}a_{33} - a_{23}a_{32}) - a_{12}(a_{21}a_{33} - a_{23}a_{31})$$
$$+ a_{13}(a_{21}a_{32} - a_{22}a_{31}).$$

Since each expression in parentheses can be expressed as a second-order determinant, we can write

$$A = a_{11}\begin{vmatrix} a_{22} & a_{23} \\ a_{32} & a_{33} \end{vmatrix} - a_{12}\begin{vmatrix} a_{21} & a_{23} \\ a_{31} & a_{33} \end{vmatrix} + a_{13}\begin{vmatrix} a_{21} & a_{22} \\ a_{31} & a_{32} \end{vmatrix}.$$

Expressing each second-order determinant as a minor yields

$$|A| = a_{11}M_{11} - a_{12}M_{12} + a_{13}M_{13},$$

or

$$|A| = a_{11}C_{11} + a_{12}C_{12} + a_{13}C_{13}.$$

The last expression for $|A|$ is called the *cofactor expansion* of the determinant of A about the first row. The cofactor expansion of $|A|$ about the first row is a sum of terms formed by multiplying each element in the first row by its cofactor.

The terms in the equation defining $|A|$ in Definition 8-1 could also have been regrouped and factored in such a way as to have a cofactor expansion of $|A|$ about any certain row or any certain column. In general, the *cofactor expansion* of $|A|$ about row i ($i = 1, 2,$ or 3) is

$$|A| = a_{i1}C_{i1} + a_{i2}C_{i2} + a_{i3}C_{i3},$$

and the cofactor expansion of $|A|$ about column j ($j = 1, 2,$ or 3) is

$$|A| = a_{1j}C_{1j} + a_{2j}C_{2j} + a_{3j}C_{3j}.$$

EXAMPLE 4   Evaluate $|A|$ below by expanding about the second column.

$$|A| = \begin{vmatrix} 3 & 2 & 1 \\ -4 & -1 & 5 \\ -3 & -4 & -2 \end{vmatrix}$$

SOLUTION   Using

$$|A| = a_{12}C_{12} + a_{22}C_{22} + a_{32}C_{32},$$

we have

$$|A| = (2)(-1)^3 \begin{vmatrix} -4 & 5 \\ -3 & -2 \end{vmatrix} + (-1)(-1)^4 \begin{vmatrix} 3 & 1 \\ -3 & -2 \end{vmatrix}$$
$$+ (-4)(-1)^5 \begin{vmatrix} 3 & 1 \\ -4 & 5 \end{vmatrix}$$
$$= (-2)[8 - (-15)] + (-1)[-6 - (-3)] + (4)[15 - (-4)]$$
$$= (-2)(23) + (-1)(-3) + (4)(19)$$
$$= 33.$$

This agrees with the value obtained in Example 2 for the same determinant.

Determinants of order $n > 3$ can also be evaluated by the method of cofactor expansion. The cofactor expansion of an nth-order determinant is given in the following theorem.

THEOREM 8-3   Let $A = [a_{ij}]_{(n, n)}$ be a square matrix of order n, and let C_{ij} be the cofactor of element a_{ij}, for $i = 1, 2, \ldots, n$ and $j = 1, 2, \ldots, n$. Then the cofactor expansion of $|A|$ about row i is

$$|A| = a_{i1}C_{i1} + a_{i2}C_{i2} + \cdots + a_{in}C_{in},$$

and the cofactor expansion of $|A|$ about column j is

$$|A| = a_{1j}C_{1j} + a_{2j}C_{2j} + \cdots + a_{nj}C_{nj}.$$

Theorem 8-3 is proved in the study of linear algebra. That is, it is shown that $|A|$ does not depend on the choice of the row number i, or on the choice of the column number j. This proof is very complicated, and does not properly belong in this text.

EXAMPLE 5   Evaluate the following determinant.

$$|A| = \begin{vmatrix} -2 & 1 & 0 & 1 \\ 0 & -2 & 1 & 4 \\ -3 & 1 & 0 & -2 \\ 1 & 0 & -2 & 1 \end{vmatrix}$$

SOLUTION   We expand $|A|$ about the third column since it contains more zeros than any other column or row, and consequently this cofactor expansion will have the fewest nonzero terms. We obtain

$$|A| = a_{13}C_{13} + a_{23}C_{23} + a_{33}C_{33} + a_{43}C_{43}$$

$$= 0 \cdot C_{13} + (1)(-1)^5 \begin{vmatrix} -2 & 1 & 1 \\ -3 & 1 & -2 \\ 1 & 0 & 1 \end{vmatrix} + 0 \cdot C_{33}$$

$$+ (-2)(-1)^7 \begin{vmatrix} -2 & 1 & 1 \\ 0 & -2 & 4 \\ -3 & 1 & -2 \end{vmatrix}$$

$$= -\begin{vmatrix} -2 & 1 & 1 \\ -3 & 1 & -2 \\ 1 & 0 & 1 \end{vmatrix} + 2\begin{vmatrix} -2 & 1 & 1 \\ 0 & -2 & 4 \\ -3 & 1 & -2 \end{vmatrix}.$$

Expanding the first third-order determinant about its third row, and the second third-order determinant about its first column, we have

$$|A| = -\left[(1)(-1)^4 \begin{vmatrix} 1 & 1 \\ 1 & -2 \end{vmatrix} + 0 + (1)(-1)^6 \begin{vmatrix} -2 & 1 \\ -3 & 1 \end{vmatrix} \right]$$

$$+ 2\left[(-2)(-1)^2 \begin{vmatrix} -2 & 4 \\ 1 & -2 \end{vmatrix} + 0 + (-3)(-1)^4 \begin{vmatrix} 1 & 1 \\ -2 & 4 \end{vmatrix} \right]$$

$$= -[(1)(-2-1) + 0 + (1)(-2-(-3))]$$
$$+ 2[(-2)(4-4) + 0 + (-3)(4-(-2))]$$
$$= -[-3+0+1] + 2[0+0-18]$$
$$= 2 - 36$$
$$= -34.$$

EXERCISES 8-1

Evaluate the following determinants. Letters represent real numbers.

1. $\begin{vmatrix} -3 & 2 \\ 1 & 1 \end{vmatrix}$

2. $\begin{vmatrix} 0 & -3 \\ -1 & 3 \end{vmatrix}$

3. $\begin{vmatrix} 5 & 8 \\ 2 & 3 \end{vmatrix}$

4. $\begin{vmatrix} 4 & -3 \\ 3 & -2 \end{vmatrix}$

5. $\begin{vmatrix} 7 & 0 \\ 0 & -3 \end{vmatrix}$

6. $\begin{vmatrix} 0 & 0 \\ -1 & 1 \end{vmatrix}$

7. $\begin{vmatrix} -1 & -1 \\ 1 & 1 \end{vmatrix}$

8. $\begin{vmatrix} 2 & 9 \\ -4 & 10 \end{vmatrix}$

9. $\begin{vmatrix} -\frac{1}{2} & \frac{1}{3} \\ \frac{1}{6} & \frac{1}{9} \end{vmatrix}$

10. $\begin{vmatrix} \frac{1}{4} & -1 \\ 0 & -4 \end{vmatrix}$

11. $\begin{vmatrix} x & -x \\ 2 & 3 \end{vmatrix}$

12. $\begin{vmatrix} x & y \\ 2x & 3y \end{vmatrix}$

Use a cofactor expansion to evaluate the following determinants.

13. $\begin{vmatrix} 0 & 3 \\ -2 & 1 \end{vmatrix}$

14. $\begin{vmatrix} 8 & 0 \\ 1 & -5 \end{vmatrix}$

15. $\begin{vmatrix} 1 & 11 \\ -2 & 4 \end{vmatrix}$

16. $\begin{vmatrix} 1 & 4 \\ -2 & 1 \end{vmatrix}$

17. $\begin{vmatrix} 2 & -1 & 2 \\ 0 & 2 & 0 \\ -3 & 1 & 0 \end{vmatrix}$

18. $\begin{vmatrix} -2 & 6 & 1 \\ 0 & 0 & 4 \\ 1 & -4 & 0 \end{vmatrix}$

19. $\begin{vmatrix} 2 & 0 & -1 \\ -1 & 2 & 0 \\ 0 & -1 & 2 \end{vmatrix}$

20. $\begin{vmatrix} 2 & 0 & 1 \\ 0 & 1 & 1 \\ 4 & -1 & 1 \end{vmatrix}$

21. $\begin{vmatrix} 1 & 4 & 0 \\ -1 & 2 & -1 \\ 0 & 3 & -1 \end{vmatrix}$

22. $\begin{vmatrix} 1 & 0 & 1 \\ 0 & 1 & 3 \\ -1 & 2 & 1 \end{vmatrix}$

23. $\begin{vmatrix} 1 & 3 & -2 \\ 1 & 4 & 0 \\ 2 & 5 & -7 \end{vmatrix}$

24. $\begin{vmatrix} 1 & 0 & 3 \\ -1 & 1 & -3 \\ 1 & -3 & 2 \end{vmatrix}$

25. $\begin{vmatrix} -1 & 1 & 1 \\ 1 & -1 & 1 \\ 1 & 1 & 1 \end{vmatrix}$

26. $\begin{vmatrix} -2 & 3 & 3 \\ 3 & 1 & -2 \\ 1 & 1 & -1 \end{vmatrix}$

27. $\begin{vmatrix} 1 & -2 & 3 \\ 2 & 1 & 2 \\ 3 & -3 & 6 \end{vmatrix}$

28. $\begin{vmatrix} 2 & 1 & 1 \\ 9 & 4 & 3 \\ 6 & 3 & 4 \end{vmatrix}$

29. $\begin{vmatrix} 4 & -5 & 1 \\ 5 & -9 & 2 \\ 2 & -4 & 1 \end{vmatrix}$

30. $\begin{vmatrix} 4 & -3 & -8 \\ 1 & 3 & -2 \\ 2 & 1 & -3 \end{vmatrix}$

31. $\begin{vmatrix} 2 & 4 & 3 \\ 5 & -9 & -2 \\ -1 & 11 & 5 \end{vmatrix}$

32. $\begin{vmatrix} -3 & -1 & 2 \\ 1 & 6 & -3 \\ -5 & 4 & 1 \end{vmatrix}$

33. $\begin{vmatrix} 1 & 0 & 1 & 2 \\ -2 & 1 & -1 & -4 \\ 1 & 0 & 0 & 1 \\ 0 & -2 & 1 & 5 \end{vmatrix}$

34. $\begin{vmatrix} 6 & -1 & -1 & 3 \\ 8 & 0 & -3 & 2 \\ -9 & 2 & -2 & 1 \\ 2 & 0 & -1 & 1 \end{vmatrix}$

35. $\begin{vmatrix} 0 & 4 & -2 & 1 \\ -1 & -2 & -1 & 1 \\ -2 & -1 & -3 & 3 \\ 1 & 0 & 2 & -1 \end{vmatrix}$

36. $\begin{vmatrix} 1 & 3 & -1 & 1 \\ 2 & 9 & -2 & 3 \\ 2 & 10 & -3 & 2 \\ 3 & -8 & 0 & 1 \end{vmatrix}$

Solve for x in each of the following equations.

37. $\begin{vmatrix} 3 & x \\ -2 & 1 \end{vmatrix} = 1$

38. $\begin{vmatrix} -x & 3 \\ -4 & -1 \end{vmatrix} = -3$

39. $\begin{vmatrix} 2x-1 & 2 \\ 1 & 4 \end{vmatrix} = 10$

40. $\begin{vmatrix} 1 & 3 \\ 2-x & -1 \end{vmatrix} = -5$

41. $\begin{vmatrix} 2 & 0 & 1 \\ 4 & x & 5 \\ 1 & 4 & -3 \end{vmatrix} = -10$

42. $\begin{vmatrix} -2 & 0 & 1 \\ 1 & -3 & -2 \\ x & 1 & 1 \end{vmatrix} = 0$

43. $\begin{vmatrix} 3 & x & x \\ -2 & 1 & 0 \\ 4 & 1 & -1 \end{vmatrix} = 21$

44. $\begin{vmatrix} 1 & 3 & 2 \\ 2x & 1 & x \\ 5 & 1 & -4 \end{vmatrix} = 0$

The values of x which satisfy the equation $|A - xI| = 0$ are called the *eigenvalues*, or the *characteristic values*, of the matrix A. Find the eigenvalues of the following matrices.

*45. $\begin{bmatrix} 2 & 0 \\ 0 & -3 \end{bmatrix}$

*46. $\begin{bmatrix} 5 & 6 \\ 4 & 0 \end{bmatrix}$

*47. $\begin{bmatrix} 1 & -2 \\ 3 & -4 \end{bmatrix}$

*48. $\begin{bmatrix} 1 & 3 \\ 9 & 7 \end{bmatrix}$

*49. $\begin{bmatrix} 5 & 1 & -1 \\ 0 & -3 & 2 \\ 0 & 0 & 2 \end{bmatrix}$

*50. $\begin{bmatrix} 1 & 3 & 15 \\ -2 & 0 & -2 \\ 1 & 0 & 1 \end{bmatrix}$

*51. $\begin{bmatrix} 1 & -2 & 0 \\ 0 & 0 & -3 \\ 2 & -4 & 0 \end{bmatrix}$

*52. $\begin{bmatrix} 1 & -1 & 1 \\ 1 & -1 & -1 \\ -1 & 1 & -3 \end{bmatrix}$

*53. Use Definition 8-1 to show that a determinant $|A|$ of order 3 can be evaluated by any one of the following equations.

(a) $|A| = a_{11}C_{11} + a_{21}C_{21} + a_{31}C_{31}$

(b) $|A| = a_{21}C_{21} + a_{22}C_{22} + a_{23}C_{23}$

(c) $|A| = a_{12}C_{12} + a_{22}C_{22} + a_{32}C_{32}$

(d) $|A| = a_{31}C_{31} + a_{32}C_{32} + a_{33}C_{33}$

(e) $|A| = a_{13}C_{13} + a_{23}C_{23} + a_{33}C_{33}$

8-2

Evaluation of Determinants

In Section 7-3, we used three types of row operations on augmented matrices to solve linear systems by Gaussian elimination. These operations are listed below.

Row Operations

1. Interchange any two rows.

2. Multiply (or divide) every element in a row by the same nonzero number.

3. Add (or subtract) a multiple of one row to another row.

Corresponding to each of these row operations, there is a similar *column operation*. These column operations are described below.

Column Operations

1. Interchange any two columns.

2. Multiply (or divide) every element in a column by the same non-zero number.

3. Add (or subtract) a multiple of one column to another column.

When a row or column operation is performed on a square matrix A, the value of $|A|$ is sometimes (but not always) changed. The effect of performing an operation of each type is described in the next three theorems. Each theorem can be stated in terms of rows or in terms of columns. We indicate this by stating each theorem in terms of rows and by inserting the word "column" in parentheses. This indicates that a dual theorem is obtained simply by replacing the word "row" by the word "column." As it was with Theorem 8-3, the proofs of these theorems for the general case are more appropriate in a linear algebra course, and they are not presented here.

THEOREM 8-4 If the matrix B is obtained from a matrix A by interchanging any two rows (columns) of A, then $|B| = -|A|$.

EXAMPLE 1 Some uses of Theorem 8-4 are presented below.

(a) If $|A| = \begin{vmatrix} a & b & c \\ d & e & f \\ g & h & i \end{vmatrix}$, then $\begin{vmatrix} a & b & c \\ g & h & i \\ d & e & f \end{vmatrix} = -|A|$.

(b) Every time two rows or two columns are interchanged, the sign of the determinant changes. Thus

$$\begin{vmatrix} 1 & -2 & 3 \\ 4 & -1 & 0 \\ -1 & 1 & 2 \end{vmatrix} = -\begin{vmatrix} -1 & 1 & 2 \\ 4 & -1 & 0 \\ 1 & -2 & 3 \end{vmatrix},$$ since rows 1 and 3 have been interchanged,

$$= \begin{vmatrix} -1 & 2 & 1 \\ 4 & 0 & -1 \\ 1 & 3 & -2 \end{vmatrix},$$ since columns 2 and 3 have been interchanged.

THEOREM 8-5 If the matrix B is obtained from a matrix A by multiplying each element of a row (column) of A by the same number c, then $|B| = c|A|$.

EXAMPLE 2 The determinant

$$\begin{vmatrix} 12 & -4 \\ 9 & 2 \end{vmatrix}$$

can be expressed as

$$12\begin{vmatrix} 1 & -1 \\ 3 & 2 \end{vmatrix}$$

by applying Theorem 8-5 twice:

$$\begin{vmatrix} 12 & -4 \\ 9 & 2 \end{vmatrix} = 4\begin{vmatrix} 3 & -1 \\ 9 & 2 \end{vmatrix}$$

$$= (4)(3)\begin{vmatrix} 1 & -1 \\ 3 & 2 \end{vmatrix}.$$

Multiplication of a determinant by a number c has the same effect as multiplying one column or one row of the determinant by c. By contrast, in multiplication of a matrix by the number c, every element in the matrix is multiplied by c. This difference between multiplication of a matrix by a number and multiplication of a determinant by a number is illustrated in the next example.

EXAMPLE 3 Let

$$A = \begin{bmatrix} -7 & -4 \\ 2 & 5 \end{bmatrix}.$$

Then

$$3A = \begin{bmatrix} -21 & -12 \\ 6 & 15 \end{bmatrix},$$

and $|3A| = -243 = 9(-27) = 9|A|$. But

$$3|A| = \begin{vmatrix} -21 & -4 \\ 6 & 5 \end{vmatrix} = \begin{vmatrix} -7 & -12 \\ 2 & 15 \end{vmatrix}$$

$$= \begin{vmatrix} -21 & -12 \\ 2 & 5 \end{vmatrix} = \begin{vmatrix} -7 & -4 \\ 6 & 15 \end{vmatrix} = -81.$$

THEOREM 8-6 If the matrix B is obtained from a matrix A by adding a multiple of one row (column) to another row (column), then $|B| = |A|$.

EXAMPLE 4 Adding 2 times the first row to the third row in

$$\begin{vmatrix} 1 & 3 & -1 \\ 0 & 1 & 5 \\ -2 & 1 & 4 \end{vmatrix}$$

yields

$$\begin{vmatrix} 1 & 3 & -1 \\ 0 & 1 & 5 \\ 0 & 7 & 2 \end{vmatrix},$$

and these determinants are equal:

$$\begin{vmatrix} 1 & 3 & -1 \\ 0 & 1 & 5 \\ -2 & 1 & 4 \end{vmatrix} = \begin{vmatrix} 1 & 3 & -1 \\ 0 & 1 & 5 \\ 0 & 7 & 2 \end{vmatrix}$$

since each determinant is -33.

Theorem 8-6 proves to be most useful in evaluating determinants. By introducing zeros into a row (or column), the cofactor expansion can be reduced to only one term. This is illustrated in Example 5. In the example, we use a notation for row and column operations that is similar to that used in Section 7-3:

$R_i \leftrightarrow R_j$ indicates that row i and row j are interchanged;
$C_i \leftrightarrow C_j$ indicates that column i and column j are interchanged;
$cR_i + R_j$ indicates that row j is replaced by the sum of row j and c times row i;
$cC_i + C_j$ indicates that column j is replaced by the sum of column j and c times column i.

EXAMPLE 5 Evaluate the following determinant by introducing zeros into column 2.

$$|A| = \begin{vmatrix} 3 & 2 & -2 \\ -1 & -1 & 4 \\ 2 & 4 & -1 \end{vmatrix}$$

SOLUTION We use row operations to introduce two zeros into column 2 as follows:

$$|A| = \begin{vmatrix} 3 & 2 & -2 \\ -1 & -1 & 4 \\ 2 & 4 & -1 \end{vmatrix}$$

$$\overset{2R_2 + R_1}{=} \begin{vmatrix} 1 & 0 & 6 \\ -1 & -1 & 4 \\ 2 & 4 & -1 \end{vmatrix}$$

$$\overset{4R_2 + R_3}{=} \begin{vmatrix} 1 & 0 & 6 \\ -1 & -1 & 4 \\ -2 & 0 & 15 \end{vmatrix}.$$

The cofactor expansion of this determinant about the second column reduces to one term.

$$|A| = (0) \cdot C_{12} + (-1)C_{22} + (0) \cdot C_{32}$$
$$= (-1)\begin{vmatrix} 1 & 6 \\ -2 & 15 \end{vmatrix}$$
$$= (-1)(15 + 12)$$
$$= -27.$$

The use of column operations is illustrated in the next example.

EXAMPLE 6 Evaluate the following determinant by introducing zeros before expanding about a row or column.

$$|A| = \begin{vmatrix} 3 & 1 & -5 & 0 \\ 2 & 1 & 4 & 1 \\ -1 & 2 & -4 & 1 \\ 0 & -3 & 1 & 5 \end{vmatrix}$$

SOLUTION Since there is already one zero in the first row, we choose to introduce two more zeros in this row by using column operations.

$$|A| = \begin{vmatrix} 3 & 1 & -5 & 0 \\ 2 & 1 & 4 & 1 \\ -1 & 2 & -4 & 1 \\ 0 & -3 & 1 & 5 \end{vmatrix}$$

$$\underset{-3C_2 + C_1}{=} \begin{vmatrix} 0 & 1 & -5 & 0 \\ -1 & 1 & 4 & 1 \\ -7 & 2 & -4 & 1 \\ 9 & -3 & 1 & 5 \end{vmatrix}$$

$$\underset{5C_2 + C_3}{=} \begin{vmatrix} 0 & 1 & 0 & 0 \\ -1 & 1 & 9 & 1 \\ -7 & 2 & 6 & 1 \\ 9 & -3 & -14 & 5 \end{vmatrix}.$$

The cofactor expansion about the first row is

$$|A| = (1)(-1) \begin{vmatrix} -1 & 9 & 1 \\ -7 & 6 & 1 \\ 9 & -14 & 5 \end{vmatrix}.$$

Next we choose to introduce zeros into the third column. (Any column or row can be used.) Using row operations, we have

$$|A| = - \begin{vmatrix} -1 & 9 & 1 \\ -7 & 6 & 1 \\ 9 & -14 & 5 \end{vmatrix}$$

$$\underset{-R_1 + R_2}{=} - \begin{vmatrix} -1 & 9 & 1 \\ -6 & -3 & 0 \\ 9 & -14 & 5 \end{vmatrix}$$

$$\underset{-5R_1 + R_3}{=} - \begin{vmatrix} -1 & 9 & 1 \\ -6 & -3 & 0 \\ 14 & -59 & 0 \end{vmatrix}.$$

The cofactor expansion about the third column is

$$|A| = -(1)\begin{vmatrix} -6 & -3 \\ 14 & -59 \end{vmatrix}.$$

Factoring -3 out of the first row, we have

$$|A| = -(1)(-3)\begin{vmatrix} 2 & 1 \\ 14 & -59 \end{vmatrix}$$

$$= (3)(-118 - 14)$$

$$= -396.$$

EXERCISES 8-2

Without evaluating the determinants, determine the value of the variable which makes each of the following statements true.

1. $\begin{vmatrix} 3 & -4 \\ 1 & 5 \end{vmatrix} = -\begin{vmatrix} 1 & 5 \\ 3 & x \end{vmatrix}$

2. $\begin{vmatrix} 11 & -2 \\ 7 & 3 \end{vmatrix} = -\begin{vmatrix} -2 & x \\ 3 & 7 \end{vmatrix}$

3. $\begin{vmatrix} 5 & -1 \\ 2 & -3 \end{vmatrix} = a\begin{vmatrix} 5 & 1 \\ 2 & 3 \end{vmatrix}$

4. $\begin{vmatrix} 3 & -2 \\ 7 & 0 \end{vmatrix} = y\begin{vmatrix} -3 & 2 \\ -7 & 0 \end{vmatrix}$

5. $\begin{vmatrix} 1 & 0 & 2 \\ -2 & 1 & 4 \\ 0 & 1 & 5 \end{vmatrix} = x\begin{vmatrix} 0 & 2 & 1 \\ 1 & 4 & -2 \\ 1 & 5 & 0 \end{vmatrix}$

6. $\begin{vmatrix} -2 & 1 & 11 \\ 3 & 1 & 4 \\ 0 & -1 & 0 \end{vmatrix} = t\begin{vmatrix} 3 & 1 & 4 \\ 0 & -1 & 0 \\ -2 & 1 & 11 \end{vmatrix}$

7. $\begin{vmatrix} -3 & 4 \\ 9 & -2 \end{vmatrix} = x\begin{vmatrix} -1 & -2 \\ 3 & 1 \end{vmatrix}$

8. $\begin{vmatrix} 1 & -1 & 3 \\ 4 & 20 & -12 \\ 5 & -25 & 30 \end{vmatrix} = y\begin{vmatrix} 1 & -1 & 1 \\ 1 & 5 & -1 \\ 1 & -5 & 2 \end{vmatrix}$

9. $\begin{vmatrix} 3 & 2 & 1 \\ 5 & -4 & -3 \\ 1 & 1 & 2 \end{vmatrix} = \begin{vmatrix} 3 & 2 & 1 \\ 14 & x & 0 \\ 1 & 1 & 2 \end{vmatrix}$

10. $\begin{vmatrix} -1 & -3 & 1 \\ 1 & -2 & 1 \\ 4 & -2 & 2 \end{vmatrix} = \begin{vmatrix} x & -5 & 3 \\ 1 & -2 & 1 \\ 4 & -2 & 2 \end{vmatrix}$

11. $\begin{vmatrix} -1 & 11 & -1 & 4 \\ 0 & 3 & 0 & 1 \\ 1 & 1 & -2 & 1 \\ 4 & 1 & 5 & -2 \end{vmatrix} = \begin{vmatrix} -1 & -1 & -1 & 4 \\ 0 & 0 & 0 & 1 \\ 1 & x & -2 & 1 \\ 4 & 7 & 5 & -2 \end{vmatrix}$

12. $\begin{vmatrix} 1 & -1 & 2 & 3 \\ -2 & 1 & 1 & 4 \\ 1 & 5 & 0 & 2 \\ 1 & 1 & -2 & 3 \end{vmatrix} = \begin{vmatrix} 1 & -1 & 2 & 3 \\ -1 & 0 & x & 7 \\ 1 & 5 & 0 & 2 \\ 1 & 1 & -2 & 3 \end{vmatrix}$

Evaluate the following determinants by introducing zeros before expanding about a row or column.

13. $\begin{vmatrix} 2 & -1 & 3 \\ 1 & 2 & -1 \\ 0 & 2 & 1 \end{vmatrix}$

14. $\begin{vmatrix} 1 & 3 & -2 \\ 0 & 1 & 3 \\ 2 & 0 & 5 \end{vmatrix}$

15. $\begin{vmatrix} 7 & 4 & -2 \\ 1 & 2 & -1 \\ 4 & 2 & 0 \end{vmatrix}$

16. $\begin{vmatrix} 4 & -2 & 1 \\ -2 & 2 & -1 \\ 5 & 0 & 2 \end{vmatrix}$

17. $\begin{vmatrix} 1 & 3 & -1 \\ 3 & -2 & 4 \\ 2 & 1 & 3 \end{vmatrix}$

18. $\begin{vmatrix} 13 & 3 & 1 \\ 11 & 4 & -2 \\ 4 & -1 & 3 \end{vmatrix}$

19. $\begin{vmatrix} 1 & 2 & 3 \\ 1 & 1 & 2 \\ 1 & 1 & 3 \end{vmatrix}$

20. $\begin{vmatrix} 2 & -1 & -2 \\ 1 & 1 & -5 \\ -1 & 3 & 4 \end{vmatrix}$

21. $\begin{vmatrix} 2 & -1 & 2 \\ 3 & -5 & 1 \\ 2 & 1 & 3 \end{vmatrix}$

22. $\begin{vmatrix} 8 & 2 & -3 \\ 5 & 3 & -1 \\ 2 & -7 & 3 \end{vmatrix}$

23. $\begin{vmatrix} -2 & 1 & -2 \\ 2 & -1 & 3 \\ -4 & 2 & 1 \end{vmatrix}$

24. $\begin{vmatrix} -1 & 1 & 7 \\ 3 & -3 & -1 \\ 1 & 3 & -3 \end{vmatrix}$

25. $\begin{vmatrix} 5 & 0 & 0 & 3 \\ 2 & 4 & -3 & 1 \\ 0 & 1 & 0 & -1 \\ 1 & -1 & 2 & -1 \end{vmatrix}$

26. $\begin{vmatrix} 4 & -3 & 1 & 0 \\ 2 & 1 & -1 & 0 \\ -1 & 0 & 1 & 2 \\ 2 & 1 & 1 & 1 \end{vmatrix}$

27. $\begin{vmatrix} 1 & 0 & 1 & 0 \\ 0 & 1 & 2 & 1 \\ 4 & 0 & 0 & -2 \\ 1 & 2 & 3 & 0 \end{vmatrix}$

28. $\begin{vmatrix} -3 & 1 & 2 & 0 \\ 1 & 0 & 1 & 2 \\ 0 & 1 & 0 & 1 \\ 4 & 5 & -2 & 2 \end{vmatrix}$

29. $\begin{vmatrix} 5 & -2 & 2 & 3 \\ 0 & 1 & 1 & 3 \\ 1 & 0 & 1 & 1 \\ -2 & -1 & 0 & 6 \end{vmatrix}$

30. $\begin{vmatrix} 24 & 10 & 4 & 1 \\ 21 & 3 & 0 & -2 \\ 3 & 2 & 3 & 0 \\ 1 & 1 & 1 & -4 \end{vmatrix}$

31. $\begin{vmatrix} 2 & 2 & -1 & 3 \\ 1 & 1 & -1 & 1 \\ 1 & -1 & 1 & 1 \\ 4 & 2 & 1 & 5 \end{vmatrix}$

32. $\begin{vmatrix} 3 & -2 & -1 & 2 \\ 4 & 1 & 2 & -3 \\ -9 & -5 & 7 & -8 \\ 1 & 5 & 3 & -2 \end{vmatrix}$

*33. Show that

$$\begin{vmatrix} 1 & 1 & 1 \\ x & y & z \\ x^2 & y^2 & z^2 \end{vmatrix} = (x - y)(y - z)(z - x).$$

*34. Show that

$$\begin{vmatrix} -x & 1 & 0 & 0 \\ 0 & -x & 1 & 0 \\ 0 & 0 & -x & 1 \\ -c_0 & -c_1 & -c_2 & -c_3 - x \end{vmatrix} = x^4 + c_3 x^3 + c_2 x^2 + c_1 x + c_0.$$

*35. Show that

$$\begin{vmatrix} a & b & c \\ d & e & f \\ h & i & j \end{vmatrix} + \begin{vmatrix} a & b & c \\ k & m & n \\ h & i & j \end{vmatrix} = \begin{vmatrix} a & b & c \\ d + k & e + m & f + n \\ h & i & j \end{vmatrix}.$$

*36. Prove that if a matrix has all zero elements in one row, then its determinant is zero.

*37. If the rows of a 3×3 matrix A are the columns of B in the same order, prove that $|A| = |B|$.

*38. Show that if A is a square matrix of order 3 and c is a number, then $|cA| = c^3 |A|$.

*39. Prove Theorem 8-4 for square matrices of order 3.

*40. Prove Theorem 8-5 for square matrices of order 3.

8-3
Cramer's Rule

In Sections 7-3 and 7-5, we studied two matrix methods for solving systems of linear equations. The first method was that of Gaussian elimination, and the second was the method which used inverses. Determinants can also be used to solve linear systems. The method which uses determinants is called *Cramer's Rule*. This method is presented in this section.

We begin with a linear system of two equations in two variables.

$$a_1 x + b_1 y = c_1$$
$$a_2 x + b_2 y = c_2$$

Suppose first that $a_1 b_2 - a_2 b_1 \neq 0$. We shall use the elimination method to solve for x, and then formulate the solution in terms of determinants. To eliminate y, we multiply the first equation by b_2, the second equation by b_1, and then subtract.

$$a_1 b_2 x + b_1 b_2 y = c_1 b_2$$
$$\underline{a_2 b_1 x + b_1 b_2 y = c_2 b_1}$$
$$(a_1 b_2 - a_2 b_1)x \qquad = c_1 b_2 - c_2 b_1$$

Since $a_1b_2 - a_2b_1 \neq 0$, this gives

$$x = \frac{c_1b_2 - c_2b_1}{a_1b_2 - a_2b_1} = \frac{\begin{vmatrix} c_1 & b_1 \\ c_2 & b_2 \end{vmatrix}}{\begin{vmatrix} a_1 & b_1 \\ a_2 & b_2 \end{vmatrix}}.$$

Similarly, x can be eliminated by multiplying the first equation by a_2 and the second equation by a_1. The solution for y is found to be

$$y = \frac{a_1c_2 - a_2c_1}{a_1b_2 - a_2b_1} = \frac{\begin{vmatrix} a_1 & c_1 \\ a_2 & c_2 \end{vmatrix}}{\begin{vmatrix} a_1 & b_1 \\ a_2 & b_2 \end{vmatrix}}.$$

The two fractions for x and y have the same denominator, which is the determinant of the coefficients. We shall denote this determinant by D:

$$D = \begin{vmatrix} a_1 & b_1 \\ a_2 & b_2 \end{vmatrix}.$$

The expressions that we have obtained for x and y constitute the first part of Cramer's Rule, as given below. The second and third parts describe the possibilities when $D = 0$, and we omit the proofs of these statements.

Cramer's Rule

1. If $D = \begin{vmatrix} a_1 & b_1 \\ a_2 & b_2 \end{vmatrix} \neq 0$ in the system

$$a_1x + b_1y = c_1$$
$$a_2x + b_2y = c_2,$$

the solution is given by

$$x = \frac{\begin{vmatrix} c_1 & b_1 \\ c_2 & b_2 \end{vmatrix}}{\begin{vmatrix} a_1 & b_1 \\ a_2 & b_2 \end{vmatrix}}, \quad y = \frac{\begin{vmatrix} a_1 & c_1 \\ a_2 & c_2 \end{vmatrix}}{\begin{vmatrix} a_1 & b_1 \\ a_2 & b_2 \end{vmatrix}}.$$

2. If $D = 0$ and either $\begin{vmatrix} c_1 & b_1 \\ c_2 & b_2 \end{vmatrix}$ or $\begin{vmatrix} a_1 & c_1 \\ a_2 & c_2 \end{vmatrix}$ is not zero, there is no solution. In this case, the system is called *inconsistent*.

3. If $D = 0$ and both $\begin{vmatrix} c_1 & b_1 \\ c_2 & b_2 \end{vmatrix}$ and $\begin{vmatrix} a_1 & c_1 \\ a_2 & c_2 \end{vmatrix}$ are zero, there are many solutions. In this case the system is called *dependent*.

Before considering an example, we make some observations about the determinants in part (1) of Cramer's Rule. If the first column of the coefficient matrix is replaced by the column of constants, the determinant of the resulting matrix is the numerator in the expression for x (the first variable in the system). If the second column of the coefficient matrix is replaced by the column of constants, the determinant of the resulting matrix is the numerator in the expression for y (the second variable in the system). The standard notations for these determinants are

$$D_x = \begin{vmatrix} c_1 & b_1 \\ c_2 & b_2 \end{vmatrix} \quad \text{and} \quad D_y = \begin{vmatrix} a_1 & c_1 \\ a_2 & c_2 \end{vmatrix}.$$

With this notation, the solutions are given by

$$x = \frac{D_x}{D}, \qquad y = \frac{D_y}{D}.$$

EXAMPLE 1 Use Cramer's Rule to solve the following system.

$$3x + y = -1$$
$$4x - y = -13$$

SOLUTION We first calculate D:

$$D = \begin{vmatrix} 3 & 1 \\ 4 & -1 \end{vmatrix} = -3 - 4 = -7.$$

Since $D \neq 0$, the solution is given by the formulas in part (1) of Cramer's Rule. The determinants D_x and D_y are given by

$$D_x = \begin{vmatrix} -1 & 1 \\ -13 & -1 \end{vmatrix} = 1 + 13 = 14,$$

$$D_y = \begin{vmatrix} 3 & -1 \\ 4 & -13 \end{vmatrix} = -39 + 4 = -35.$$

Thus

$$x = \frac{D_x}{D} = \frac{14}{-7} = -2,$$

$$y = \frac{D_y}{D} = \frac{-35}{-7} = 5.$$

These values can easily be checked in the original system.

EXAMPLE 2 Use Cramer's Rule to solve the following system.

$$x - 4y = 2$$
$$2x - 8y = 2$$

Since

$$D = \begin{vmatrix} 1 & -4 \\ 2 & -8 \end{vmatrix} = -8 + 8 = 0$$

and

$$D_y = \begin{vmatrix} 1 & 2 \\ 2 & 2 \end{vmatrix} = 2 - 4 \neq 0,$$

the system has no solution. In other words, the system is inconsistent.

Let us consider now a general system of n linear equations in n unknowns.

$$a_{11}x_1 + a_{12}x_2 + \cdots + a_{1n}x_n = b_1$$
$$a_{21}x_1 + a_{22}x_2 + \cdots + a_{2n}x_n = b_2$$
$$\cdot \qquad\qquad \cdot$$
$$\cdot \qquad\qquad \cdot$$
$$\cdot \qquad\qquad \cdot$$
$$a_{n1}x_1 + a_{n2}x_2 + \cdots + a_{nn}x_n = b_n$$

We let D denote the determinant of the coefficient matrix, and let D_{x_i} denote the determinant of the matrix formed by replacing the elements in the ith column of the coefficient matrix by the column of constants. Using these notations, Cramer's Rule may be stated as follows.

Cramer's Rule

1. If $D \neq 0$, the solution is given by

$$x_1 = \frac{D_{x_1}}{D}, \quad x_2 = \frac{D_{x_2}}{D}, \quad \cdots, \quad x_n = \frac{D_{x_n}}{D}.$$

2. If $D = 0$ and one or more D_{x_i} is not zero, there is no solution. In this case, the system is called *inconsistent*.

3. If $D = 0$ and all $D_{x_i} = 0$, the system is *dependent*, and there are many solutions.

EXAMPLE 3 Use Cramer's Rule to solve the following system.

$$2x + 3y - z = 4$$
$$-x + y + 2z = -2$$
$$3x - y - 2z = 1$$

SOLUTION　Evaluation of the four determinants involved yields

$$D = \begin{vmatrix} 2 & 3 & -1 \\ -1 & 1 & 2 \\ 3 & -1 & -2 \end{vmatrix} = 14, \qquad D_x = \begin{vmatrix} 4 & 3 & -1 \\ -2 & 1 & 2 \\ 1 & -1 & -2 \end{vmatrix} = -7,$$

$$D_y = \begin{vmatrix} 2 & 4 & -1 \\ -1 & -2 & 2 \\ 3 & 1 & -2 \end{vmatrix} = 15, \qquad D_z = \begin{vmatrix} 2 & 3 & 4 \\ -1 & 1 & -2 \\ 3 & -1 & 1 \end{vmatrix} = -25.$$

Thus the solution is given by

$$x = \frac{D_x}{D} = \frac{-7}{14} = -\frac{1}{2}, \qquad y = \frac{D_y}{D} = \frac{15}{14}, \qquad z = \frac{D_z}{D} = -\frac{25}{14}.$$

EXERCISES 8-3

Use Cramer's Rule to find the solution, if the determinant of the coefficients is not zero. If the system is dependent or inconsistent, state so.

1. $2x + 9y = 3$
    $3x - 2y = -11$

2. $x + 2y = 4$
    $3x - 2y = -12$

3. $7x - y = 9$
    $3x + 5y = -7$

4. $4x + 3y = -6$
    $5x - y = 27$

5. $4a + 3b = 1$
    $2a - 5b = -19$

6. $a + 2b = 4$
    $3a - 2b = -12$

7. $3x - 12y = 9$
    $-x + 4y = -3$

8. $5x + 2y = 8$
    $15x + 6y = 24$

9. $7x - y = 1$
    $-14x + 2y = -1$

10. $-5x + 4y = 0$
    $10x - 8y = 2$

11. $x + y = 3$
    $x + 2y - z = 5$
    $2y + z = 1$

12. $4x + z = -6$
    $y + 2z = -3$
    $2x + 5y + z = 1$

13. $3x + 2y + z = -1$
    $x - 2z = -3$
    $y + 2z = -2$

14. $-2x - y + 3z = 1$
    $2x + y - z = -2$
    $-x + 3y + 2z = 4$

15. $3x + y + 2z = 3$
    $x - 5y - z = 0$
    $2x + 3y + 2z = 0$

16. $x + 3y - z = 4$
    $2x + y + 3z = 11$
    $3x - 2y + 4z = 11$

17. $a + 3b - c = 4$
    $3a - 2b + 4c = 11$
    $2a + b + 3c = 13$

18. $2r - s + 2t = 2$
    $2r - s + t = -5$
    $r - 2s + 3t = 4$

19. $2x - y + 3z = 17$
    $5x - 2y + 4z = 28$
    $3x + 3y - z = -1$

20. $x + 3y - 2z = 1$
    $-4x + 2y - 2z = 5$
    $2x - y + z = 3$

21. $2x + y - z = 5$
    $x - y + 2z = -3$
    $-3y + 5z = -11$

22. $4x - y + 2z = -7$
    $x + y + z = 4$
    $5x - 5y + z = -26$

23. $2x_1 - x_2 + 3x_3 = 17$
    $5x_1 - 2x_2 + 4x_3 = 28$
    $3x_1 + 3x_2 - x_3 = 1$

24. $5x_1 + 3x_2 - x_3 = 4$
    $2x_1 - 7x_2 + 3x_3 = -36$
    $3x_1 - x_2 - 2x_3 = -13$

25. $6x - 4y + 3z = 1$
    $12x - 8y - 9z = -3$
    $12x - 4y + 9z = 4$

26. $x - 2y + 3z = 4$
    $2x + y - 4z = 3$
    $-x + 5y - 5z = 1$

27. $3x - y + 2z = 0$
    $x + 2y + 4z = -1$
    $7x - 7y - 2z = 1$

28. $2x + 2y - 3z = 9$
    $x - y + 2z = -1$
    $x - 5y + 9z = 1$

29. $x - y - 2z = 2$
    $2x + 3y - z = 4$
    $3x - y - 2z = 1$

30. $x + y + z = 1$
    $x - 3y - 5z = -1$
    $4x - 5y + 2z = -35$

31. $x + 2y + 3z = 0$
    $y + z = 0$
    $x + y + 3z = 1$

32. $3x - y + 2z = 4$
    $x + 2y - z = -2$
    $x + 2y - 3z = -2$

33. $3x - y - 2z = -13$
    $5x + 3y - z = 4$
    $2x - 7y + 3z = -36$

34. $x + y - 2z = -5$
    $3x + y + z = 3$
    $4x + y - 2z = -2$

35. $x - y - z - w = 2$
    $x + y - w = 0$
    $3y + 2z + w = -2$
    $y + w = 0$

36. $x - 2y + 3z = -3$
    $x + 2z + w = 0$
    $-2x + 4y + 5w = 10$
    $2x - 4z - 3w = -2$

37. $x + 3y - 2z + w = 2$
    $2x + 4y + 2z - 3w = -1$
    $x - y - z - 2w = -1$
    $2x + 5y - z - w = 1$

38. $4x - 2y - 5z - 6w = 1$
    $2x + y + 3w = 3$
    $7x + 2y - 3z + 2w = 4$
    $4x - y + 7z + 5w = 2$

*39. Find m and b so that the line with equation $y = mx + b$ passes through the points $(1, -2)$ and $(3, 2)$.

*40. Find a and b so that the line with equation $ax + by = 2$ passes through the points $(-3, -4)$ and $(2, 6)$.

*41. Find $a, b,$ and c so that the parabola with equation $y = ax^2 + bx + c$ passes through the points $(2, 10)$, $(-1, -5)$, and $(-3, 5)$.

*42. Find $a, b,$ and c so that the circle with equation $x^2 + ax + y^2 + by = c$ passes through the points $(1, 1)$, $(3, 3)$, and $(5, 1)$.

1. Evaluate the determinant: $\begin{vmatrix} 4 & -2 \\ -1 & -3 \end{vmatrix}$.

2. Use a cofactor expansion to evaluate the determinant.
$$\begin{vmatrix} 3 & -2 & 1 \\ 4 & 0 & 1 \\ 2 & 5 & -2 \end{vmatrix}$$

3. Solve for x in the equation
$$\begin{vmatrix} -2 & x & -2 \\ 3 & 1 & 0 \\ 1 & 2x & 1 \end{vmatrix} = 0.$$

Without evaluating the determinants, determine the value of the variable which makes each statement true.

4. $\begin{vmatrix} 2 & 1 & 1 \\ -3 & 5 & -3 \\ 4 & 2 & 1 \end{vmatrix} = \begin{vmatrix} 2 & 1 & 1 \\ -3 & 5 & -3 \\ 0 & 0 & x \end{vmatrix}$

5. $\begin{vmatrix} 1 & -2 & -1 \\ 3 & 0 & 2 \\ 2 & 1 & 1 \end{vmatrix} = x \begin{vmatrix} -1 & -2 & 1 \\ 2 & 0 & 3 \\ 1 & 1 & 2 \end{vmatrix}$

6. $\begin{vmatrix} -2 & 4 & 8 \\ -4 & 2 & -6 \\ 2 & 2 & -4 \end{vmatrix} = x \begin{vmatrix} -1 & 2 & 4 \\ -2 & 1 & -3 \\ 1 & 1 & -2 \end{vmatrix}$

7. Evaluate the determinant by introducing zeros before expanding about a row or column.
$$\begin{vmatrix} -1 & 0 & 2 & 0 \\ 2 & 3 & 1 & 2 \\ 1 & 1 & 1 & 0 \\ -1 & 2 & -1 & 3 \end{vmatrix}$$

Use Cramer's Rule to find the solution if the determinant of the coefficients is not zero. If the system is dependent or inconsistent, state so.

8. $3x - y = -2$
$6x - 2y = -4$

9. $5x + 2y = 4$
$-x + 2y = -8$

10. $x + y + 2z = 1$
$-x + 2y + z = 5$
$3x + 3z = 1$

**Theory
of Polynomials**

9-1
Complex Numbers In Section 4-6, the quadratic formula was derived for the solution of the general quadratic equation

$$ax^2 + bx + c = 0,$$

where a, b, and c are real numbers. This formula states that the solutions are given by

$$x = \frac{-b \pm \sqrt{b^2 - 4ac}}{2a},$$

provided that $b^2 - 4ac \geq 0$. The derivation in Section 4-6 breaks down in step 3 if $b^2 - 4ac$ is negative. The fundamental difficulty is that a negative number does not have a square root in the set of real numbers. That is, the square of a real number is never negative.

This situation is very unsatisfactory to a mathematician. Simple equations such as $x^2 + 1 = 0$ and $x^2 + x + 2 = 0$ have no solutions in the real numbers. This inadequacy of the real numbers is the primary reason for the construction of the system of complex numbers, which begins with the introduction of a number i such that $i^2 = -1$. The formal definition is as follows.

> **DEFINITION 9-1**   The number i is, by definition, a number such that
>
> $$i^2 = -1.$$

That is, $i \cdot i = -1$. A *complex number* is a number of the form

$$a + bi,$$

where a and b are real numbers. The set of all complex numbers is denoted by $\mathcal{C}$.

EXAMPLE 1   Examples of complex numbers are

$$3 + 7i, \quad \pi + 4i, \quad 1 + (-2)i, \quad -5 + (-\tfrac{3}{7})i, \quad 0 + 3i, \quad 4 + 0i.$$

If b is negative, as it is in $1 + (-2)i$, we drop the "$+$" sign, and simply write $1 - 2i$. In this set $\mathcal{C}$, we identify the set $\mathcal{R}$ of real numbers as being the set of all numbers of the form $a + 0i$. More precisely, each real number a is identified as being the same as $a + 0i$. For example, we write 4 for $4 + 0i$. Also, those numbers of the special form $0 + bi$ are called *pure imaginary numbers*, and are written as bi instead of $0 + bi$. We write $3i$ for $0 + 3i$, and so on. Thus the list of complex numbers above can be written as

$$3 + 7i, \quad \pi + 4i, \quad 1 - 2i, \quad -5 - \tfrac{3}{7}i, \quad 3i, \quad 4.$$

> **DEFINITION 9-2**   In a given complex number $a + bi$, where a and b are real, a is called the *real part* of the complex number, and b is called the *imaginary part* of the complex number.

Our goal in this section is to define the operations of addition, subtraction, multiplication, and division in the set of complex numbers, and to do this so that the field properties hold (see Section 1-2), and so that the calculations are consistent with those of the real numbers (i.e., numbers of the form $a + 0i$). It is appropriate to begin with the definition of equality.

> **DEFINITION 9-3**   For two complex numbers $a + bi$ and $c + di$,
>
> $$a + bi = c + di$$
>
> if and only if
>
> $$a = c \quad \text{and} \quad b = d.$$

That is, *equality* of complex numbers requires that they have equal real parts and equal imaginary parts.

EXAMPLE 2   Some uses of the definition of equality are given below.

(a) If x and y are real numbers, then

$$x - 3i = 2 - yi$$

if and only if

$$x = 2 \quad \text{and} \quad y = 3.$$

(b) If a, b, x, and y are real numbers, then

$$a - bi = x + 2yi$$

if and only if

$$a = x \quad \text{and} \quad b = -2y.$$

DEFINITION 9-4   For arbitrary complex numbers $a + bi$ and $c + di$, *addition* is defined by

$$(a + bi) + (c + di) = (a + c) + (b + d)i.$$

That is, to add two complex numbers, we simply add their real parts and their imaginary parts. This is the same pattern as one follows with binomials:

$$(a + bx) + (c + dx) = (a + c) + (b + d)x.$$

It is easy to see that $0 = 0 + 0i$ is the *additive identity*, and that the *additive inverse* of $a + bi$ is given by

$$-(a + bi) = -a + (-b)i.$$

Subtraction is then defined by

$$(a + bi) - (c + di) = [a + bi] + [-(c + di)]$$
$$= (a - c) + (b - d)i.$$

EXAMPLE 3   Addition and subtraction are illustrated by the following examples.

(a) $(6 + i) + (7 + 3i) = (6 + 7) + (1 + 3)i$
$$= 13 + 4i$$

(b) $(2 - 3i) + (7 + i) = (2 + 7) + (-3 + 1)i$
$$= 9 - 2i$$

(c) $(9 - 5i) - (6 + 4i) = (9 - 6) + (-5 - 4)i$
$$= 3 - 9i$$

The set $\mathcal{C}$ of complex numbers is *closed* under the operations of addition and subtraction. That is, the result of performing either operation on two complex numbers is always another complex number.[1] We shall see that $\mathfrak{C}$ is closed under multiplication and division, when these operations are defined.

To motivate the definition of multiplication, consider the product of two binomials:

$$(a + bx)(c + dx) = ac + (ad + bc)x + bdx^2.$$

This result is a consequence of the field properties, which hold for real numbers. If the same properties are to hold for complex numbers, this equation should be valid when x is replaced by i:

$$(a + bi)(c + di) = ac + (ad + bc)i + bdi^2$$
$$= (ac - bd) + (ad + bc)i,$$

where the last equality follows from the fact that $i^2 = -1$. Thus it is natural to make the following definition.

DEFINITION 9-5   For arbitrary complex numbers $a + bi$ and $c + di$, multiplication is defined by

$$(a + bi)(c + di) = (ac - bd) + (ad + bc)i.$$

EXAMPLE 4   Illustrating the definition of multiplication, we have

$$(2 - 3i)(3 + 4i) = [6 - (-12)] + [8 + (-9)]i$$
$$= 18 - i,$$

and

$$(4 + 3i)(4 - 3i) = (16 + 9) + (-12 + 12)i$$
$$= 25 + 0i$$
$$= 25.$$

Although we do not prove it here, multiplication of complex numbers has the associative and commutative properties. The real number 1 is the multiplicative identity, and we shall see presently that the other field properties, as described in Section 1-2, also hold for the complex numbers.[2]

In connection with division of complex numbers, it is convenient to have available the notion of the conjugate of a complex number.

---

[1]Note that if a and b are added as real numbers or as complex numbers $a + 0i$ and $b + 0i$, corresponding results are obtained.

[2]Note that if a and b are multiplied as real numbers or as complex numbers $a + 0i$ and $b + 0i$, corresponding results are obtained.

DEFINITION 9-6   For any complex number $z = a + bi$, the *conjugate* of z is the complex number

$$\bar{z} = a - bi.$$

EXAMPLE 5   Some examples of conjugates are given below.

(a) If $z = 3 + 4i$, then $\bar{z} = 3 - 4i$. That is,
$$\overline{3 + 4i} = 3 - 4i.$$
(b) The conjugate of $2 - 3i$ is $2 + 3i$. That is,
$$\overline{2 - 3i} = 2 + 3i.$$
(c) $\bar{4} = 4$.

Two important facts concerning conjugate complex numbers are listed below.

---

1. The product of z and its conjugate $\bar{z}$ is always a nonnegative real number. In fact,
$$(a + bi)(a - bi) = a^2 + b^2,$$
which is positive if $a + bi \neq 0$.
2. If $z = \bar{z}$, then z is a real number.

However, the most important use of conjugates is in connection with division, as we shall see shortly.

If a complex number $z = a + bi$ is not zero, then the multiplicative inverse of z exists, and is given by

$$\frac{1}{z} = \frac{a}{a^2 + b^2} - \frac{b}{a^2 + b^2}i.$$

This is easily verified by direct multiplication.

At this point, it can be verified that the field properties, as listed in Section 1-2, are valid for the complex numbers. This verification is routine, but lengthy, and it is omitted here for this reason.

The operation of division by a nonzero complex number can be described now. If $z_1 = a + bi$ and $z_2 = c + di$ are complex numbers with $z_2 \neq 0$, then

$$\begin{aligned}
\frac{z_1}{z_2} &= \frac{a + bi}{c + di} \\
&= \frac{(a + bi)(c - di)}{(c + di)(c - di)} \\
&= \frac{ac + bd}{c^2 + d^2} + \frac{bc - ad}{c^2 + d^2}i.
\end{aligned}$$

This expression for the quotient is *not* a formula to be memorized. Its purpose is to show that the quotient of two complex numbers can be found by multiplying both numerator and denominator by the conjugate of the denominator. This is illustrated in Example 6.

EXAMPLE 6 Perform the following divisions, and express the result in the form $a + bi$, where a and b are real numbers. (This form is called the *standard form* for complex numbers.)

(a) $\dfrac{6 - 2i}{3 + i}$

(b) $\dfrac{3 + 5i}{4 - 2i}$

(c) $\dfrac{1}{a + bi}$, where $a + bi \neq 0$

SOLUTION

(a) $\dfrac{6 - 2i}{3 + i} = \dfrac{(6 - 2i)(3 - i)}{(3 + i)(3 - i)}$

$= \dfrac{16 - 12i}{10}$

$= \dfrac{(2)(8 - 6i)}{(2)(5)}$

$= \dfrac{8 - 6i}{5}$

$= \dfrac{8}{5} - \dfrac{6}{5}i$

(b) $\dfrac{3 + 5i}{4 - 2i} = \dfrac{3 + 5i}{4 - 2i} \cdot \dfrac{4 + 2i}{4 + 2i}$

$= \dfrac{2 + 26i}{20}$

$= \dfrac{(2)(1 + 13i)}{(2)(10)}$

$= \dfrac{1 + 13i}{10}$

$= \dfrac{1}{10} + \dfrac{13}{10}i$

(c) $\dfrac{1}{a + bi} = \dfrac{(1)(a - bi)}{(a + bi)(a - bi)}$

$= \dfrac{a - bi}{a^2 + b^2}$

$= \dfrac{a}{a^2 + b^2} - \dfrac{b}{a^2 + b^2}i$

We note that this last result agrees with the expression previously given for the multiplicative inverse.

The powers of i under multiplication are of some interest. We first note that -1 has a square root in the set of complex numbers, since

$$i^2 = -1.$$

Using Definition 9-5, we find that -1 has another square root. Since

$$(-bi)(-ci) = -bc$$

we have

$$(-i)(-i) = -1.$$

Thus i and $-i$ are both square roots of -1. Similarly, $2i$ and $-2i$ are square roots of -4. Every negative real number is of the form $-a$, where

a is a positive real number, and $-a$ has two square roots, $i\sqrt{a}$ and $-i\sqrt{a}$. The square root of $-a$ which has the positive imaginary part is designated as the principal square root, and is denoted by $\sqrt{-a}$. Thus

$$\sqrt{-1} = i, \quad \sqrt{-4} = 2i, \quad \sqrt{-25} = 5i, \quad \sqrt{-7} = \sqrt{7}\,i,$$

and so on.

Beginning with $i^2 = -1$, the next several powers of i are given by

$$i^2 = -1,$$
$$i^3 = i^2 \cdot i = (-1)i = -i,$$
$$i^4 = i^2 \cdot i^2 = (-1)(-1) = 1,$$
$$i^5 = i^4 \cdot i = 1 \cdot i = i,$$
$$i^6 = i^4 \cdot i^2 = 1 \cdot i^2 = -1,$$
$$i^7 = i^4 \cdot i^3 = 1 \cdot i^3 = -i,$$
$$i^8 = i^4 \cdot i^4 = 1 \cdot 1 = 1.$$

At this point, a pattern of repetition is clear. We see that, for any positive integer n, i^n can be reduced to one of the numbers i, -1, $-i$, or 1. In fact, if division of n by 4 gives $n = 4q + r$, where r is 0, 1, 2, or 3, then

$$i^n = i^{4q+r}$$
$$= i^{4q} \cdot i^r$$
$$= (i^4)^q \cdot i^r$$
$$= (1)^q \cdot i^r$$
$$= i^r.$$

This proves the following theorem.

THEOREM 9-7 Suppose that the positive integer n is divided by 4 to obtain

$$n = 4q + r,$$

where r is 0, 1, 2, or 3. Then

$$i^n = i^r,$$

and consequently i^n has one of the values 1, i, -1, $-i$.

EXAMPLE 7 Reduce each power of i to one of the values 1, i, -1, $-i$.

(a) i^{47}
(b) i^{97}

(a) Since $47 = (4)(11) + 3$, we have

$$i^{47} = i^{(4)(11)+3}$$
$$= (i^4)^{11} \cdot i^3$$
$$= (1)^{11} \cdot i^3$$
$$= i^3$$
$$= -i.$$

(b) Since $97 = 4(24) + 1$, we have

$$i^{97} = i^{(4)(24)+1}$$
$$= (i^4)^{24} \cdot i^1$$
$$= (1)^{24} \cdot i$$
$$= i.$$

EXERCISES 9-1

Use the definition of equality of complex numbers to solve for x and y, where x and y are real numbers.

1. $2 - yi = x + 6i$
2. $x - yi = 4 + 2i$
3. $x - 7i = yi$
4. $2x - 3yi = 6 + 9i$
5. $x^2 - yi = 4 + y^2i$
6. $8 - y^2i = x^2 - 1 - 2yi$
7. $(2x - y) - (5y + 4)i = 5 - 4xi$
8. $(3x + 1) + (x - y)i = (2 - 4y) - (1 - 2x)i$

Evaluate each expression.

9. $\sqrt{-16}$
10. $\sqrt{-25}$
11. $-\sqrt{-49}$
12. $-\sqrt{-64}$
13. $-\sqrt{-5}$
14. $-\sqrt{-3}$

Perform the indicated operations, and leave your result in standard form. All letters represent real numbers. (See Examples 3, 4, and 6.)

15. $(3 + 2i) + (6 - 3i)$
16. $(7 - i) - (3 - 6i)$
17. $(65 - 3i) - (50 - 70i)$
18. $(64 + 32i) + (-59 - 75i)$
19. $(2 - 5i) - (6 - 3i)$
20. $(6 - 19i) + (32 - 7i)$
21. $(2x - 3i) - (7 - 6yi)$
22. $(4x - 7zi) + (16 - 5zi)$
23. $(x^2 - 7i) + (y^2 - i)$
24. $(16x - 3yi) - 7i$
25. $(2 + 3i)(3 - i)$
26. $(11 - 5i)(2 - 3i)$
27. $(6 + 4i)(-7 + 3i)$
28. $(2 - 3i)(2 + 3i)$

29. $(6 - 3i)^2$

30. $(3 - 2i)^2$

31. $i(12 - 4i)(3 + i)$

32. $i(2 - 5i)(3 - 7i)$

33. $(x - 2yi)(3 - yi)$

34. $(x - 2i)(y - 7xi)$

35. $(6 - 7i) \div 3$

36. $(3 - 9i) \div 9$

37. $6 \div (3 - i)$

38. $4 \div (2 - i)$

39. $\dfrac{5 - 3i}{6 + i}$

40. $\dfrac{9 - i}{3 + 2i}$

41. $\dfrac{14 - 7i}{4 - 2i}$

42. $\dfrac{7 - 5i}{3 - 7i}$

43. $(4 - 3i) \div (3 + 4i)$

44. $(5 - 10i) \div (2 + i)$

Find the multiplicative inverse of the given complex number.

45. $3 + 4i$

46. $8 + 6i$

47. $5 - 12i$

48. $8 - 15i$

49. $15 + 8i$

50. $12 + 5i$

Reduce each power of i to one of the values $1, i, -1, -i$.

51. i^{72}

52. i^{56}

53. i^{29}

54. i^{13}

55. i^{91}

56. i^{39}

57. i^{-15}

58. $\dfrac{1}{i^{95}}$

59. i^{62}

60. i^{78}

61. i^{915}

62. i^{231}

*63. For any two complex numbers $z_1 = a_1 + b_1 i$ and $z_2 = a_2 + b_2 i$, in standard form, prove that
   (a) $\overline{z_1 + z_2} = \bar{z}_1 + \bar{z}_2$,   (b) $\overline{z_1 \cdot z_2} = \bar{z}_1 \cdot \bar{z}_2$.

*64. Extend the results of problem 63 to any number of terms:
   (a) $\overline{z_1 + z_2 + \cdots + z_n} = \bar{z}_1 + \bar{z}_2 + \cdots + \bar{z}_n$
   (b) $\overline{z_1 \cdot z_2 \cdots z_n} = \bar{z}_1 \cdot \bar{z}_2 \cdots \bar{z}_n$

*65. Use problem 64(b) to show that $\overline{(z^n)} = (\bar{z})^n$ for any complex number z.

*66. Use the definitions of this section to show that the system of complex numbers is a field. That is, verify the field properties listed in Section 1-2.

9-2
The Factor
and Remainder
Theorems

Prior to this chapter, all variables have been restricted to the set of real numbers. For the remainder of this chapter, this restriction is often removed. Unless otherwise stated, the domain of a variable is the set $\mathbb{C}$ of all complex numbers. In particular, the quadratic formula can be used to find the complex numbers which are solutions to a quadratic equation $ax^2 + bx + c = 0$.

EXAMPLE 1 Solve the quadratic equation

$$x^2 + 2x + 2 = 0.$$

SOLUTION Using the quadratic formula with $a = 1$, $b = 2$, and $c = 2$, we have

$$x = \frac{-2 \pm \sqrt{4 - 8}}{2}$$

$$= \frac{-2 \pm \sqrt{-4}}{2}$$

$$= \frac{-2 \pm 2i}{2}$$

$$= -1 \pm i.$$

Thus the solution set is $\{-1 + i, -1 - i\}$.

When the coefficients a, b, and c in

$$ax^2 + bx + c = 0, \qquad a \neq 0,$$

are allowed to be complex numbers, the quadratic formula still yields the solutions to the equation. This is true because the steps used in Section 4-6 to derive the quadratic formula remain valid when a, b, and c are complex numbers. However, the solution of the general situation is too complicated for us to consider here.

EXAMPLE 2 Solve the quadratic equation

$$2ix^2 - 3x + 3i = 0.$$

SOLUTION We use $a = 2i$, $b = -3$, and $c = 3i$ in the quadratic formula to obtain

$$x = \frac{3 \pm \sqrt{9 - 4(2i)(3i)}}{4i}$$

$$= \frac{3 \pm \sqrt{9 - 24i^2}}{4i}$$

$$= \frac{3 \pm \sqrt{33}}{4i}$$

$$= \frac{(3 \pm \sqrt{33})(-i)}{(4i)(-i)}$$

$$= \frac{-3 \pm \sqrt{33}}{4} i.$$

The two solutions are $x = (-3 + \sqrt{33})i/4$ and $x = (-3 - \sqrt{33})i/4$.

We recall from Section 2-4 that, in dividing a polynomial $P(x)$ by

250

a nonzero polynomial $D(x)$, there is a quotient $Q(x)$ and remainder $R(x)$ such that

$$\frac{P(x)}{D(x)} = Q(x) + \frac{R(x)}{D(x)},$$

or

$$P(x) = D(x)Q(x) + R(x),$$

where the remainder $R(x)$ is either 0, or has degree less than the degree of $D(x)$. This is valid whether the coefficients of $P(x)$ and $Q(x)$ are real or complex numbers. In particular, when the divisor is a linear polynomial of the form $x - c$, the remainder is a constant:

$$P(x) = (x - c) \cdot Q(x) + r,$$

where r is a constant. This equality of polynomials holds for all values of x, and, in particular, for $x = c$. If we assign the value c to x, we obtain

$$
\begin{aligned}
P(c) &= (c - c) \cdot Q(c) + r \\
&= 0 \cdot Q(c) + r \\
&= 0 + r \\
&= r.
\end{aligned}
$$

This means that the remainder, r, is the same as the value of $P(x)$ when $x = c$. This important result is known as the Remainder Theorem.

THEOREM 9-8 (REMAINDER THEOREM) If a real or complex polynomial $P(x)$ is divided by $x - c$, for c real or complex, the remainder is $P(c)$.

EXAMPLE 3 Use the Remainder Theorem to find the remainder in each of the following divisions.

(a) $(x^2 + 3x + 2) \div (x - 2)$

(b) $\dfrac{x^{40} - 2x^{10} + 1}{x + 1}$

(c) $(x^3 - x^2 - x + 2) \div (x + i)$

SOLUTION (a) By the Remainder Theorem, when $P(x) = x^2 + 3x + 2$ is divided by $x - 2$, the remainder is

$$
\begin{aligned}
r &= P(2) \\
&= (2)^2 + 3(2) + 2 \\
&= 12.
\end{aligned}
$$

(b) With $P(x) = x^{40} - 2x^{10} + 1$ and $x - c = x + 1 = x - (-1)$, the remainder is

$$r = P(-1)$$
$$= (-1)^{40} - 2(-1)^{10} + 1$$
$$= 1 - 2 + 1$$
$$= 0.$$

(c) With $P(x) = x^3 - x^2 - x + 2$ and $x + i$ as divisor, the remainder is

$$r = P(-i)$$
$$= (-i)^3 - (-i)^2 - (-i) + 2$$
$$= i + 1 + i + 2$$
$$= 3 + 2i.$$

The Remainder Theorem can also be used in the other direction: the value of $P(c)$ can be found by obtaining the remainder when $P(x)$ is divided by $x - c$. As the next example shows, this procedure is very useful when it is combined with synthetic division (see Section 2-4).

EXAMPLE 4 Use the Remainder Theorem and synthetic division to find the value of $P(5)$ if

$$P(x) = x^4 - 2x^3 + 2x^2 - 1.$$

SOLUTION To find $P(5)$, we divide by $x - 5$ as follows:

$$
\begin{array}{r|rrrrr}
5 & 1 & -2 & 2 & 0 & -1 \\
  &   & 5 & 15 & 85 & 425 \\
\hline
  & 1 & 3 & 17 & 85 & 424
\end{array}
$$

Thus $P(5) = 424$. This method is easier than the direct calculation $P(5) = (5)^4 - 2(5)^3 + 2(5)^2 - 1$.

For an arbitrary complex number c, the Remainder Theorem states that

$$P(x) = (x - c)Q(x) + P(c).$$

Now $x - c$ is a factor of $P(x)$ if and only if the remainder is 0 when $P(x)$ is divided by $x - c$, and therefore the equation above shows that $x - c$ is a factor of $P(x)$ if and only if $P(c) = 0$. This establishes the following Factor Theorem.

THEOREM 9-9 (FACTOR THEOREM) A polynomial $P(x)$ has a factor $x - c$ if and only if $P(c) = 0$.

A *zero* of the polynomial $P(x)$ is a complex number c such that $P(c) = 0$. The Factor Theorem states that $x - c$ is a factor of $P(x)$ if and only if c is a zero of $P(x)$.

EXAMPLE 5   For each polynomial $P(x)$ below, use the Factor Theorem to decide whether or not the given number c is a zero of $P(x)$.

(a) $P(x) = x^4 - x^3 - 5x^2 + 3x + 2$, $c = -2$
(b) $P(x) = x^3 - 2x^2 + 9x - 27$, $c = 3$

SOLUTION   (a) By the Factor Theorem, -2 is a zero of $P(x) = x^4 - x^3 - 5x^2 + 3x + 2$ if and only if $x + 2$ is a factor of $P(x)$. Synthetic division can be used to check if the remainder is 0 when $P(x)$ is divided by $x + 2 = x - (-2)$.

$$
\begin{array}{r|rrrrr}
-2 & 1 & -1 & -5 & 3 & 2 \\
   &   & -2 & 6 & -2 & -2 \\
\hline
   & 1 & -3 & 1 & 1 & 0
\end{array}
$$

Thus $x + 2$ is a factor of $P(x)$, and -2 is a zero of $P(x)$.
(b) The number 3 is a zero of $P(x) = x^3 - 2x^2 + 9x - 27$ if and only if $x - 3$ is a factor of $P(x)$. Dividing $P(x)$ by $x - 3$, we have

$$
\begin{array}{r|rrrr}
3 & 1 & -2 & 9 & -27 \\
  &   & 3 & 3 & 36 \\
\hline
  & 1 & 1 & 12 & 9
\end{array}
$$

The remainder is 9, so $x - 3$ is not a factor of $P(x)$, and 3 is not a zero of $P(x)$.

Sometimes it is easier to compute $P(c)$ than to divide $P(x)$ by $x - c$. In these cases, the Factor Theorem is useful in deciding whether or not $x - c$ is a factor of $P(x)$.

EXAMPLE 6   Use the Factor Theorem to decide whether or not $D(x)$ is a factor of $P(x)$.

(a) $P(x) = x^4 - 3x + 2$, $D(x) = x + 2$
(b) $P(x) = x^3 - 3x^2 + 9x - 27$, $D(x) = x - 3$

SOLUTION   (a) By direct computation,

$$P(-2) = (-2)^4 - 3(-2) + 2$$
$$= 16 + 6 + 2$$
$$= 24.$$

Thus $P(-2) \neq 0$, and $x + 2$ is not a factor of $x^4 - 3x + 2$.

(b) We have

$$P(3) = (3)^3 - 3(3)^2 + 9(3) - 27$$
$$= 27 - 27 + 27 - 27$$
$$= 0,$$

and therefore $x - 3$ is a factor of $x^3 - 3x^2 + 9x - 27$.

In certain instances, the Factor Theorem can be useful in solving equations. This is illustrated in the next example.

EXAMPLE 7 Given that 1 is a zero of $P(x) = x^3 - 3x^2 + x + 1$, solve the equation

$$x^3 - 3x^2 + x + 1 = 0.$$

SOLUTION Since it is given that 1 is a zero, we know that $x - 1$ is a factor of $P(x)$. To find the quotient, we divide by $x - 1$.

$$
\begin{array}{r|rrrr}
1 & 1 & -3 & 1 & 1 \\
  &   & 1 & -2 & -1 \\
\hline
  & 1 & -2 & -1 & 0
\end{array}
$$

Thus the quotient is $x^2 - 2x - 1$, and the given equation is equivalent to

$$(x - 1)(x^2 - 2x - 1) = 0.$$

Since a product is zero if and only if one of the factors is zero, the other solutions of the given equation are the same as the solutions to

$$x^2 - 2x - 1 = 0.$$

These solutions are easily found to be

$$x = \frac{2 \pm \sqrt{4 + 4}}{2}$$
$$= \frac{2 \pm 2\sqrt{2}}{2}$$
$$= 1 \pm \sqrt{2}.$$

Thus the solution set for the equation

$$x^3 - 3x^2 + x + 1 = 0$$

is $\{1, 1 + \sqrt{2}, 1 - \sqrt{2}\}$.

EXERCISES 9-2

Solve each quadratic equation, and write the solutions in standard form $a + bi$.

1. $2x^2 + x + 1 = 0$

2. $2x^2 + 2x + 3 = 0$

3. $x^2 + 4x + 5 = 0$

4. $x^2 + 4x + 13 = 0$

5. $4x^2 - 3x + 1 = 0$

6. $4x^2 - 5x + 2 = 0$

7. $x^2 - 3ix + 4 = 0$

8. $x^2 + ix + 2 = 0$

9. $2ix^2 + 3x + 2i = 0$

10. $3ix^2 - x + 4i = 0$

11. $(1 + i)x^2 + ix + 1 - i = 0$

12. $(1 - i)x^2 - ix + 1 + i = 0$

Use the Remainder Theorem to find the remainder when $P(x)$ is divided by $D(x)$.

13. $P(x) = x^4 - 2x^3 + 2x^2 - 1, \quad D(x) = x - 2$

14. $P(x) = 2x^4 - 3x^3 + 4x - 5, \quad D(x) = x - 1$

15. $P(x) = 2x^3 - 2x^2 + 4, \quad D(x) = x - \sqrt{2}$

16. $P(x) = x^3 + 2x^2 - 5x + 1, \quad D(x) = x + 3$

17. $P(x) = x^2 - 3x + 1, \quad D(x) = x - i$

18. $P(x) = x^3 + x^2 + x + 1, \quad D(x) = x - 2i$

19. $P(x) = x^{1023} - 3x^{15} + 7, \quad D(x) = x + 1$

20. $P(x) = x^{95} - 17x^4 + 9, \quad D(x) = x - 1$

Use the Remainder Theorem and synthetic division to find $P(c)$.

21. $P(x) = 4x^3 - 2x + 6, \quad c = \sqrt{2}$

22. $P(x) = 5x^3 + 2x^2 - \sqrt{3}, \quad c = \sqrt{3}$

23. $P(x) = 3x^4 - 2x^2 + x - 5, \quad c = 1 - i$

24. $P(x) = 2x^4 - 4x^2 + 2x - 3, \quad c = 1 + i$

25. $P(x) = x^4 + 2x^2 - 3x, \quad c = 1 + 2i$

26. $P(x) = x^4 - 3x^2 + 4x, \quad c = 2 - i$

Use the Factor Theorem to determine whether or not $D(x)$ is a factor of $P(x)$.

27. $P(x) = x^2 - 3x + 4, \quad D(x) = x + 1$

28. $P(x) = x^3 - 4x^2 + 9, \quad D(x) = x - 3$

29. $P(x) = x^{19} - x^{17} + x^2 - 1, \quad D(x) = x - 1$

30. $P(x) = x^{30} - x^5 - 2, \quad D(x) = x + 1$

31. $P(x) = x^4 - 3x^2 + 7x - 1, \quad D(x) = x + 2$

32. $P(x) = 7x^3 - 3x^2 - 19, \quad D(x) = x - 3$

33. $P(x) = x^3 + 5x^2 + x + 5, \quad D(x) = x - i$

34. $P(x) = x^2 + x + 4, \quad D(x) = x - (1 - 2i)$

35. $P(x) = x^2 - (3 - i)x + 8 + i, \quad D(x) = x - (1 + 2i)$

36. $P(x) = x^3 - 3ix + i + 7, \quad D(x) = x - (1 - i)$

Use the Factor Theorem to decide whether or not the given number c is a zero of $P(x)$.

37. $P(x) = x^3 - x^2 - x - 5, \quad c = 2$

38. $P(x) = x^3 - 2x^2 - 2x + 7, \quad c = 3$

39. $P(x) = x^3 - 3x^2 + 4x - 2, \quad c = 1 + i$

40. $P(x) = x^3 - 3x^2 + x - 3, \quad c = i$

41. Given that 3 is a zero of $P(x) = x^3 - 3x^2 + x - 3$, solve the equation

$$x^3 - 3x^2 + x - 3 = 0.$$

42. Given that -1 is a zero of $P(x) = x^3 - x^2 + 2$, solve the equation

$$x^3 - x^2 + 2 = 0.$$

*43. Determine k so that $P(x) = x^3 - kx^2 + 3x + 7k$ is divisible by $x + 2$.

*44. Determine k so that $P(x) = x^4 + kx^3 - 3kx + 9$ is divisible by $x - 3$.

9-3

The Fundamental Theorem of Algebra and Descartes' Rule of Signs

We recall from the last section that a *zero* of a polynomial $P(x)$ is a complex number c such that $P(c) = 0$. If c is a real number and $P(c) = 0$, then c is called a *real zero* of $P(x)$. We have seen that any second-degree polynomial $P(x) = ax^2 + bx + c$ has two zeros in the complex numbers, and the zeros are given by the quadratic formula.

In this chapter, we are concerned primarily with the problem of finding the zeros of a given polynomial

$$P(x) = a_n x^n + a_{n-1} x^{n-1} + \cdots + a_1 x + a_0.$$

This is the same problem as that of solving the polynomial equation

$$a_n x^n + a_{n-1} x^{n-1} + \cdots + a_1 x + a_0 = 0.$$

The solutions to an equation are frequently referred to as the *roots* of the equation. We are especially interested in finding the real roots of a polynomial equation which has real coefficients.

Since the problem of finding the roots of a polynomial equation of degree two is completely resolved by the quadratic formula, it might be expected that there are similar formulas for the roots of polynomial equations with higher degree. This is true, up to a point. In the period 1500–1550, Italian mathematicians named Tartaglia, Cardan, and Ferrari obtained formulas which could be used to solve the general equations of the third and fourth degrees. For over 200 years afterward, mathematicians searched for similar formulas for equations of degree greater than four. It was not until 1824 that a Norwegian mathematician, N. H. Abel, proved that it is impossible to express the roots of a general equation of degree greater than four by a formula involving only the four fundamental operations and the extraction of roots.

Thus the problem that we are dealing with in this chapter is far from simple. Our development begins with the following theorem, which was first proved in 1799 by the German mathematician J. K. F. Gauss (1777–1855).

THEOREM 9-10 (THE FUNDAMENTAL THEOREM OF ALGEBRA)   Let

$$P(x) = a_n x^n + a_{n-1} x^{n-1} + \cdots + a_1 x + a_0$$

denote a polynomial of degree $n \geq 1$ with coefficients which are real or complex numbers. Then $P(x)$ has a zero in the field of complex numbers.

In other words, the conclusion of Theorem 9-10 states that there is a complex number r_1 such that

$$P(r_1) = 0.$$

This complex number r_1 may happen to be a real number; that is, r_1 may be a real zero of $P(x)$. In any case, $x - r_1$ is a factor of $P(x)$, by the Factor Theorem. Thus we can write

$$P(x) = (x - r_1)Q_1(x),$$

where $Q_1(x)$ has degree $n - 1$. If the quotient $Q_1(x)$ has degree ≥ 1, then by Theorem 9-10, $Q_1(x)$ has a zero r_2, and a corresponding factor $x - r_2$. That is,

$$Q_1(x) = (x - r_2)Q_2(x),$$

and

$$P(x) = (x - r_1)(x - r_2)Q_2(x).$$

If the quotient $Q_2(x)$ has degree ≥ 1, the procedure can be repeated again, obtaining a factor $x - r_3$ of $P(x)$. Each time another factor is obtained, the degree of the new quotient is one less than the degree of the previous quotient. After n applications of the Fundamental Theorem and the Factor Theorem, we arrive at the factorization

$$P(x) = (x - r_1)(x - r_2) \cdots (x - r_n)(a_n).$$

The last quotient must be a_n, since this is the coefficient of x^n in $P(x)$ in the beginning. Thus we have the following theorem.

THEOREM 9-11   Let $P(x)$ be a polynomial of degree $n \geq 1$, with coefficients which are real or complex numbers. Then $P(x)$ can be factored as

$$P(x) = a_n(x - r_1)(x - r_2) \cdots (x - r_n),$$

where $r_1, r_2, \ldots, r_n$ are n complex numbers which are zeros of $P(x)$ and a_n is the leading coefficient of $P(x)$.

The zeros $r_1, r_2, \ldots, r_n$ in Theorem 9-11 are not necessarily distinct. That is, a given factor $x - r$ may be repeated in the factorization of $P(x)$. If $x - r$ occurs as a factor exactly k times in the factorization, we say that the factor $x - r$ is of *multiplicity* k, or that r is a zero of multiplicity k. Thus a polynomial of degree $n \geq 1$ has exactly n zeros in the complex numbers if a zero of multiplicity k is counted as a zero k times.

EXAMPLE 1   Find a polynomial $P(x)$ of least degree which has 2, -1, and $2i$ as zeros.

SOLUTION   From Theorem 9-11, we know that $P(x)$ must have degree 3, and must factor as
$$P(x) = a_3(x - 2)(x + 1)(x - 2i),$$
where a_3 is the leading coefficient in $P(x)$. The choice $a_3 = 1$ makes $P(x)$ a monic polynomial:
$$P(x) = (x - 2)(x + 1)(x - 2i)$$
$$= x^3 - (1 + 2i)x^2 - (2 - 2i)x + 4i.$$

EXAMPLE 2   State the factors $x - r$ and zeros of the polynomial
$$P(x) = 3(x - 2)^3(x + 1)^2(x - 3).$$

SOLUTION   According to our discussion above, $P(x)$ has six factors of the form $x - r$ and six zeros, counting repetitions. The six factors are:
$$x - 2, \quad \text{with multiplicity 3;}$$
$$x - (-1), \quad \text{with multiplicity 2;}$$
and
$$x - 3.$$
The six zeros are 2, 2, 2, -1, -1, and 3.

Although Theorem 9-11 assures us that a polynomial of degree $n \geq 1$ has exactly n zeros in the complex numbers, it gives us no help at all in finding these zeros. As was indicated earlier, we are especially interested in finding the real zeros of a polynomial which has real coefficients. The next theorem is often useful along these lines when it is used in combination with results which are presented in the next two sections. This theorem is known as "Descartes' rule of signs."

Descartes' rule of signs allows certain predictions to be made about the number of positive real zeros, or about the number of negative real zeros, of a polynomial which has real coefficients. The predictions are based on the number of "variations in sign" which occur when the terms of the polynomial are arranged in the usual order of descending powers of x:

$$P(x) = a_n x^n + a_{n-1} x^{n-1} + \cdots + a_1 x + a_0.$$

After any powers of x with zero coefficients are deleted, a *variation in sign* is said to occur when two consecutive coefficients are opposite in sign. For example,

$$P(x) = x^4 - 3x^2 + 7x + 11$$

has two variations in sign, as indicated by the arrows.

THEOREM 9-12 (DESCARTES' RULE OF SIGNS)   Let

$$P(x) = a_n x^n + a_{n-1} x^{n-1} + \cdots + a_1 x + a_0$$

be a polynomial with real coefficients. The number of positive real zeros of $P(x)$ is either equal to the number of variations of sign occuring in $P(x)$, or is less than this number by an even positive integer. The number of negative real zeros of $P(x)$ is either equal to the number of variations in sign occurring in $P(-x)$, or is less than this number by an even positive integer.

The proof of Descartes' rule of signs, like the proof of the Fundamental Theorem of Algebra, is much beyond the level of this text, and so is not included. Some applications of the rule are given in the example below, but its full usefulness is not illustrated until Section 9-4.

EXAMPLE 3   Use Descartes' rule of signs to discuss the nature of the zeros of each polynomial. That is, describe the possibilities as to the number of positive real zeros, the number of negative real zeros, and the number of complex zeros which are not real.

(a) $P(x) = x^3 - x^2 - 3$
(b) $P(x) = 2x^4 + 4x^2 - 6x - 5$
(c) $P(x) = x^4 - 3x^2 - 7x + 11$

SOLUTION   (a) Counting the number of variations of sign in

$$P(x) = x^3 - x^2 - 3,$$

we see that there is only one variation, so $P(x)$ has one positive real zero. In

$$P(-x) = (-x)^3 - (-x)^2 - 3$$
$$= -x^3 - x^2 - 3,$$

there are no variations in sign, so $P(x)$ has no negative real zeros. Thus we know that $P(x)$ has one positive real zero and therefore two complex zeros which are not real.

(b) There is one variation of sign in

$$P(x) = 2x^4 + 4x^2 - 6x - 5,$$

so $P(x)$ has one positive real zero. Also,

$$P(-x) = 2x^4 + 4x^2 + 6x - 5$$

has one variation in sign, so $P(x)$ has one negative real zero. Thus $P(x)$ has one positive real zero, one negative real zero, and two complex zeros which are not real.

(c) There are two variations of sign in

$$P(x) = x^4 - 3x^2 - 7x + 11,$$

and also two variations of sign in

$$P(-x) = x^4 - 3x^2 + 7x + 11.$$

Thus any one of the following is a possibility as to the nature of the zeros of $P(x)$:

   (i) two positive zeros, two negative zeros;

  (ii) two positive zeros, two complex zeros;

 (iii) two negative zeros, two complex zeros;

 (iv) four complex zeros.

There is one more basic fact that we shall need concerning the zeros of a polynomial which has real coefficients. This fact, which is stated in Theorem 9-13, involves the conjugates of the zeros of $P(x)$.

We recall from Section 9-1 that if $z = a + bi$ is a complex number in standard form, the conjugate of z is the complex number $\bar{z} = a - bi$. Problem 64 in Exercises 9-1 stated the following facts regarding conjugates:

$$\overline{z_1 + z_2 + \cdots + z_n} = \bar{z}_1 + \bar{z}_2 + \cdots + \bar{z}_n,$$

and

$$\overline{z_1 \cdot z_2 \cdots z_n} = \bar{z}_1 \cdot \bar{z}_2 \cdots \bar{z}_n.$$

In words, the conjugate of a sum is the sum of the conjugates, and the conjugate of a product is the product of the conjugates. A special case of the last equation is

$$(\overline{z^n}) = (\bar{z})^n.$$

These facts about conjugates have an interesting implication concerning polynomials with real coefficients. Let

$$P(x) = a_n x^n + a_{n-1} x^{n-1} + \cdots + a_1 x + a_0$$

represent a polynomial that has real coefficients. For any complex number z,

$$
\begin{aligned}
\overline{P(z)} &= \overline{a_n z^n + a_{n-1} z^{n-1} + \cdots + a_1 z + a_0} \\
&= \overline{a_n z^n} + \overline{a_{n-1} z^{n-1}} + \cdots + \overline{a_1 z} + \overline{a_0} \\
&= \overline{a_n}(\overline{z^n}) + \overline{a_{n-1}}(\overline{z^{n-1}}) + \cdots + \overline{a_1}\bar{z} + \overline{a_0} \\
&= \overline{a_n}(\bar{z})^n + \overline{a_{n-1}}(\bar{z})^{n-1} + \cdots + \overline{a_1}\bar{z} + \overline{a_0} \\
&= a_n(\bar{z})^n + a_{n-1}(\bar{z})^{n-1} + \cdots + a_1\bar{z} + a_0,
\end{aligned}
$$

where the last equality follows from the fact that $\overline{a_i} = a_i$ because each a_i is real. Now the last expression above is the same as that obtained when $P(x)$ is evaluated at $\bar{z}$. That is,

$$
\overline{P(z)} = P(\bar{z}),
$$

for any complex number z. The main use that we have for this result is when z is a zero of $P(x)$. For when $P(z) = 0$, then $\overline{P(z)} = \bar{0} = 0$, and therefore $P(\bar{z}) = 0$. That is, if z is a zero of $P(x)$, then $\bar{z}$ is also a zero. This is stated in the following theorem.

THEOREM 9-13   Let

$$
P(x) = a_n x^n + a_{n-1} x^{n-1} + \cdots + a_1 x + a_0
$$

be a polynomial with real coefficients. If $z = a + bi$ is a zero of $P(x)$, then $\bar{z} = a - bi$ is also a zero of $P(x)$.

This means that, if $P(x)$ is a polynomial with real coefficients, the zeros of $P(x)$ always occur in conjugate pairs.

EXAMPLE 4   Given that $1 - i$ is a zero of

$$
P(x) = x^3 - 4x^2 + 6x - 4,
$$

find all zeros of $P(x)$.

SOLUTION   Since $1 - i$ is a zero of $P(x)$, $x - (1 - i)$ is a factor of $P(x)$. We use synthetic division to divide $P(x)$ by $x - (1 - i)$.

$$
\begin{array}{r|rrrr}
1 - i & 1 & -4 & 6 & -4 \\
 & & 1 - i & -4 + 2i & 4 \\
\hline
 & 1 & -3 - i & 2 + 2i & 0
\end{array}
$$

Thus

$$
P(x) = [x - (1 - i)][x^2 + (-3 - i)x + (2 + 2i)].
$$

By Theorem 9-13, $\overline{1 - i} = 1 + i$ is also a zero of $P(x)$. Since $1 + i$ is not a zero of the factor $x - (1 - i)$, it must be a zero of the quotient. We have

$$1 + i \,\overline{\big)\, 1 \quad -3 - i \quad 2 + 2i}$$
$$\underline{ 1 + i \quad -2 - 2i}$$
$$1 \quad -2 \qquad 0$$

and the new quotient is $x - 2$. This means that

$$P(x) = [x - (1 - i)][x - (1 + i)](x - 2),$$

so the zeros of $P(x)$ are $1 - i$, $1 + i$, and 2.

EXAMPLE 5 Find a polynomial $Q(x)$ of least degree with *real* coefficients which has $3i$ and 4 as zeros.

SOLUTION Since $Q(x)$ is to have *real* coefficients and is to have $3i$ as a zero, it must also have $\overline{3i} = -3i$ as a zero. The product

$$Q(x) = (x - 3i)(x + 3i)(x - 4)$$
$$= (x^2 + 9)(x - 4)$$
$$= x^3 - 4x^2 + 9x - 36$$

is a monic polynomial with the required properties.

   In connection with Example 5, we note that the monic polynomial of least degree which has $3i$ and 4 as zeros is

$$P(x) = (x - 3i)(x - 4)$$
$$= x^2 - (4 + 3i)x + 12i,$$

but this polynomial does not have the real coefficients which were required in Example 5.

EXERCISES 9-3

Find (a) a polynomial $P(x)$ of least degree which has the given numbers as zeros and (b) a polynomial $Q(x)$ of least degree with real coefficients which has the given numbers as zeros.

1. $3, -5$ 2. $2, -2$

3. $2i$ 4. $-3i$

5. $3, 2i$ 6. $2, 1 + i$

7. $3, 1 - i, 3 + 2i$ 8. $1 - 2i, 3i, -2$

Use Descartes' rule of signs to discuss the nature of the zeros of each polynomial. (See Example 3.)

9. $P(x) = 2x^4 + 3x^3 - 2x + 1$

10. $P(x) = 4x^4 - 7x^3 + 2x^2 + 3x + 2$

11. $P(x) = x^6 + x^3 + 2x + 3$ 12. $P(x) = 4x^4 + 2x^3 + 4x - 2$

13. $P(x) = 4x^4 + 1$ 14. $P(x) = 2x^7 - 3$

15. $P(x) = x^6 + 4x^4 + x^2 + 5$ 16. $P(x) = x^5 + 3x^3 + 7x$

In problems 17–26, some of the zeros of the polynomial are given. Find the other zeros.

17. $P(x) = x^2 + 9$; $-3i$ is a zero.

18. $P(x) = x^2 + 2x + 2$; $-1 + i$ is a zero.

19. $Q(x) = x^3 + x + 10$; -2 and $1 + 2i$ are zeros.

20. $Q(x) = x^3 + x^2 - x + 15$; -3 and $1 + 2i$ are zeros.

21. $P(x) = x^4 + 20x^2 + 64$; $-2i$ and $4i$ are zeros.

22. $P(x) = x^4 + 17x^2 + 16$; $-i$ and $4i$ are zeros.

23. $Q(x) = x^4 + x^3 + 10x^2 + 9x + 9$; $-3i$ is a zero.

24. $Q(x) = x^4 + 3x^3 + 6x^2 + 12x + 8$; $2i$ is a zero.

25. $P(x) = x^5 - 6x^4 + 16x^3 - 24x^2 + 20x - 8$; $1 + i$ is a zero of multiplicity 2.

26. $P(x) = x^5 - 9x^4 + 34x^3 - 66x^2 + 65x - 25$; $2 - i$ is a zero of multiplicity 2.

*27. Prove that if all the coefficients of $P(x)$ are positive, then $P(x)$ has no positive real zeros.

*28. Prove that if $P(x)$ has no odd powers of x and all its coefficients are of the same sign, then $P(x)$ has no real zeros different from 0.

*29. Prove that every polynomial of odd degree with real coefficients has at least one real zero.

9-4
Rational Zeros

In the practical applications of algebra, it is common to encounter a problem which calls for finding the rational numbers[3] which are zeros of a polynomial $P(x)$ with integral coefficients. In such cases, the following theorem is fundamental to the solution of the problem.

THEOREM 9-14   Let

$$P(x) = a_n x^n + a_{n-1} x^{n-1} + \cdots + a_1 x + a_0$$

[3]Recall that a rational number is a quotient p/q of integers p and q, with $q \neq 0$.

be a polynomial in which all coefficients are integers, and let p/q denote a rational number which has been reduced to lowest terms. If p/q is a zero of $P(x)$, then p is a factor of a_0 and q is a factor of a_n.

That is, if p/q is a zero of $P(x)$ which is written so that the greatest common divisor of p and q is 1, then p must be a factor of the constant term, and q must be a factor of the leading coefficient of $P(x)$.

To see why the theorem is true, suppose that p/q, in lowest terms, is a zero of $P(x)$. Then

$$a_n\left(\frac{p}{q}\right)^n + a_{n-1}\left(\frac{p}{q}\right)^{n-1} + \cdots + a_1\left(\frac{p}{q}\right) + a_0 = 0.$$

Multiplying both sides by q^n, we have

$$a_n p^n + a_{n-1} p^{n-1} q + \cdots + a_1 p q^{n-1} + a_0 q^n = 0.$$

Subtracting $a_0 q^n$ from both sides yields

$$a_n p^n + a_{n-1} p^{n-1} q + \cdots + a_1 p q^{n-1} = -a_0 q^n,$$

and

$$p(a_n p^{n-1} + a_{n-1} p^{n-2} q + \cdots + a_1 q^{n-1}) = -a_0 q^n.$$

This equation shows that p is a factor of $a_0 q^n$. Since p/q is in lowest terms, the greatest common divisor of p and q is 1, and this means that p is a factor of a_0. Similarly, the equation

$$a_{n-1} p^{n-1} q + \cdots + a_1 p q^{n-1} + a_0 q^n = -a_n p^n$$

can be used to show that q is a factor of a_n.

The special case where $P(x)$ is a monic polynomial is important enough to designate as a corollary to Theorem 9-14.

COROLLARY 9-15 Let

$$P(x) = x^n + a_{n-1} x^{n-1} + \cdots + a_1 x + a_0$$

be a monic polynomial with integral coefficients. Then any rational zero of $P(x)$ is an integral factor of a_0.

EXAMPLE 1 Find all rational zeros of

$$P(x) = 2x^3 - x^2 - 8x - 5.$$

SOLUTION By Theorem 9-14, any rational zero of $P(x)$ has the form p/q, where p is a factor of -5 and q is a factor of 2. This means that

$$p \in \{\pm 1, \pm 5\}$$

and
$$q \in \{\pm 1, \pm 2\}.$$

Thus any rational zero p/q of $P(x)$ is included in the list

$$\pm 1, \quad \pm 5, \quad \pm \tfrac{1}{2}, \quad \pm \tfrac{5}{2}.$$

It is good practice to list the possible zeros in order from left to right, and test them systematically. We rewrite the possibilities as

$$-5, \quad -\tfrac{5}{2}, \quad -1, \quad -\tfrac{1}{2}, \quad \tfrac{1}{2}, \quad 1, \quad \tfrac{5}{2}, \quad 5.$$

Descartes' rule of signs indicates there is one positive zero of

$$P(x) = 2x^3 - x^2 - 8x - 5,$$

but this positive zero is not necessarily rational. Using synthetic division and testing the positive possibilities in order, we have

$$
\begin{array}{r|rrrr}
\tfrac{1}{2} & 2 & -1 & -8 & -5 \\
             &   & 1 & 0 & -4 \\
\hline
             & 2 & 0 & -8 & -9 \\
\end{array}
$$

$$
\begin{array}{r|rrrr}
1 & 2 & -1 & -8 & -5 \\
  &   & 2 & 1 & -7 \\
\hline
  & 2 & 1 & -7 & -12 \\
\end{array}
$$

$$
\begin{array}{r|rrrr}
\tfrac{5}{2} & 2 & -1 & -8 & -5 \\
             &   & 5 & 10 & 5 \\
\hline
             & 2 & 4 & 2 & 0 \\
\end{array}
$$

Thus $\tfrac{5}{2}$ is a rational zero of $P(x)$.

Before continuing, let us observe that the three synthetic divisions above could have been tabulated as follows, where the bottom row of each synthetic division is written to the right of the possibility that is being tested.

	2	-1	-8	-5
$\tfrac{1}{2}$	2	0	-8	-9
1	2	1	-7	-12
$\tfrac{5}{2}$	2	4	2	0

A table such as this is efficient when several divisions are to be performed in succession. With a little practice, the necessary arithmetic can be done mentally.

It would be straightforward to continue testing the remaining possible rational zeros, which are $5, -\tfrac{1}{2}, -1, -\tfrac{5}{2}, -5$. However, we know

from the last division performed that $x - \frac{5}{2}$ is a factor of $P(x)$, and
$$P(x) = (x - \tfrac{5}{2})(2x^2 + 4x + 2).$$
Thus any remaining zeros of $P(x)$ are zeros of the quotient $2x^2 + 4x + 2$ $= 2(x^2 + 2x + 1)$. We need only solve
$$x^2 + 2x + 1 = 0$$
to finish the problem. We have
$$(x + 1)^2 = 0,$$
so the other zero of $P(x)$ is -1, with a multiplicity of two. That is, the rational zeros of $P(x)$ are given by
$$\{\tfrac{5}{2}, -1, -1\}.$$

As Example 1 shows, it may happen that the testing for rational zeros leads to a situation where the quotient is a quadratic polynomial. In such a case, it is easier to find the zeros of the quotient than to continue checking rational zero possibilities. Sometimes this method will produce all the zeros of $P(x)$, not just the rational zeros.

EXAMPLE 2   Find all zeros of
$$P(x) = x^3 + x^2 - x + 2.$$

SOLUTION   According to Corollary 9-15, any rational zero of $P(x)$ is a factor of 2. Thus the possible rational zeros are
$$-2, \quad -1, \quad 1, \quad 2.$$
There are two variations of sign in
$$P(x) = x^3 + x^2 - x + 2,$$
and one in
$$P(-x) = -x^3 + x^2 + x + 2.$$
Thus the number of positive zeros is either 2 or 0, and there is one negative zero. Using synthetic division, we obtain the following table:

	1	1	-1	2
1	1	2	1	3
2	1	3	5	12
-1	1	0	-1	3
-2	1	-1	1	0

The last division shows that -2 is a zero, and that
$$P(x) = (x + 2)(x^2 - x + 1).$$
To find the other zeros of $P(x)$, we set
$$x^2 - x + 1 = 0$$
and obtain
$$x = \frac{1 \pm i\sqrt{3}}{2}$$
The zeros of $P(x)$ are given by
$$\left\{ -2, \frac{1 + i\sqrt{3}}{2}, \frac{1 - i\sqrt{3}}{2} \right\}.$$

It is important to note that a polynomial may have a large set of possible rational zeros, and yet, in fact, not have any rational zeros. For example, the polynomial
$$P(x) = 4x^4 + 7x^2 + 3$$
has, as possible rational zeros, the set
$$\{\pm 1, \pm 3, \pm \tfrac{1}{2}, \pm \tfrac{3}{2}, \pm \tfrac{1}{4}, \pm \tfrac{3}{4}\}.$$
But Descartes' rule of signs shows that $P(x)$ has no real zeros, and certainly no rational zeros.

When there is a large number of possible rational zeros, the following theorem can be extremely useful. We accept the theorem without proof.

THEOREM 9-16   Let
$$P(x) = a_n x^n + a_{n-1} x^{n-1} + \cdots + a_1 x + a_0$$
be a polynomial with real coefficients and $a_n > 0$, and suppose that synthetic division is used to divide $P(x)$ by $x - r$, where r is real. The last row in the synthetic division can be used in the following manner.

---

1. If $r > 0$ and all numbers in the last row are nonnegative, then $P(x)$ has no zero greater than r.
2. If $r < 0$ and the numbers in the last row alternate in sign (with 0 written as $+0$ or -0), then $P(x)$ has no zero less than r.

In case $P(x)$ has no zero less than the number a, we say that a is a *lower bound* for the zeros. Similarly, if $P(x)$ has no zeros greater than b, then b is called an *upper bound* for the zeros. Theorem 9-16 can frequently be used to obtain a positive upper bound and a negative lower bound for the zeros of $P(x)$.

EXAMPLE 3   Find all rational zeros of

$$P(x) = 2x^4 + x^3 - 8x^2 + x - 10.$$

SOLUTION   Any rational zero of $P(x)$ has the form p/q, where p is a factor of -10 and q is a factor of 2. That is,

$$p \in \{\pm 1, \pm 2, \pm 5, \pm 10\},$$

and

$$q \in \{\pm 1, \pm 2\}.$$

This gives the following set of possible rational zeros:

$$\{\pm 1, \pm 2, \pm 5, \pm 10, \pm \tfrac{1}{2}, \pm \tfrac{5}{2}\}.$$

Since there are three variations of sign in

$$P(x) = 2x^4 + x^3 - 8x^2 + x - 10,$$

the number of positive zeros is either 3 or 1. We arrange the positive possibilities in order of size,

$$\tfrac{1}{2}, 1, 2, \tfrac{5}{2}, 5, 10,$$

and begin a systematic check.

	2	1	-8	1	-10
$\frac{1}{2}$	2	2	-7	$-\frac{5}{2}$	$-\frac{45}{4}$
1	2	3	-5	-4	-14
2	2	5	2	5	0

The last division shows two things: 2 is a rational zero of $P(x)$, and 2 is an upper bound of the zeros of $P(x)$. There is no need to try the other positive possibilities for rational zeros. Since

$$P(x) = (x - 2)(2x^3 + 5x^2 + 2x + 5),$$

we concentrate now on the negative zeros of

$$2x^3 + 5x^2 + 2x + 5.$$

Systematically checking the possibilities

$$-\tfrac{1}{2}, \quad -1, \quad -2, \quad -\tfrac{5}{2}, \quad -5, \quad -10,$$

we obtain the following table:

	2	5	2	5
$-\frac{1}{2}$	2	4	0	5
-1	2	3	-1	6
-2	2	1	0	5
$-\frac{5}{2}$	2	-0	2	-0

From the last row of the table, we see that $-\frac{5}{2}$ is a rational zero, and $-\frac{5}{2}$ is a lower bound of the zeros. Since $-\frac{5}{2}$ is a lower bound for the zeros, there is no need to test -5 and -10. We have

$$P(x) = (x - 2)(x + \tfrac{5}{2})(2x^2 + 2)$$
$$= 2(x - 2)(x + \tfrac{5}{2})(x^2 + 1).$$

The complete set of zeros of $P(x)$ is $\{2, -\frac{5}{2}, i, -i\}$, and the rational zeros of $P(x)$ are given by

$$\{2, \quad -\tfrac{5}{2}\}.$$

EXERCISES 9-4

In problems 1–8, (a) find the smallest positive integer which Theorem 9-16 detects as an upper bound for the zeros of the given polynomial, and (b) find the negative integer nearest 0 which Theorem 9-16 detects as a lower bound for the zeros of the given polynomial.

1. $2x^3 + x^2 - 10x - 4$ 2. $3x^3 - 2x^2 - 21x + 15$

3. $2x^3 - 3x^2 + 8x - 13$ 4. $3x^3 - 7x^2 + 15x - 35$

5. $x^4 + x^3 + 2x^2 + 4x - 9$ 6. $x^4 - x^3 - 4x^2 - 2x - 13$

7. $x^4 + x^3 - 5x^2 + x - 5$ 8. $x^4 + x^3 - 11x^2 + x - 10$

Find all zeros of the given polynomial.

9. $x^3 - 3x^2 + 4x - 12$ 10. $x^3 + 2x^2 + 6x + 12$

11. $2x^3 + 7x^2 + 2x - 3$ 12. $2x^3 - 3x^2 - 7x - 6$

13. $3x^3 - 5x^2 - 4$ 14. $2x^3 + x^2 - 2x - 6$

15. $x^4 - x^3 - 2x^2 + 6x - 4$ 16. $x^4 + x^3 - 2x^2 - 6x - 4$

17. $2x^4 - x^3 - 13x^2 + 5x + 15$ 18. $2x^4 + x^3 - 13x^2 - 5x + 15$

19. $2x^4 + 5x^3 - 7x^2 - 10x + 6$ 20. $2x^4 - x^3 - x^2 - x - 3$

Find all rational zeros of the given polynomial.

21. $x^3 + 7x - 6$

22. $x^3 - 2x^2 + 10$

23. $2x^4 + 3x^3 + 2x^2 + 11x + 12$

24. $3x^4 - 2x^3 + 3x^2 + 16x - 12$

25. $2x^5 - x^4 - 8x^3 + 3x^2 + 5x - 6$

26. $3x^5 + 4x^4 - 7x^3 - x^2 + 8x - 4$

*27. Show that $\sqrt{3}$ is irrational by applying Theorem 9-14 to $P(x) = x^2 - 3$.

*28. Show that $\sqrt{2}$ is irrational by applying Theorem 9-14 to $P(x) = x^2 - 2$.

9-5
Approximation
of Real Zeros

In the preceding section, we saw how the rational zeros of a polynomial with integral coefficients can be found. The other real zeros (i.e., the irrational zeros) of such a polynomial can be found to any desired degree of accuracy by using the Location Theorem, which is stated below.

THEOREM 9-17 (THE LOCATION THEOREM)   Let $P(x)$ be a polynomial with real coefficients. If a and b are real numbers such that $P(a)$ and $P(b)$ have opposite signs, then $P(x)$ has at least one zero between a and b.

The essence of this theorem is that the values of $P(x)$ cannot change from positive to negative, or from negative to positive, without assuming the value 0 between. This is one aspect of the continuity of polynomial functions, a topic which is related to Section 9-6 as well as this one. The sketches in Figure 9.1 show how the graph of $y = P(x)$ might look over

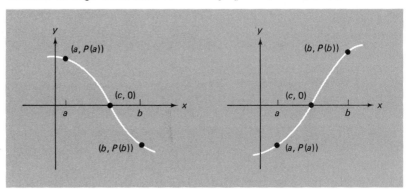

FIGURE 9.1

the interval from $x = a$ to $x = b$. A point c where $P(c) = 0$ is indicated in each case.

Theorem 9-17 is the basis of our method for approximating the real zeros of a polynomial with real coefficients to any desired degree of accuracy. The fundamental idea is to locate the zeros of $P(x)$ in intervals having length small enough to give the desired accuracy. The method is illustrated in the following example.

EXAMPLE 1 Find the real zeros of the polynomial
$$P(x) = x^4 + 4x^2 - 6x - 5,$$
correct to the nearest tenth.

SOLUTION Since there is one variation of sign in $P(x)$ and also one variation of sign in
$$P(-x) = x^4 + 4x^2 + 6x - 5,$$
the polynomial has one positive real zero and one negative real zero, by Descartes' rule of signs. The set of possible rational zeros is
$$\{-5, -1, 1, 5\},$$
by Corollary 9-15. Using synthetic division, we have

		1	0	4	-6	-5
1		1	1	5	-1	-6
5		1	5	29	139	690
-1		1	-1	5	-11	6

The alternating signs in the last row show that -1 is a lower bound, so we do not try -5 as a rational zero. There are no rational zeros of $P(x)$, but the work above shows that $P(1) = -6$ and $P(5) = 690$, so the positive real zero is between 1 and 5, by the Location Theorem. The values $P(1) = -6$ and $P(5) = 690$ indicate that the zero is probably much closer to 1 than 5, so we try 2 next in synthetic division.

$$
\begin{array}{r|rrrrr}
2 & 1 & 0 & 4 & -6 & -5 \\
  &   & 2 & 4 & 16 & 20 \\
\hline
  & 1 & 2 & 8 & 10 & 15 \\
\end{array}
$$

We have $P(1) = -6$ and $P(2) = 15$, so the zero is between 1 and 2. We next evaluate $P(1.5)$, since this will locate the zero in one half of the interval or the other.

$$
\begin{array}{r|rrrrr}
1.5 & 1 & 0 & 4 & -6 & -5 \\
    &   & 1.5 & 2.25 & 9.375 & 5.0625 \\
\hline
    & 1 & 1.5 & 6.25 & 3.375 & 0.0625 \\
\end{array}
$$

Since $P(1.5) = 0.0625$ is positive, the zero is between 1 and 1.5, probably nearer 1.5. We next find $P(1.4)$.

$$
\begin{array}{r|rrrr}
1.4 & 1 & 0 & 4 & -6 & -5 \\
 & & 1.4 & 1.96 & 8.344 & 3.2816 \\
\hline
 & 1 & 1.4 & 5.96 & 2.344 & -1.7184
\end{array}
$$

Thus the zero is between 1.4 and 1.5, probably closer to 1.5. Computing $P(1.45)$, we have

$$
\begin{array}{r|rrrr}
1.45 & 1 & 0 & 4 & -6 & -5 \\
 & & 1.45 & 2.1025 & 8.8486 & 4.1305 \\
\hline
 & 1 & 1.45 & 6.1025 & 2.8486 & -0.8695
\end{array}
$$

Since $P(1.5) = 0.0625$ and $P(1.45) = -0.8695$, the positive zero of $P(x)$ is between 1.45 and 1.5. To the nearest tenth, then, the zero is 1.5.

Looking for the negative real zero, we find $P(0) = -5$ and $P(-1) = 6$, so the negative real zero of $P(x)$ is between -1 and 0. Following the same sort of procedure as with the positive zero, we obtain the following synthetic division table:

	1	0	4	-6	-5
0	1	0	4	-6	-5
-1	1	-1	5	-11	6
-0.5	1	-0.5	4.25	-8.125	-0.9375
-0.6	1	-0.6	4.36	-8.616	0.1696
-0.55	1	-0.55	4.3025	-8.3664	-0.3985

The negative real zero is between -0.55 and -0.6, and has the value -0.6, to the nearest tenth. Thus the set of real zeros, to the nearest tenth, of $P(x)$ is

$$\{1.5, -0.6\}.$$

The real (rational or irrational) zeros of a polynomial with real (not necessarily rational) coefficients can be approximated to the nearest hundredth, or to any needed degree of accuracy, by the method used in Example 1. The numerical calculations, of course, become more and more tedious as accuracy increases.

EXERCISES 9-5

Use the Location Theorem to approximate, to the nearest tenth, the zero of the given polynomial which is in the indicated interval.

1. $2x^3 - 11x^2 + 15x - 1$, between 2 and 3

2. $x^3 - 3x^2 + x + 1$, between 0 and 1

3. $x^3 + x^2 - 9x + 4$, between 0 and 1

4. $x^3 - 9x + 7$, between 2 and 3

5. $2x^4 - 5x^3 + 6x^2 - 22x + 7$, between 2 and 3

6. $x^4 + 2x^3 - 3x^2 + 7x - 4$, between 0 and 1

Each of the following polynomials has exactly one real zero. Find the value of the zero, correct to the nearest tenth.

7. $4x^3 + 4x^2 + 2x + 1$ 8. $x^3 + x^2 + x - 2$

9. $x^3 + 3x^2 + 3x - 10$ 10. $x^3 + x^2 - x + 1$

Find all real zeros of the given polynomial, correct to the nearest tenth.

11. $x^3 - x^2 - 3x + 1$ 12. $x^3 - 3x + 1$

13. $3x^3 - 9x^2 + 7x - 1$ 14. $2x^3 - 2x^2 - 3x - 2$

15. $x^3 - x^2 - 15x - 17$ 16. $x^3 + 2x^2 - 14x - 32$

17. $x^4 - 2x^3 + 3x^2 - 12x + 10$ 18. $x^4 - 2x^3 + 2x^2 - 6x + 3$

9-6
Graphs
of Polynomial
Functions

We restrict our attention in this section to polynomial functions defined by $y = P(x)$, where $P(x)$ is a polynomial with real coefficients.

In Chapter 5, we have seen that the graph of a first-degree polynomial function is always a straight line, and the graph of a second-degree polynomial function is always a parabola. In this section, we study the graphs of polynomial functions which have degree greater than two.

The number of possibilities as to the shape of the graph increases as the degree of the polynomial function increases. Even for third-degree polynomials, there are several possibilities. Some (not all) of these are indicated in Figure 9.2, where attention is called to the number of real zeros of the polynomial.

The graphs shown in Figure 9.2 represent third-degree polynomials

$$y = a_3 x^3 + a_2 x^2 + a_1 x + a_0,$$

with a leading coefficient a_3 which is positive. Similar graphs could be drawn for the case where $a_3 < 0$. These graphs would simply correspond to reversing the directions on the y-axis, since that is the effect of multiplying a polynomial by -1.

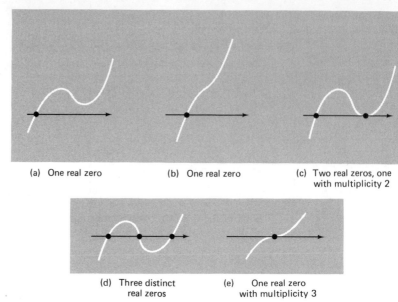

(a) One real zero

(b) One real zero

(c) Two real zeros, one with multiplicity 2

(d) Three distinct real zeros

(e) One real zero with multiplicity 3

FIGURE 9.2

The point here is that there is no single characteristic shape, such as a parabola, for the graphs of higher-degree polynomials. As the degree increases, the number of possible shapes increases. However, some general remarks can be made. First, the graph of a polynomial function is a continuous, smooth curve (no breaks, gaps, or sharp corners). As a usual rule, the number of turning points on the graph is one less than the degree (note the exceptions in Figure 9.2). Also, a polynomial of odd degree always has a real zero (see problem 29 of Exercises 9-3).

For maximum efficiency in graphing a polynomial function, the use of calculus is necessary. For certain simpler functions, however, a reasonably good graph can be drawn by plotting suitably selected points on the curve which have integral x-coordinates, and connecting these points with a smooth curve. Synthetic division, Descartes' rule of signs, the Location Theorem, and Theorem 9-16 are frequently very useful. An example is provided below.

EXAMPLE 1 Sketch the graph of the polynomial function defined by

$$y = -x^4 + 24x^2 - 12x + 4.$$

SOLUTION Since there are three variations of sign in

$$P(x) = -x^4 + 24x^2 - 12x + 4,$$

the graph crosses the positive x-axis in either one or three places. Also, there is one variation of sign in

$$P(-x) = -x^4 + 24x^2 + 12x + 4,$$

so the graph crosses the negative x-axis at one place. Using synthetic division, we obtain the following table:

	−1	0	24	−12	4
0	−1	0	24	−12	4
1	−1	−1	23	11	15
2	−1	−2	20	28	60
3	−1	−3	15	33	103
4	−1	−4	8	20	84
5	−1	−5	−1	−17	−81

From the last two lines, we have $P(4) = 84$ and $P(5) = -81$, so there is a zero between 4 and 5. With a negative leading coefficient, all negative numbers in the last row tells us that 5 is an upper bound for the zeros (see Theorem 9-16). Thus it appears that there is only one positive real zero. This is not certain, however, for two more zeros of $P(x)$ may lie between consecutive integers. The methods of calculus are needed for checking possibilities such as this.

Before plotting our points, we compute the function values for some negative integers.

	−1	0	24	−12	4
−1	−1	1	23	−35	39
−2	−1	2	20	−52	108
−3	−1	3	15	−57	175
−4	−1	4	8	−44	180
−5	−1	5	−1	−7	39
−6	−1	6	−12	60	−356

The last two lines show that the negative zero is between −5 and −6. Because the function values vary over a very large range as compared to the x-values, it is desirable to use a smaller scale on the y-axis when plotting our points. Even then, it is convenient to omit the point corresponding to $P(-6) = -356$. A plausible sketch of the graph is shown in Figure 9.3.

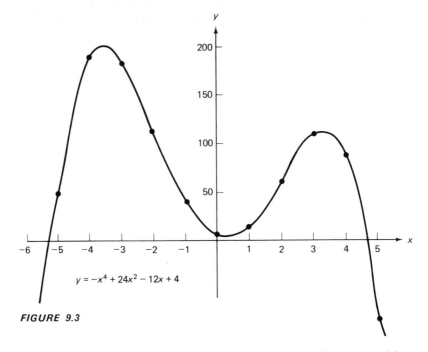

$$y = -x^4 + 24x^2 - 12x + 4$$

FIGURE 9.3

In Example 1, it is reasonably clear that, as x increases without bound, the values of y decrease without bound. Also, as x decreases without bound, y decreases without bound. This behavior is due to the fact that the leading coefficient is negative. The behavior of y, as x increases or decreases without bound, is controlled by the sign of the leading coefficient. This is true because the highest-degree term in the polynomial dominates the other terms for values of x which are sufficiently large in absolute value.

When a polynomial is presented as a product of linear factors, the multiplicity of each zero is apparent, and this makes the graphing much easier. It is easy to see that if $x - r$ is a factor of $P(x)$ with *even* multiplicity, then $P(x)$ does not change sign as x increases through r, and the graph of $y = P(x)$ is tangent to the x-axis at $x = r$. However, if $x - r$ is a factor of $P(x)$ with *odd* multiplicity, then $P(x)$ changes sign as x increases through r, and the graph of $y = P(x)$ crosses the x-axis at $x = r$.

The remarks above are illustrated in the following examples.

EXAMPLE 2   Sketch the graph of the polynomial function

$$P(x) = x^3.$$

SOLUTION   It is clear that $P(x) = x^3$ has a zero of multiplicity 3 at $x = 0$, and the graph crosses the x-axis at $x = 0$. A table of values and the graph are given in Figure 9.4.

Theory of Polynomials

x	y
0	0
1	1
2	8
3	27
-1	-1
-2	-8
-3	-27

$P(x) = x^3$

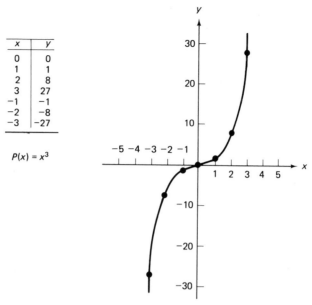

FIGURE 9.4

EXAMPLE 3 Sketch the graph of

$$y = (x - 1)(x - 2)^2.$$

SOLUTION Since $x - 2$ is a factor with even multiplicity, the graph is tangent to the x-axis at $x = 2$. Also, since $x - 1$ is a factor with odd multiplicity, the graph crosses the x-axis at $x = 1$. A table of values and the graph are shown in Figure 9.5.

FIGURE 9.5

x	y
0	-4
1/2	-9/8
1	0
3/2	1/8
2	0
3	2

$y = (x - 1)(x - 2)^2$

EXAMPLE 4   Sketch the graph of

$$y = (x - 3)(x - 1)^2(x + 1).$$

SOLUTION   Observing that -1 and 3 are zeros of multiplicity 1 and that 1 is a zero of multiplicity 2, we know that the graph crosses the x-axis at -1 and 3, and is tangent to the x-axis at $x = 1$. A table of values and the graph are shown in Figure 9.6.

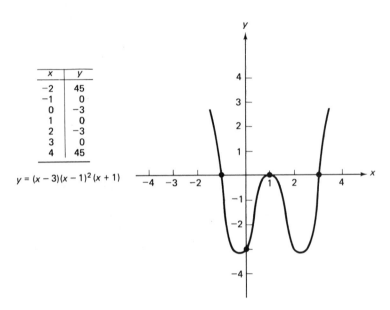

x	y
−2	45
−1	0
0	−3
1	0
2	−3
3	0
4	45

$y = (x - 3)(x - 1)^2 (x + 1)$

FIGURE 9.6

EXERCISES 9-6

Sketch the graph of the polynomial function defined by each equation.

1. $y = (x - 1)(x + 2)(x - 3)$

2. $y = (x + 1)(x - 2)(x - 4)$

3. $y = (x - 1)^2(x + 4)^2$

4. $y = (x + 3)^2(x - 2)^2$

5. $y = (-3x + 2)(2x - 1)(x + 3)$

6. $y = (x + 2)(-2x + 3)(2x + 1)$

7. $y = -2(x + 1)^2(1 - x)$

8. $y = (x - 2)^2(x + 3)$

9. $y = -2(x^2 + 1)(x - 1)(x + 2)$

10. $y = -2(x^2 + 2)(x^2 - 1)$

11. $y = (2 - x)(x^2 - 3x + 2)$

12. $y = (1 - x)(x^2 - 3x + 2)$

13. $y = (x - 2)^2(x + 3) + 4$

14. $y = (x + 1)^2(x - 2) + 2$

15. $y = (2x - 3)^2(3x - 4)(2x + 1)$

16. $y = (2x + 1)^2(4x - 3)(2x - 5)$

17. $y = (x - 2)(x^3 + 1)$

18. $y = (x + 3)(x^3 - 8)$

19. $y = (x^2 + x + 1)(x - 2)$

20. $y = (x^2 + x + 2)(x + 1)$

21. $y = x^4 - 8x^2$

22. $y = x^4 - 27x^2$

23. $y = x^3 - 2x^2 - 15x$

24. $y = x^3 + 4x^2 + 4x$

25. $y = x^3 - 3x + 4$

26. $y = x^3 - 3x - 4$

27. $y = x^3 - x^2 - 3x + 1$

28. $y = x^3 - 3x + 1$

29. $y = x^4 - 4x^3 + 4x^2 - 1$

30. $y = x^4 + 3x^3 + 3x^2 - 1$

31. $y = x^4 - 2x^3 + 3x^2 - 12x + 10$

32. $y = x^4 - 2x^3 + 2x^2 - 6x + 3$

Perform the indicated operations and leave the result in standard form.

1. $\dfrac{3 - i}{2 + 5i}$ 2. i^{67}

3. Solve the equation $2x^2 - 3ix + 3 = 0$.

4. Use the Remainder Theorem to find $P(-2)$ if
$$P(x) = 5x^4 - 2x^3 - 4x + 1.$$

5. Find a polynomial of least degree with real coefficients that has -1, -2, and $3i$ as zeros.

6. Use Descartes' rule of signs to discuss the nature of the zeros of the polynomial
$$P(x) = 2x^4 + x^3 - 4x - 3.$$

7. (a) Find a positive integer, as small as possible, which is an upper bound for the zeros of the polynomial
$$2x^3 + 4x^2 - 3x - 6.$$

   (b) Find a negative integer, as large as possible, which is a lower bound for the zeros of the polynomial in part (a).

8. Find all rational zeros of the polynomial in problem 7.

9. The polynomial below has one negative real zero. Find the value of the zero, correct to the nearest tenth.
$$P(x) = 2x^4 + x^2 - 4x - 3$$

10. Sketch the graph of the polynomial function defined by the equation
$$y = (x - 1)^2(x + 2).$$

Exponential
and Logarithmic
Functions

10-1
Exponential
Functions

In this section, we consider a type of function called an exponential function. These functions are indispensable in working with problems which involve population growths, decay of radioactive materials, and other processes which occur in nature.

In an exponential function, the function value is obtained by raising a fixed number, called the *base*, to a power. Recall that, for $a \neq 0$, a^x was defined for integral values of x in Section 2-1. In Sections 3-1 and 3-2, this definition was extended to include values $x = m/n$ which are rational numbers. However, if a is negative, it may happen that there is no real number $a^{m/n}$. [Such an example is provided by $(-4)^{1/2}$.] For this reason, *we restrict our attention in this chapter to the case where a is positive*. Once this restriction is made, a^x is defined by Definition 3-3 for all rational values of x. It is possible to extend the definition of a^x to include irrational values of x, but a rigorous treatment of this extension requires a degree of mathematical sophistication which is beyond this text. A complete treatment of this sort of topic belongs to the area of mathematics known as *analysis*.

In order to get some intuitive feeling for the situation involving an irrational exponent, we consider a specific example, say $2^{\sqrt{5}}$. The reasoning that we use is this: if $\sqrt{5}$ is approximated by a rational number m/n, then $2^{\sqrt{5}}$ should be approximated by $2^{m/n}$. More specifically, if $\sqrt{5}$ is approximated by successively closer rational numbers such as

$$2.2, \quad 2.23, \quad 2.236,$$

then successively closer approximations to $2^{\sqrt{5}}$ should be provided by

$$2^{2.2}, \quad 2^{2.23}, \quad 2^{2.236}.$$

Each of these approximations is meaningful, since the exponents are rational. For example,

$$2^{2.2} = 2^{22/10} = 2^{11/5} = (\sqrt[5]{2})^{11}.$$

This intuitive procedure is justified in more advanced courses. For our work, we accept the procedure without justification. In the same spirit, we accept the following theorem, which is a generalization of Theorem 2-2.

THEOREM 10-1   Let a and b denote positive real numbers. For any real numbers x and y,

(a) $a^x \cdot a^y = a^{x+y}$;

(b) $\dfrac{a^x}{a^y} = a^{x-y}$;

(c) $(a^x)^y = a^{xy}$;

(d) $(ab)^x = a^x \cdot b^x$;

(e) $\left(\dfrac{a}{b}\right)^x = \dfrac{a^x}{b^x}$.

With the preceding facts in mind, we make the following definition.

DEFINITION 10-2   If $a > 0$ and $a \neq 1$, then the function defined by

$$f(x) = a^x$$

is the *exponential function* with base a.

The value $a = 1$ is excluded in the definition because $1^x = 1$ for all real numbers x. That is, the value $a = 1$ would define a constant function, and it is excluded from the exponential functions for this reason.

As we shall see in the examples that follow, it is not difficult to sketch the graph of an exponential function.

EXAMPLE 1   Sketch the graph of the exponential function
$$f(x) = 2^x.$$

SOLUTION   To locate some points on the graph conveniently, we assign a few integral values to x, and compute the corresponding function values. Note that the graph approaches the negative x-axis asymptotically.

x	-3	-2	-1	0	1	2	3
2^x	$\frac{1}{8}$	$\frac{1}{4}$	$\frac{1}{2}$	1	2	4	8

After plotting these points, we sketch a smooth curve as shown in Figure 10.1.

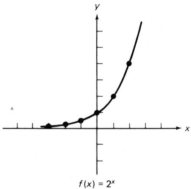

$f(x) = 2^x$

FIGURE 10.1

EXAMPLE 2   Sketch the graph of the exponential function defined by
$$f(x) = (\tfrac{1}{2})^x.$$

SOLUTION   Following the same procedure as in Example 1, we obtain the following table. Note that the x-axis is an asymptote for large values of x.

x	-3	-2	-1	0	1	2	3
$(\tfrac{1}{2})^x$	8	4	2	1	$\frac{1}{2}$	$\frac{1}{4}$	$\frac{1}{8}$

The graph is sketched in Figure 10.2.

The graphs shown in Figures 10.1 and 10.2 are typical, and they illustrate the following observations concerning exponential functions.

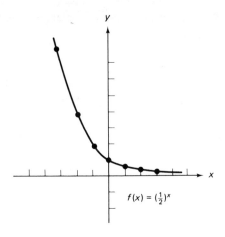

$f(x) = (\frac{1}{2})^x$

FIGURE 10.2

Important Properties of the Exponential Functions

$$f(x) = a^x, \qquad a > 0 \text{ and } a \neq 1$$

1. The *domain* of f is the set of all real numbers.

2. The *range* of f is the set of all positive real numbers.

3. If $a > 1$, the graph of $f(x) = a^x$ resembles that of $f(x) = 2^x$. (See Figure 10.1.)

4. If $0 < a < 1$, the graph of $f(x) = a^x$ resembles that of $f(x)$ $= (\frac{1}{2})^x$. (See Figure 10.2.)

5. $a^u = a^v$ if and only if $u = v$.

In connection with property 3 in the list, we note that if $a > 1$, then a^x increases as x increases. For this reason, we say that a^x is an *increasing function* when $a > 1$. Similarly, if $0 < a < 1$, then a^x decreases as x increases, and a^x is called a *decreasing function* when $0 < a < 1$.

Property 5 in the list allows us to solve certain types of equations for an unknown which appears in an exponent. A more general method of solution is presented in Section 10-5.

EXAMPLE 3   Solve for x in the following equations.

(a) $3^x = 81$

(b) $2^x = \frac{1}{32}$

(c) $4^x = 32$

(d) $9^{-x} = 27$

(a) We recognize that $81 = 3^4$, so the given equation can be rewritten as

$$3^x = 81$$
$$= 3^4.$$

By property 5, $x = 4$ is the only solution.

(b) Since $2^5 = 32$, we have

$$2^x = \frac{1}{2^5}$$
$$= 2^{-5}.$$

Therefore, $x = -5$ is the only solution.

(c) In this case, we can write each side of the equation as a power of 2.

$$4^x = 32,$$
$$(2^2)^x = 2^5,$$
$$2^{2x} = 2^5.$$

Therefore, $2x = 5$, and $x = \frac{5}{2}$ is the solution.

(d) Each member of the given equation can be written as a power of 3.

$$9^{-x} = 27,$$
$$(3^2)^{-x} = 3^3,$$
$$3^{-2x} = 3^3.$$

Therefore, $-2x = 3$, and $x = -\frac{3}{2}$ is the solution.

A more general type of exponential function is one defined by an equation of the form $y = a^{P(x)}$, where $P(x)$ is a polynomial in x. The graphs of two such functions are shown in Examples 4 and 5.

EXAMPLE 4 Sketch the graph of

$$f(x) = 2^{-x^2}.$$

SOLUTION We have the following table:

x	-2	-1	0	1	2
$f(x)$	$\frac{1}{16}$	$\frac{1}{2}$	1	$\frac{1}{2}$	$\frac{1}{16}$

Noting that

$$f(x) = 2^{-x^2} = \frac{1}{2^{x^2}},$$

we see that the largest possible function value of f is $f(0) = 1$, and that the function values approach 0 as $|x|$ increases without bound. The graph is shown in Figure 10.3.

Exponential and
Logarithmic Functions

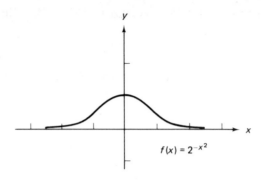

$$f(x) = 2^{-x^2}$$

FIGURE 10.3

EXAMPLE 5   Sketch the graph of

$$y = 3^{2x-1}.$$

SOLUTION   A table of values is shown below, and the graph is drawn in Figure 10.4.

x	-2	-1	0	$\frac{1}{2}$	1	2
y	$\frac{1}{243}$	$\frac{1}{27}$	$\frac{1}{3}$	1	3	27

Here it is important to single out the value that makes the exponent zero, namely, $x = \frac{1}{2}$.

FIGURE 10.4

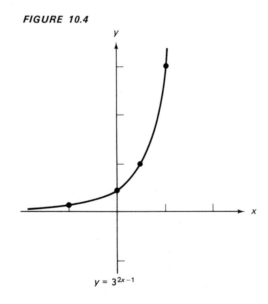

$$y = 3^{2x-1}$$

Exponential and
Logarithmic Functions

If $f(x) = (\frac{3}{2})^x$, find each function value.

1. $f(0)$ 2. $f(1)$

3. $f(2)$ 4. $f(3)$

5. $f(-1)$ 6. $f(-2)$

7. $f(-\frac{1}{2})$ 8. $f(\frac{1}{2})$

Solve for x in each equation.

9. $3^x = 27$ 10. $2^x = 32$

11. $2^x = \frac{1}{16}$ 12. $3^x = \frac{1}{81}$

13. $2^x = 1$ 14. $3^x = 1$

15. $(\frac{1}{2})^x = 8$ 16. $(\frac{1}{3})^x = 9$

17. $(\frac{2}{3})^x = \frac{9}{4}$ 18. $(\frac{3}{2})^x = \frac{8}{27}$

19. $8^x = \frac{1}{16}$ 20. $(27)^x = \frac{1}{9}$

21. $5^{-x} = 625$ 22. $4^{-x} = 128$

23. $3^{2x-1} = \frac{1}{81}$ 24. $2^{3x+1} = 64$

25. $(81)^{2x+1} = \frac{1}{3}$ 26. $(32)^{3x-2} = \frac{1}{4}$

Sketch the graph of each function.

27. $f(x) = 3^x$ 28. $f(x) = 4^x$

29. $f(x) = (10)^x$ 30. $f(x) = (10)^{-x}$

31. $f(x) = 3^{-x}$ 32. $f(x) = 4^{-x}$

33. $f(x) = -2^x$ 34. $f(x) = -(\frac{1}{2})^x$

35. $f(x) = -3^x$ 36. $f(x) = -4^x$

37. $f(x) = -3^{-x}$ 38. $f(x) = -4^{-x}$

39. $f(x) = (\frac{1}{3})^x$ 40. $f(x) = (\frac{1}{4})^x$

41. $f(x) = (\frac{2}{3})^x$ 42. $f(x) = (\frac{3}{2})^x$

Sketch the graph of each equation.

43. $y = 2^{2x-1}$ 44. $y = 3^{2x+1}$

45. $y = 2 \cdot 5^x$ 46. $y = 2 \cdot 3^x$

47. $y = 5^{-x} + 1$ 48. $y = 2^x - 1$

49. $y = 4^{-x^2}$ 50. $y = (2)^{1-x^2}$

51. The population of a city is now 16,000. The population t years from now
is given by the formula $P = 16,000 \cdot 2^{t/10}$. What will the population be
40 years from now? How often does the population double?

52. A woman's savings grow according to the formula $P = P_0(3)^{0.08t}$, where P_0 is the amount of her original investment, and t is the time in years since the investment. If $P_0 = \$10,000$, how much will she have in 25 years?

10-2
Logarithmic
Functions

Exponential functions provide a basis for defining logarithmic functions. More specifically, two of the properties listed for exponential functions in the preceding section are fundamental in defining the term *logarithm*. For $a > 0$ and $a \neq 1$, these properties are

2. the range of a^x is the set of all positive real numbers,

and

5. $a^u = a^v$ if and only if $u = v$.

Property 2 means that every positive real number can be obtained by raising a to an appropriate power, and property 5 means that there is only one power of a that will give the desired number. These two facts are easily seen from the graphs in Figures 10.1 and 10.2.

DEFINITION 10-3   Let a denote a real number such that $a > 0$ and $a \neq 1$. For any positive number x, the exponent y to which a must be raised to have $x = a^y$ is called the *logarithm of x to the base a*, and is abbreviated as

$$y = \log_a x.$$

In other words, for $x > 0$,

$$y = \log_a x \quad \text{if and only if} \quad x = a^y.$$

The last statement in Definition 10-3 sets up an equivalence between logarithmic statements $y = \log_a x$ and exponential statements $x = a^y$, so that a statement in exponential form can be translated into an equivalent statement in logarithmic form, and vice versa.

EXAMPLE 1   The following statements illustrate changes from exponential form to logarithmic form.

(a) $2^3 = 8$ means that $\log_2 8 = 3$.
(b) $(27)^{2/3} = 9$ means that $\log_{27} 9 = \frac{2}{3}$.
(c) $4^{-2} = \frac{1}{16}$ means that $\log_4 \frac{1}{16} = -2$.
(d) $6^0 = 1$ means that $\log_6 1 = 0$.

EXAMPLE 2   The following statements illustrate changes from logarithmic form to exponential form.

(a) $\log_2 \frac{1}{16} = -4$ since $2^{-4} = \frac{1}{16}$
(b) $\log_{3/4} \frac{9}{16} = 2$ since $(\frac{3}{4})^2 = \frac{9}{16}$
(c) $\log_4 \frac{1}{8} = -\frac{3}{2}$ since $4^{-3/2} = \frac{1}{8}$

For a fixed number $a > 0$ and $a \neq 1$, the equation

$$y = \log_a x$$

defines a function, the *logarithmic function* with base a. This defining equation is equivalent to $x = a^y$, and this equation can be obtained from the defining equation $y = a^x$ for the exponential function by interchanging x and y. Thus the logarithmic function

$$f(x) = \log_a x$$

and the exponential function

$$g(x) = a^x$$

are *inverse functions* of one another. This implies that $f(g(x)) = x$ for all real numbers x. That is,

$$\log_a a^x = x,$$

for all real numbers x. Similarly, $g(f(x)) = x$, or

$$a^{\log_a x} = x$$

for all positive real numbers x. As special cases of the first of these important formulas, we note that

$$\log_a a = 1$$

and

$$\log_a 1 = 0.$$

If two of the variables a, x, and y are specified in the equation $y = \log_a x$, the unknown variable can frequently be found by converting to an exponential statement.

EXAMPLE 3   Solve for a, x, or y.

    (a) $y = \log_5 \frac{1}{125}$
    (b) $-2 = \log_3 x$
    (c) $\log_a \frac{16}{81} = -4$

SOLUTION   (a) Since $125 = 5^3$, we have

$$y = \log_5 \frac{1}{5^3}$$
$$= \log_5 5^{-3}$$
$$= -3.$$

(b) From Definition 10-3, the equation

$$-2 = \log_3 x$$

is equivalent to

$$x = 3^{-2} = \tfrac{1}{9}.$$

(c) By Definition 10-3,

$$\log_a \tfrac{16}{81} = -4$$

means that

$$a^{-4} = \tfrac{16}{81}.$$

Therefore,

$$(a^{-4})^{-1/4} = (\tfrac{16}{81})^{-1/4},$$

and

$$a = \left(\frac{2^4}{3^4}\right)^{-1/4}$$
$$= (\tfrac{2}{3})^{-1}$$
$$= \tfrac{3}{2}.$$

We recall that the graph of the inverse of a function f can be obtained from the graph of f by reflecting the graph through the line $y = x$. This fact is used in the next two examples.

EXAMPLE 4   Sketch the graph of

$$y = \log_2 x.$$

SOLUTION   Since $y = \log_2 x$ is the inverse function of $y = 2^x$, the graph of $y = \log_2 x$ may be obtained by reflecting the graph of $y = 2^x$ through the line $y = x$. The graph of $y = 2^x$ was drawn in Figure 10.1 and is sketched as a dashed curve in Figure 10.5. The graph of $y = \log_2 x$ can be obtained by reflecting through the line $y = x$.

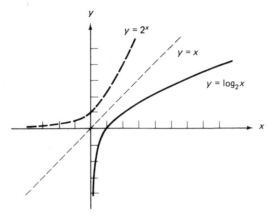

FIGURE 10.5

EXAMPLE 5   Sketch the graph of

$$y = \log_{1/2}x.$$

SOLUTION   The graph of $y = (\frac{1}{2})^x$ is shown in Figure 10.2 and is reproduced as a dashed curve in Figure 10.6. The graph of $y = \log_{1/2}x$ can be obtained by reflecting through the line $y = x$. However, the graph of $y = \log_{1/2}x$

x	$\log_{1/2}x$
1	0
2	−1
4	−2
8	−3
1/2	1
1/4	2
1/8	3

FIGURE 10.6

can also be obtained by plotting a few points and sketching in a smooth curve. To illustrate this, a table of values is also included in Figure 10.6.

The graphs shown in Figures 10.5 and 10.6 are typical, and they illustrate the following properties of logarithmic functions.

Important Properties of the Logarithmic Functions

$$f(x) = \log_a x, \quad a > 0 \text{ and } a \neq 1$$

1. The *domain* of f is the set of all positive real numbers.
2. The *range* of f is the set of all real numbers.
3. If $a > 1$, the graph of $f(x) = \log_a x$ resembles that of $f(x) = \log_2 x$. (See Figure 10.5.)
4. If $0 < a < 1$, the graph of $f(x) = \log_a x$ resembles that of $f(x) = \log_{1/2} x$. (See Figure 10.6.)
5. $\log_a u = \log_a v$ if and only if $u = v$.

Since the exponential and logarithmic functions with base a are inverses of each other, the properties stated in Theorem 10-1 have some important implications concerning logarithms. These are stated in the next theorem.

THEOREM 10-4   Let a denote a real number such that $a > 0$ and $a \neq 1$. If u, v, and r are real numbers with u and v positive, then

(a) $\log_a(u \cdot v) = \log_a u + \log_a v$;

(b) $\log_a \dfrac{u}{v} = \log_a u - \log_a v$;

(c) $\log_a u^r = r \log_a u$.

All three parts of Theorem 10-4 can be proved by using Theorem 10-1. As an illustration, we shall show how part (c) can be obtained. Let u be an arbitrary positive number, and let $x = \log_a u$. Then $a^x = u$. To find $\log_a u^r$, we express u^r as a power of a:

$$u^r = (a^x)^r$$
$$= a^{rx},$$

by Theorem 10-1(c). Therefore,

$$\log_a u^r = rx$$
$$= r \log_a u.$$

Before hand-held calculators came into common use, logarithms were frequently used as an aid in numerical computations. Their usefulness in this respect depended on the properties listed in Theorem 10-4. Although this usefulness has greatly diminished, they are still of value in solving equations where the unknown appears in an exponent. The computational aspect of logarithms is presented in the remaining sections of this chapter. The use of Theorem 10-4 along these lines is illustrated in the next example.

EXAMPLE 6   Given $\log_{10}3 = 0.4771$ and $\log_{10}4 = 0.6020$, use Theorem 10-4 to find the value of each logarithm.

(a) $\log_{10}12$
(b) $\log_{10}2$
(c) $\log_{10}25$

SOLUTION   Using the fact that $\log_{10}3^x4^y = x\log_{10}3 + y\log_{10}4$, we can find $\log_{10}N$ for any number of the form $N = 3^x4^y$.

(a) $\log_{10}(3)(4) = \log_{10}3 + \log_{10}4 = 0.4771 + 0.6020 = 1.0791$
(b) $\log_{10}2 = \log_{10}(4)^{1/2} = \frac{1}{2}\log_{10}4 = \frac{1}{2}(0.6020) = 0.3010$
(c) $\log_{10}25 = \log_{10}\frac{100}{4} = \log_{10}100 - \log_{10}4 = 2 - 0.6020 = 1.3980$

Theorem 10-4 can be used to change from one form of logarithmic expression to another. Examples are provided below.

EXAMPLE 7   (a) Express $\log_a\sqrt{\dfrac{x^2y^3}{z^4}}$ in terms of logarithms of x, y, and z.

(b) Write the expression
$$\log_a6xy^3 - 2\log_a3xy^2z$$
as a single logarithm.

SOLUTION   (a) We have
$$\log_a\sqrt{\frac{x^2y^3}{z^4}} = \log_a\left(\frac{x^2y^3}{z^4}\right)^{1/2}$$
$$= \tfrac{1}{2}\log_a\left(\frac{x^2y^3}{z^4}\right)$$
$$= \tfrac{1}{2}(\log_a x^2y^3 - \log_a z^4)$$
$$= \tfrac{1}{2}(\log_a x^2 + \log_a y^3 - \log_a z^4)$$
$$= \tfrac{1}{2}(2\log_a x + 3\log_a y - 4\log_a z)$$
$$= \log_a x + \tfrac{3}{2}\log_a y - 2\log_a z.$$

(b) By Theorem 10-4,
$$\log_a6xy^3 - 2\log_a3xy^2z = \log_a6xy^3 - \log_a(3xy^2z)^2$$
$$= \log_a\frac{6xy^3}{(3xy^2z)^2}$$
$$= \log_a\frac{2}{3xyz^2}.$$

Express each equation in logarithmic form.

1. $4^2 = 16$

2. $5^3 = 125$

3. $3^4 = 81$

4. $2^4 = 16$

5. $3^{-2} = \frac{1}{9}$

6. $4^{-3} = \frac{1}{64}$

7. $(10)^{-2} = \frac{1}{100}$

8. $8^{-1/3} = \frac{1}{2}$

9. $5^0 = 1$

10. $2^1 = 2$

11. $4^{-1/2} = \frac{1}{2}$

12. $(64)^{1/6} = 2$

Express each equation in exponential form.

13. $\log_2 64 = 6$

14. $\log_2 \frac{1}{8} = -3$

15. $\log_3 \frac{1}{27} = -3$

16. $\log_{64} \frac{1}{2} = -\frac{1}{6}$

17. $\log_{3/4} \frac{9}{16} = 2$

18. $\log_{1/3} 9 = -2$

19. $\log_{10}(0.01) = -2$

20. $\log_{10} 100 = 2$

Solve for a, x, or y.

21. $y = \log_2 1$

22. $y = \log_5 5$

23. $y = \log_2 8$

24. $y = \log_7 49$

25. $y = \log_{1/3} \frac{1}{9}$

26. $y = \log_{10} 1000$

27. $2 = \log_3 x$

28. $3 = \log_2 x$

29. $\log_3 x = -4$

30. $\log_4 x = -3$

31. $y = \log_{1/10} 100$

32. $y = \log_{1/3} 81$

33. $y = \log_{1/3} 9$

34. $y = \log_{1/2} 4$

35. $3 = \log_a 8$

36. $4 = \log_a 625$

37. $\log_a 125 = -3$

38. $\log_a 100 = -2$

39. $y = \log_5 \frac{1}{125}$

40. $y = \log_5 \frac{1}{25}$

41. $y = \log_2 4$

42. $y = \log_{1/3} 9$

Given that $\log_{10} 3 = 0.4771$ and $\log_{10} 5 = 0.6990$, use Theorem 10-4 to find the values of the following logarithms. (*Hint:* First write each number as $3^x \cdot 5^y$ for suitable x and y.)

43. $\log_{10} 15$

44. $\log_{10} 45$

45. $\log_{10} 27$

46. $\log_{10} 25$

47. $\log_{10}(5)^{1/2}$

48. $\log_{10} \sqrt{3}$

49. $\log_{10} \frac{3}{5}$

50. $\log_{10} \frac{5}{3}$

51. $\log_{10} \frac{9}{25}$

52. $\log_{10} \frac{125}{27}$

Graph each function.

53. $y = \log_3 x$ 54. $y = \log_5 x$

55. $y = \log_{1/3} x$ 56. $y = \log_{1/5} x$

57. $y = \log_8 x$ 58. $y = \log_{10} x$

59. $y = \log_3(x + 4)$ 60. $y = \log_{10}(x - 3)$

Express each of the following in terms of logarithms of x, y, and z.

61. $\log_a xy$ 62. $\log_a xyz$

63. $\log_a \dfrac{x^2 y^3}{z^4}$ 64. $\log_a \dfrac{x^3 y^4}{z^2}$

65. $\log_a \dfrac{\sqrt[5]{xz^2}}{z}$ 66. $\log_a \dfrac{\sqrt[3]{x^4 y^2}}{\sqrt{z^5}}$

Write each of the following as a single logarithm.

67. $\log_a x - 2 \log_a y + \frac{1}{2} \log_a z$

68. $\log_a 18x^2 + 3 \log_a z - \frac{1}{3} \log_a 6y$

69. $\log_a x^2 y + 2 \log_a 5xy^3 - \log_a 10x^2 y^2$

70. $\log_a x^3 y^4 - 3 \log_a 4y^2 z + \log_a 8x^2 yz$

71. $\frac{2}{3} \log_a 8x^2 z^3 + \log_a 3 + \frac{1}{3} \log_a 27x^4 y^6 z^9$

72. $\frac{3}{4} \log_a 8x^2 y^4 + \log_a 5 + \frac{1}{4} \log_a 8x^6 z^4$

*73. Use Theorem 10-1 to prove part (a) of Theorem 10-4.

*74. Use Theorem 10-1 to prove part (b) of Theorem 10-4.

*75. Sometimes it is desirable to change the base of a logarithm. Show that

$$\log_b x = \frac{1}{\log_a b} \cdot \log_a x.$$

(*Hint:* $b^y = x$ implies that $y \log_a b = \log_a x$.)

*76. Use Problem 75 to show that

$$\log_a b = \frac{1}{\log_b a}$$

whenever the expressions involved are defined.

10-3
Common
Logarithms
(**Optional**)

As mentioned in the last section, we concern ourselves for the remainder of this chapter with the use of logarithms as an aid in numerical computations. Since our number system uses the base 10, logarithms to base 10 are the most useful for computational purposes, and they are referred to as *common logarithms*.

For the rest of our work with logarithms, we deal primarily with logarithms to the base 10. For this reason, we adopt the convention that log x indicates $\log_{10}x$. That is, the base is understood to be 10 unless it is specified otherwise.

In order to use common logarithms in performing computations, one must be able to find the value of log x for any positive number x. As a starting point, we observe that

$$\log 0.01 = \log 10^{-2} = -2,$$
$$\log 0.1 = \log 10^{-1} = -1,$$
$$\log 1 = \log 10^{0} = 0,$$
$$\log 10 = \log 10^{1} = 1,$$
$$\log 100 = \log 10^{2} = 2.$$

When we recall (see Figure 10.5) the characteristic shape of the graph of $y = \log_a x$ for $a > 1$, it is clear that

$$\text{for } 0.1 < x < 1, \qquad -1 < \log x < 0;$$
$$\text{for } 1 < x < 10, \qquad 0 < \log x < 1;$$
$$\text{for } 10 < x < 100, \qquad 1 < \log x < 2;$$

and so on. Now any positive number N can be written as the product of a power of 10 and a number between 1 and 10: $N = a \times 10^n$, where $1 \le a < 10$. For example,

$$5520 = 5.52 \times 10^3$$

and

$$0.0436 = 4.36 \times 10^{-2}.$$

The expression $a \times 10^n$, where $1 \le a < 10$, is called the *scientific notation* for the number N. The equality $N = a \times 10^n$ means that log N differs from log a by an integer, since

$$\log N = \log (a \times 10^n)$$
$$= \log a + \log 10^n$$
$$= \log a + n$$
$$= n + \log a.$$

With the same numbers as above, we would have

$$\log 5520 = \log (5.52 \times 10^3)$$
$$= \log 5.52 + \log 10^3$$
$$= 3 + \log 5.52,$$

and

$$\log 0.0436 = -2 + \log 4.36.$$

The significance of this is that the logarithms of numbers between 1 and 10 can be used to find the logarithms of all positive numbers. The Table of Four-Place Common Logarithms (printed on the back end papers) gives the values of logarithms of numbers from 1 to 9.99, at intervals of 0.01. A portion of this table is reproduced in Figure 10.7. The logarithm of a three-digit number[1] between 1 and 10 can be found in the table by locating the first two digits of the number in the column under N, and the last digit in the column headings at the top of the table. The logarithm of the number is located in the row of the table which starts with the first two digits, and in the column which has the last digit at the top.

N	0	1	2	3	4	5	6	7	8	9
5.5	.7404	.7412	.7419	.7427	.7435	.7443	.7451	.7459	.7466	.7474
5.6	.7482	.7490	.7497	.7505	.7513	.7520	.7528	.7536	.7543	.7551
5.7	.7559	.7566	.7574	.7582	.7589	.7597	.7604	.7612	.7619	.7627
5.8	.7634	.7642	.7649	.7657	.7664	.7672	.7679	.7686	.7694	.7701
5.9	.7709	.7716	.7723	.7731	.7738	.7745	.7752	.7760	.7767	.7774

FIGURE 10.7

EXAMPLE 1   Use the log table to find the value of

$$\log 5.84.$$

SOLUTION   In the log table, or in the portion of the table reproduced in Figure 10.7, log 5.84 is located by matching up the row which has 5.8 under N and the column which has 4 at the top. The appropriate row and column are shaded in Figure 10.7. The value of log 5.84 is found at their intersection:

$$\log 5.84 = 0.7664.$$

Actually, the value 0.7664 is an approximation to four decimal places of log 5.84, as are most of the values in the log table. It would be more precise to write $\log 5.84 \approx 0.7664$, but we choose to write "$=$" instead of "$\approx$" as a matter of convenience.

We have noted before that any positive number N can be written in scientific notation as $N = a \times 10^n$, where $1 \le a < 10$, and consequently

$$\log N = n + \log a,$$

where $0 \le \log a < 1$. Thus log N can be expressed as the sum of an integer and a nonnegative decimal fraction less than 1. The integral part, n, of the logarithm is called the *characteristic*, and the fractional part,

[1]By a three-digit number, we mean that there are three digits when the number is written in scientific notation. Zeros used only to place the decimal are not counted.

log a, is called the *mantissa*. The integral part can be found by writing the number in scientific notation, and the mantissa for three-digit numbers can be read from the log table, with accuracy to four decimal places.

EXAMPLE 2   Use the log table to find the value of
$$\log 584.$$

SOLUTION   Writing 584 in scientific notation, we have
$$584 = 5.84 \times 10^2,$$
so the characteristic is 2, and the mantissa is
$$\log 5.84 = 0.7664,$$
from the log table. Thus
$$\log 584 = 2 + 0.7664$$
$$= 2.7664.$$

It may happen, of course, that the characteristic is a negative number.

EXAMPLE 3   Use the log table to find the value of
$$\log 0.00584.$$

SOLUTION   Writing the number in scientific notation, we have
$$0.00584 = 5.84 \times 10^{-3},$$
and the characteristic is -3. As in Example 1, the mantissa is 0.7664, so
$$\log 0.00584 = -3 + 0.7664.$$
The convention in a situation like this is to avoid combining the negative integer and the positive fraction, and write the logarithm in the form
$$\log 0.00584 = 7.7664 - 10.$$

As it was done here, a negative characteristic is usually written as a positive integer minus a multiple of 10. We shall see in Section 10-4 that this is very convenient in numerical computations.

In order to perform calculations by use of logarithms, one must be able to use the tables to find a number N when $\log N$ is known.

EXAMPLE 4   Find the number N, given that
$$\log N = 8.7657 - 10.$$

The digits in N are determined by the mantissa 0.7657, and the position of the decimal in N is determined by the characteristic, which is $8 - 10 = -2$. To find the digits in N, we search through the body of the log table until we find the entry 0.7657 (note that the mantissa increases as N increases in the table). This entry is found in the row which starts with 5.8, and in the column headed by 3. Thus

$$N = 5.83 \times 10^{-2}$$
$$= 0.0583.$$

When a number N is found by using log N, the number N is referred to as the *antilogarithm* of log N (abbreviated antilog). Thus, in Example 4, we would say that

$$\text{antilog } (8.7657 - 10) = 0.0583.$$

As mentioned earlier, the log table directly furnishes logarithms only for numbers with three digits. Tables exist which give logarithms for numbers with more than three digits However, the log table can be used to approximate the logarithm of a four-digit number with very good accuracy, employing a procedure called *linear interpolation*.

To illustrate the procedure of linear interpolation, let us consider the problem of finding the value of log 10.37. The two numbers nearest 10.37 which have logarithms in the log table are 10.30 and 10.40. Their logarithms are given by

$$\log 10.30 = 1.0128$$

and

$$\log 10.40 = 1.0170.$$

Thinking geometrically, this means that the points $P(10.30, 1.0128)$ and $Q(10.40, 1.0170)$ are on the graph of $y = \log x$. A portion of the graph which contains these points is shown in Figure 10.8. The coordinates at R on the curve are $(10.37, \log 10.37)$. The idea behind linear interpolation is this: use the straight-line segment PQ as an approximation to the curve $y = \log x$, and use the ordinate at point S as an approximation to log 10.37. The ordinate at S can be found by adding the difference d to the ordinate at P. Since 10.37 is $\frac{7}{10}$ of the distance from 10.30 to 10.40, the difference d is $\frac{7}{10}$ of the difference between the ordinates at P and Q. These differences can be set up as follows.

$$10\left[7\left[\begin{array}{l} \log 10.30 = 1.0128 \\ \log 10.37 = \underline{\hspace{2cm}} \\ \log 10.40 = 1.0170 \end{array}\right] d \right] 0.0042$$

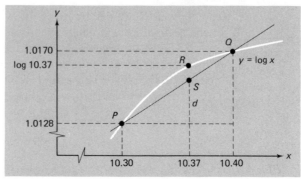

FIGURE 10.8

To find d, we use the proportion

$$\frac{d}{0.0042} = \frac{7}{10}$$

and obtain

$$d = (\tfrac{7}{10})(0.0042)$$
$$= 0.00294$$
$$= 0.0029,$$

where d is rounded to the number of decimal places in the table. Adding d to the ordinate at P, we have

$$\log 10.37 = 1.0128 + 0.0029$$
$$= 1.0157.$$

(A hand calculator gives $\log 10.37 = 1.0157788$.)

EXAMPLE 5 Use linear interpolation to find a value for $\log 0.3864$.

SOLUTION Since $0.3864 = 3.864 \times 10^{-1}$, the characteristic is -1. The nearest numbers with logarithms which can be read from the log table are 0.3860 and 0.3870. Using the same procedure as before, we obtain the following arrangement:

$$10\begin{bmatrix}4\begin{bmatrix}\log 0.3860 = 9.5866 - 10 \\ \log 0.3864 = \underline{\hspace{2cm}}\end{bmatrix}d \\ \log 0.3870 = 9.5877 - 10\end{bmatrix}0.0011$$

$$\frac{d}{0.0011} = \frac{4}{10}$$
$$d = (\tfrac{4}{10})(0.0011)$$
$$= 0.00044$$
$$= 0.0004.$$

Adding 0.0004 to log 0.3860, we have

$$\log 0.3864 = (9.5866 - 10) + 0.0004$$
$$= 9.5870 - 10.$$

(A hand calculator gives $\log 0.3864 = -0.41296288$.)

The next example shows how linear interpolation can be used in finding antilogarithms.

EXAMPLE 6 If $\log N = 8.4089 - 10$, use interpolation to approximate N to four digits.

SOLUTION The mantissa .4089 is located between .4082 and .4099 in the tables, and these mantissas correspond to the digits 256 and 257. Thus we have the following arrangement:

$$10\left[x\left[\begin{matrix} \log 0.02560 = 8.4082 - 10 \\ \log N \quad\quad = 8.4089 - 10 \\ \log 0.02570 = 8.4099 - 10 \end{matrix}\right.0.0007\right]0.0017\right.$$

$$\frac{x}{10} = \frac{0.0007}{0.0017}$$

$$x = \tfrac{70}{17}$$

$$= 4.1$$

$$= 4.$$

The number x represents the last digit in N, and is rounded to the nearest whole number. Thus we have

$$N = 0.02564.$$

(A hand calculator gives $N = 0.02563894$.)

EXERCISES 10-3

Find the common logarithms of each number, using the log table as necessary.

1. 4.16×10^{-9} 2. 1.73×10^3

3. 30.7   4. 17.3

5. 4.51   6. 3.01

7. 10.0   8. 10^{-4}

9. 1,070,000   10. 10,800

11. 0.00107   12. 0.000132

Find N in each equation, using the log table as necessary.

13. $\log N = 0.8561$ 14. $\log N = 0.9186$

15. $\log N = 8.4518$ 16. $\log N = 4.6385$

17. $\log N = 7.6776 - 10$ 18. $\log N = 6.9943 - 10$

19. $\log N = -3$ 20. $\log N = 4$

Use the log table and linear interpolation to find the logarithm of each number. Check the accuracy with a hand calculator if one is available.

21. 10.11   22. 243.6

23. 4.171   24. 1.017

25. 417,800   26. 7,103

27. 0.007717   28. 0.08354

Use the log table and linear interpolation to find N in each equation. Check the accuracy with a hand calculator if one is available.

29. $\log N = 0.1113$ 30. $\log N = 0.7549$

31. $\log N = 4.9433$ 32. $\log N = 5.4686$

33. $\log N = 7.6950 - 10$ 34. $\log N = 8.9963 - 10$

35. $\log N = 9.3881 - 10$ 36. $\log N = 9.5240 - 10$

10-4
Computations
with Logarithms
(Optional)

The properties of logarithms stated in Theorem 10-4 are the basis for their usefulness in computational work. For easy reference, we restate these properties for logarithms with base 10 in the following theorem.

THEOREM 10-5   If u, v, and r are real numbers with u and v positive, then

> (a) $\log (u \cdot v) = \log u + \log v$;
> (b) $\log \dfrac{u}{v} = \log u - \log v$;
> (c) $\log u^r = r \log u$.

In our first example, we illustrate the use of part (a) of the theorem with a simple computation.

EXAMPLE 1   Use logarithms to compute the value of $(3.85)(705)$ as a three-digit number.

SOLUTION   As a notational convenience, let

$$N = (3.85)(705).$$

Then
$$\log N = \log 3.85 + \log 705.$$

Using the log table, we find that
$$\log 3.85 = 0.5855,$$
$$\log 705 = 2.8482.$$

Adding these logarithms, we have
$$\log N = 3.4337.$$

The mantissa .4337 is not found in the log table, but we do not interpolate since we only need accuracy to three digits. The mantissa nearest .4337 in the table is .4330, which occurs with the digits 271. Thus we write
$$N = 2.71 \times 10^3$$
$$= 2710.$$

The zero in N is not a significant digit since it is used only to place the decimal. (The actual product is 2714.25.)

EXAMPLE 2 Use the log table and linear interpolation to compute the value of
$$\frac{\sqrt[3]{37.12}}{(2.931)^4},$$

with four-digit accuracy.

SOLUTION Let
$$N = \frac{\sqrt[3]{37.12}}{(2.931)^4}.$$

By parts (b) and (c) of Theorem 10-5, we have
$$\log N = \log \sqrt[3]{37.12} - \log (2.931)^4$$
$$= \tfrac{1}{3} \log 37.12 - 4 \log 2.931.$$

The interpolation to find log 37.12 is as follows:
$$10 \begin{bmatrix} 2 \begin{bmatrix} \log 37.10 = 1.5694 \\ \log 37.12 = \underline{\qquad} \end{bmatrix} d \\ \log 37.20 = 1.5705 \end{bmatrix} 0.0011$$

$$\frac{d}{0.0011} = \frac{2}{10}$$

$$d = (\tfrac{2}{10})(0.0011)$$

$$= 0.0002$$

$$\log 37.12 = 1.5694 + 0.0002$$

$$= 1.5696.$$

Similarly, we find that

$$\log 2.931 = 0.4670.$$

Thus

$$\log N = \tfrac{1}{3}(1.5696) - 4(0.4670)$$

$$= 0.5232 - 1.8680.$$

At this point, we *do not* perform the subtraction as it is indicated, because this yields a number with a *negative* decimal fraction, and the log table requires *positive* decimal fractions as mantissas. To retain a logarithm with a positive decimal part, we proceed as follows:

$$\log N = 0.5232 - 1.8680$$

$$= (10.5232 - 10) - 1.8680$$

$$= (10.5232 - 1.8680) - 10$$

$$= 8.6552 - 10.$$

To find $N = $ antilog $(8.6552 - 10)$, we use interpolation. (With a little practice, interpolation can be performed mentally.)

$$10\left[x\begin{bmatrix} \log 0.04520 = 8.6551 - 10 \\ \log N \quad\;\; = 8.6552 - 10 \end{bmatrix}0.0001 \right]0.0010$$

$$\log 0.04530 = 8.6561 - 10$$

$$x = 1$$

$$N = 0.04521$$

(A calculator gives the answer 0.0452001.)

The work in Example 2 illustrates one of the "fine points" in using logarithms for computation, in that it shows how negative decimal fractions are avoided in the values of logarithms. Another "fine point" is brought out in Example 3.

EXAMPLE 3   Use logarithms to compute

$$N = \frac{(4.16)\sqrt[3]{0.0235}}{(0.0787)^{3/2}}$$

as a three-digit number.

SOLUTION   Using all three parts of Theorem 10-5, we have

$$\log N = \log [(4.16)\sqrt[3]{0.0235}] - \log (0.0787)^{3/2}$$

$$= \log 4.16 + \log \sqrt[3]{0.0235} - \log (0.0787)^{3/2}$$

$$= \log 4.16 + \tfrac{1}{3} \log 0.0235 - \tfrac{3}{2} \log 0.0787$$

$$= 0.6191 + \tfrac{1}{3}(8.3711 - 10) - \tfrac{3}{2}(8.8960 - 10).$$

The "fine point" comes up first in the multiplication
$$\tfrac{1}{3}(8.3711 - 10).$$

If each term in parentheses is multiplied by $\tfrac{1}{3}$, the resulting difference involves two decimal fractions instead of one. This trouble is avoided by adding and subtracting a multiple of 10 which is chosen so as to make the last term evenly divisible by 3. We write

$$\tfrac{1}{3}(8.3711 - 10) = \tfrac{1}{3}(28.3711 - 30)$$
$$= 9.4570 - 10.$$

Similarly,

$$\tfrac{3}{2}(8.8960 - 10) = \tfrac{3}{2}(18.8960 - 20)$$
$$= 3(9.4480 - 10)$$
$$= 28.3440 - 30$$
$$= 8.3440 - 10.$$

Substituting these values in the equation for log N, we have

$$\log N = 0.6191 + (9.4570 - 10) - (8.3440 - 10)$$
$$= (10.0761 - 10) - (8.3440 - 10)$$
$$= 1.7321.$$

To three digits, this gives

$$N = \text{antilog } (1.7321)$$
$$= 54.0.$$

(A calculator gives 53.970179.)

EXERCISES 10-4

Use logarithms to compute a three-digit value for each of the following quantities.

1. $(2.17)(30.1)$

2. $(11.7)(4.91)$

3. $\dfrac{6.01}{31.7}$

4. $\dfrac{23.1}{3.08}$

5. $(4.63)^5$

6. $(40.5)^4$

7. $\sqrt[7]{112}$

8. $\sqrt[5]{17.1}$

9. $\dfrac{13.1}{\sqrt{41.1}}$

10. $\dfrac{\sqrt[3]{3.92}}{2.46}$

11. $\dfrac{(8.12)\sqrt[3]{0.147}}{(39.1)^2}$

12. $\dfrac{(0.0417)^2}{(0.374)\sqrt[5]{0.195}}$

Use logarithms and linear interpolation to compute a four-digit value for each quantity. Check the accuracy with a calculator if one is available.

13. $(0.006213)^3(429.5)^2$

14. $(0.1735)(66.17)^2$

15. $\sqrt[6]{0.3704}$

16. $\sqrt[5]{0.6713}$

17. $\dfrac{(27.11)^{1/3}}{(14.13)^{1/2}}$

18. $\dfrac{\sqrt[5]{171.3}}{\sqrt[7]{317.1}}$

19. $\dfrac{(4.813)^2(17.13)}{(0.3612)^{1/3}}$

20. $\dfrac{(0.3179)^{1/3} \cdot (41.95)^2}{(29.71)^4}$

21. Use logarithms to approximate 2^{32} with four significant digits.

22. Use logarithms to approximate $3^{\sqrt{3}}$ with four significant digits. (Use $\sqrt{3} = 1.732$.)

In problems 23 and 24, compute a four-digit value of the reciprocal of each number.

23. 2.173

24. 31.19

25. Use logarithms to compute a three-digit value of $57.3/\log 696$.

26. Use logarithms to evaluate

$$\frac{(-3.17) \cdot (6.13)}{(-4.19) \cdot (-3.11)}$$

as a three-digit number.

27. The volume V of a right circular cone with radius r and altitude h is given by $V = \frac{1}{3}\pi r^2 h$. A certain conical funnel has radius 12.8 centimeters and altitude 48.2 centimeters. Compute the volume of the funnel with three-digit accuracy.

28. The area A of a triangle which has sides of lengths a, b, and c is given by $A = \sqrt{s(s-a)(s-b)(s-c)}$, where s is one-half of the perimeter. A triangular-shaped plot of land has sides of length 112 meters, 121 meters, and 157 meters. Find the area of the plot of land.

10-5
*Exponential
and Logarithmic
Equations*

An equation which contains an exponential function is called an *exponential equation*. Such equations can frequently be solved by using logarithms.

EXAMPLE 1   Solve for x if

$$4^x = 23.$$

SOLUTION   The given equation is equivalent to the statement that $x = \log_4 23$, but this is no help in obtaining a decimal value for x. To obtain a decimal value of x, we first take the logarithm of both sides of $4^x = 23$:

$$\log 4^x = \log 23,$$

or

$$x \log 4 = \log 23,$$

by Theorem 10-5(c). This readily yields

$$x = \frac{\log 23}{\log 4}$$

$$= \frac{1.3617}{0.6021}.$$

Using logarithms to compute x, we have

$$\log x = \log 1.3617 - \log 0.6021$$

$$= 0.1341 - (9.7797 - 10)$$

$$= 0.3544.$$

This gives

$$x = 2.262.$$

EXAMPLE 2   Solve the equation

$$2^{2x+1} = 3^{x+4}.$$

SOLUTION   Taking the logarithm of each side of the equation, we have

$$(2x + 1) \log 2 = (x + 4) \log 3,$$

or

$$2x \log 2 + \log 2 = x \log 3 + 4 \log 3.$$

Solving for x, we have

$$(2 \log 2 - \log 3)x = 4 \log 3 - \log 2,$$

and

$$x = \frac{4 \log 3 - \log 2}{2 \log 2 - \log 3}$$

$$= \frac{1.6075}{0.1249}.$$

This quotient can be evaluated by use of logarithms, yielding $x = 12.87$ as the solution. The first form of x can be used with a hand calculator.

An equation which contains logarithmic functions is called a *logarithmic equation*. Many logarithmic equations can be solved by using Theorem 10-5 and the definition of a logarithm. Such an equation is solved in Example 3.

EXAMPLE 3   Solve the equation

$$\log x - \log (2x - 1) = 2.$$

SOLUTION   Using Theorem 10-5(b), the left side can be written as a single logarithm:

$$\log \frac{x}{2x - 1} = 2.$$

By the definition of a logarithm, this means that

$$\frac{x}{2x - 1} = 10^2 = 100.$$

Therefore,

$$x = 100(2x - 1),$$

and this is equivalent to

$$199x = 100.$$

Thus $x = 100/199 = 0.503$ is the solution of the equation.

An important application of exponential equations is the formula for computing the value of an investment, or *original principal*, P, when interest is added to the principal at the end of certain periods of time, so that the accumulated interest also earns interest in the next period of time. Interest rates are usually stated at an annual rate r, and interest is converted to principal, or *compounded*, a specified number n times per year. The *rate per period* is then r/n, and the original investment P, together with accumulated interest, has the value

$$A = P\left(1 + \frac{r}{n}\right)^{tn}$$

after t years. The total amount A is called the *compound amount*.

EXAMPLE 4 If an original investment of $1000 grows to a value of $1500 in 5 years when interest is compounded quarterly at a certain annual rate r, find r.

SOLUTION Using the formula for compound amount with $A = \$1500$, $P = \$1000$, $n = 4$, and $t = 5$, we have

$$1500 = 1000\left(1 + \frac{r}{4}\right)^{20}.$$

This gives

$$1.5 = (1 + 0.25r)^{20}.$$

Taking the logarithm of each side, we have

$$\log 1.5 = 20 \log (1 + 0.25r),$$

and

$$\log (1 + 0.25r) = \frac{\log 1.5}{20}$$

$$= \frac{0.1761}{20}$$

$$= 0.0088.$$

Thus

$$1 + 0.25r = \text{antilog}\,(0.0088)$$

$$= 1.02,$$

and we have

$$r = \frac{0.02}{0.25}$$

$$= 0.08.$$

That is, the annual interest rate is 8%.

EXERCISES 10-5

Solve for x in each equation.

1. $3^x = 17$ 2. $2^x = 5$

3. $5^x = \frac{1}{10}$ 4. $6^x = 10$

5. $2^{x-1} = 64$ 6. $3^{2x-1} = 243$

7. $4^{1-x} = 19$ 8. $7^{-x+2} = 6$

9. $9^{2x^2-1} = 11$ 10. $5^{x^2+2} = 39$

11. $7 = 6 \cdot 2^{-x+2}$ 12. $15 = 8 \cdot 3^{2x-1}$

13. $9^{x+2} = 7^{3x-1}$ 14. $5^x = 6^{x-1}$

15. $\log (2x + 1) - \log 5 = \log x$ 16. $\log 2x - \log 5 = 2$

17. $\log x^2 - \log 9 = \log x$ 18. $\log x^2 - \log 4 = \log x$

19. $\log x + \log (3x - 7) = 1$ 20. $\log (3x - 16) + \log (x - 1) = 1$

21. $\log_2 x - \log_2(x - 1) = 3$ 22. $\log_3 x - \log_3(x - 1) = 2$

23. $(1 + x)^4 = 3.12$ 24. $(1 + x)^6 = 1.34$

25. $50\left(1 + \dfrac{x}{4}\right)^{60} = 75.3$ 26. $(1 - 2x)^{1.2} = 4$

27. $(1 + 0.06)^x = 3.17$ 28. $60(1 + 0.02)^{4x} = 130$

The following problems require use of the compound amount formula

$$A = P\left(1 + \frac{r}{n}\right)^{nt}.$$

29. Find the compound amount A of $P = \$1000$ compounded quarterly for 5 years at an annual rate of 5%.

30. If $P = \$1000$ amounts to $\$1250$ in 2 years and interest is compounded semiannually, find the annual interest rate r.

31. How long will it take an original principal P to double if it is invested at 5.5% compounded semiannually?

32. How much money must Leslie invest today in order to accumulate $\$3000$ for a trip to France in 4 years if she can invest her money at 8% compounded annually?

1. If $f(x) = (\frac{2}{3})^x$, find the following function values.
   (a) $f(-2)$
   (b) $f(\frac{1}{2})$

2. Solve for x in the equation
$$(81)^x = \tfrac{1}{27}.$$

3. Sketch the graph of the function $f(x) = 2^{-x}$.

4. Solve for x in each equation.
   (a) $\log_4 x = 2$
   (b) $x = \log_2 \frac{1}{8}$

5. Graph the function $y = \log_3 x$.

6. (a) Express $\log_a \left(\frac{x^2 z}{y^3}\right)$ in terms of logarithms of x, y, and z.
   (b) Write the following expression as a single logarithm.
$$2 \log_a x^2 yz + 3 \log_a 2xyz^2 - \log_a 4x^3 y^2 z^4$$

7. Use the log table and linear interpolation to find the value of x in each equation.
   (a) $x = \log 421.6$
   (b) $\log x = 8.5155 - 10$

8. Use logarithms to compute a three-digit value for
$$\frac{26.4\sqrt{593}}{(7.87)^2}.$$

Find a three-digit value for x in each equation.

9. $5^{x+1} = 192$

10. $\log x + \log (3x + 1) = 1$

Sequences and Series

11-1
Definitions
Higher-level mathematics makes extensive use of certain types of functions, called sequences. Informally, a sequence is an ordered list in which there is a first element, a second element, a third element, and so on. Some familiar sequences are the sequence of positive integers

$$1, 2, 3, 4, \ldots$$

and the sequence of even positive integers

$$2, 4, 6, 8, \ldots.$$

These are examples of infinite sequences. Sequences which have a last, or terminal, element are called "finite sequences." An example of a finite sequence could be the ordered listing

$$2, 5, 8, 11, 14, 17.$$

These ideas are formalized in Definition 11-1.

DEFINITION 11-1 A *finite sequence* is the range of a function which has a domain of the form $\{1, 2, 3, \ldots, k\}$, where k is a fixed integer. An *infinite sequence* is the range of a function which has the set of all positive integers as its domain.

According to this definition, a finite sequence has the form

$$f(1), f(2), f(3), \ldots, f(k),$$

where f is the function which has the set $\{1, 2, 3, \ldots, k\}$ as its domain. Similarly, an infinite sequence with f as its defining function has the form

$$f(1), f(2), f(3), \ldots,$$

where the dots at the end indicate that there is no terminal element. In working with sequences, it is traditional to use a subscripted letter, such as a_n, instead of $f(n)$ to indicate the function value at the positive integer n. A finite sequence is written as

$$a_1, a_2, a_3, \ldots, a_k,$$

and an infinite sequence is written as

$$a_1, a_2, a_3, \ldots.$$

With this notation, a_n stands for the nth term, or the *general term*, in the sequence. Thus a_1 denotes the first term, a_2 denotes the second term, a_{18} denotes the eighteenth term, and so on.

EXAMPLE 1 Write the first four terms of the sequence which has the given general term.

(a) $a_n = \dfrac{1}{n}$

(b) $a_n = 2n - 1$

(c) $b_n = (-1)^n \dfrac{n}{n + 1}$

SOLUTION To find the first four terms of each sequence, we assign the values 1, 2, 3, 4 to n in succession. This gives

(a) $a_1 = 1, a_2 = \frac{1}{2}, a_3 = \frac{1}{3}, a_4 = \frac{1}{4}$;

(b) $a_1 = 1, a_2 = 3, a_3 = 5, a_4 = 7$;

(c) $b_1 = -\frac{1}{2}, b_2 = \frac{2}{3}, b_3 = -\frac{3}{4}, b_4 = \frac{4}{5}$.

Occasionally, a sequence is defined by specifying the first few terms and a rule for finding the nth term from the preceding terms. Consider the following example.

EXAMPLE 2   Find the first six terms of each of the sequences below.

$\quad$ (a) $a_1 = 1, a_n = a_{n-1} + 2$
$\quad$ (b) $a_1 = 1, a_2 = 1, a_{n+1} = a_n + a_{n-1}$

SOLUTION $\qquad$ (a) We have $a_1 = 1$, and

$$a_2 = a_1 + 2 = 1 + 2 = 3,$$
$$a_3 = a_2 + 2 = 3 + 2 = 5,$$
$$a_4 = a_3 + 2 = 5 + 2 = 7,$$
$$a_5 = a_4 + 2 = 7 + 2 = 9,$$
$$a_6 = a_5 + 2 = 9 + 2 = 11.$$

This gives the first six terms of the sequence as 1, 3, 5, 7, 9, 11.

$\quad$ (b) We have $a_1 = 1$ and $a_2 = 1$. Thus

$$a_3 = a_2 + a_1 = 1 + 1 = 2,$$
$$a_4 = a_3 + a_2 = 2 + 1 = 3,$$
$$a_5 = a_4 + a_3 = 3 + 2 = 5,$$
$$a_6 = a_5 + a_4 = 5 + 3 = 8,$$

and the first six terms are 1, 1, 2, 3, 5, 8.

DEFINITION 11-2   A *series* is the sum of the terms of a sequence.

In this section, we consider only series which are formed from a finite sequence. Such a series has the form

$$a_1 + a_2 + a_3 + \cdots + a_k.$$

It is possible in some cases for the sum of an infinite series to have meaning. Some series of this type are considered in Section 11-4.

If a formula for the general term is known, it is common practice to write a series in a compact form called the *sigma notation*. Using this notation, the series

$$a_1 + a_2 + a_3 + \cdots + a_k$$

is written $\sum_{n=1}^{k} a_n$. That is,

$$\sum_{n=1}^{k} a_n = a_1 + a_2 + a_3 + \cdots + a_k.$$

The capital Greek letter Σ (sigma) is used to indicate a *sum*, and the notations at the bottom and top of the sigma give the initial and terminal values of n. The letter n is called the *index of summation*.

EXAMPLE 3 Write the following series in expanded form, and find the value of the sum.

(a) $\displaystyle\sum_{n=1}^{5} (2n - 1)$

(b) $\displaystyle\sum_{n=1}^{6} \frac{1}{2^{n-1}}$

SOLUTION (a) The given expression represents the series which has terms that are obtained by substituting, in succession, the values 1, 2, 3, 4, and 5 for n in $(2n - 1)$. Thus

$$\sum_{n=1}^{5} (2n - 1) = 1 + 3 + 5 + 7 + 9$$
$$= 25.$$

(b) We have

$$\sum_{n=1}^{6} \frac{1}{2^{n-1}} = \frac{1}{2^0} + \frac{1}{2^1} + \frac{1}{2^2} + \frac{1}{2^3} + \frac{1}{2^4} + \frac{1}{2^5}$$
$$= 1 + \tfrac{1}{2} + \tfrac{1}{4} + \tfrac{1}{8} + \tfrac{1}{16} + \tfrac{1}{32}$$
$$= \tfrac{63}{32}.$$

There are two points which need to be made in connection with the sigma notation. First, we note that the index of summation is an arbitrary symbol, or a *dummy variable*. That is,

$$\sum_{n=1}^{k} a_n = \sum_{i=1}^{k} a_i = \sum_{j=1}^{k} a_j,$$

since each of these notations represents the sum

$$a_1 + a_2 + a_3 + \cdots + a_k.$$

The second point is that the initial value of the index is not necessarily 1. For example,

$$\sum_{j=3}^{5} \frac{1}{2^j} = \frac{1}{2^3} + \frac{1}{2^4} + \frac{1}{2^5}$$
$$= \tfrac{1}{8} + \tfrac{1}{16} + \tfrac{1}{32}$$
$$= \tfrac{7}{32}.$$

EXERCISES 11-1

In problems 1–18, write the first five terms of the sequence which has the given general term.

1. $a_n = \dfrac{1}{n + 1}$

2. $a_n = 2n^2 - 1$

3. $a_n = \dfrac{n-1}{n}$

4. $a_n = \dfrac{n^2 - 1}{2n}$

5. $a_j = (-2)^j$

6. $a_j = (-1)^j$

7. $a_n = 2$

8. $a_n = 3$

9. $a_i = \dfrac{(-1)^i}{i^2}$

10. $a_i = (-1)^i i^2$

11. $a_j = \dfrac{j}{2j - 1}$

12. $a_j = \dfrac{j}{j+1}$

13. $a_i = 3i - 2$

14. $a_i = 5i + 3$

15. $a_n = (x - 1)^n$

16. $a_n = x^{n-1}, \; x \neq 0$

17. $a_j = (-1)^j x^{2j}$

18. $a_j = (-1)^j x^{2j-1}$

Write the first six terms of each sequence.

19. $a_1 = 1, \quad a_{n+1} = 2a_n$

20. $a_1 = 2, \quad a_{n+1} = a_n + 3$

21. $a_1 = -5, \quad a_{n+1} = (-1)^n a_n$

22. $a_1 = 1, \quad a_{n+1} = x \cdot a_n$

23. $a_1 = 2, \quad a_2 = 3, \quad a_{n+1} = 2a_n + a_{n-1}$

24. $a_1 = -1, \quad a_2 = 1, \quad a_{n+1} = 2a_n - a_{n-1}$

Write each series in expanded form, and find the value of the sum.

25. $\displaystyle\sum_{n=1}^{5} 2^n$

26. $\displaystyle\sum_{n=1}^{7} (-2)^n$

27. $\displaystyle\sum_{n=3}^{9} \dfrac{n-2}{n+1}$

28. $\displaystyle\sum_{n=2}^{9} \dfrac{n-1}{n}$

29. $\displaystyle\sum_{j=1}^{7} \dfrac{j^2 - j}{2}$

30. $\displaystyle\sum_{j=1}^{11} \dfrac{j+1}{2}$

31. $\displaystyle\sum_{i=1}^{7} (-1)^i$

32. $\displaystyle\sum_{i=4}^{10} 10^i$

33. $\displaystyle\sum_{n=2}^{7} 2$

34. $\displaystyle\sum_{n=2}^{6} 3$

35. $\displaystyle\sum_{j=2}^{7} j^2 (-1)^j$

36. $\displaystyle\sum_{j=-1}^{5} j$

37. $\displaystyle\sum_{i=-1}^{5} i^3$

38. $\displaystyle\sum_{i=2}^{5} 6i(-1)^i$

39. $\displaystyle\sum_{n=1}^{6} \left(\dfrac{1}{n} - \dfrac{1}{n+1} \right)$

40. $\displaystyle\sum_{n=1}^{5} \left(\dfrac{1}{2^n} - \dfrac{1}{2^{n+1}} \right)$

41. $\displaystyle\sum_{n=2}^{5} (-1)^n \cdot 2^{-n}$

42. $\displaystyle\sum_{n=2}^{7} 3^{-n}$

An arithmetic progression is a special type of sequence, defined as follows.

> **DEFINITION 11-3** An *arithmetic progression* is a sequence in which each term after the first is obtained by adding the same number, d, to the preceding term. The constant d is called the *common difference*.

Thus an arithmetic progression with first term a_1 and common difference d is given by

$$a_1, a_1 + d, a_1 + 2d, a_1 + 3d, \ldots.$$

EXAMPLE 1 Find the first five terms of the arithmetic progression which has the first term and common difference as specified below.

(a) $a_1 = -2, d = 3$
(b) $a_1 = 5, d = -2$

SOLUTION (a) The first five terms are computed as follows:

$$a_1 = -2,$$
$$a_2 = a_1 + d = -2 + 3 = 1,$$
$$a_3 = a_2 + d = 1 + 3 = 4,$$
$$a_4 = a_3 + d = 4 + 3 = 7,$$
$$a_5 = a_4 + d = 7 + 3 = 10.$$

(b) We have

$$a_1 = 5,$$
$$a_2 = a_1 + d = 5 + (-2) = 3,$$
$$a_3 = a_2 + d = 3 + (-2) = 1,$$
$$a_4 = a_3 + d = 1 + (-2) = -1,$$
$$a_5 = a_4 + d = -1 + (-2) = -3.$$

Thus the first five terms of the arithmetic progression in (b) are given by

$$5, 3, 1, -1, -3.$$

It is easy to arrive at a formula for the nth term in an arithmetic progression. Observing the pattern in

$$a_1 = a_1,$$
$$a_2 = a_1 + d,$$
$$a_3 = a_1 + 2d,$$
$$a_4 = a_1 + 3d,$$
$$a_5 = a_1 + 4d,$$

we see that the coefficient of d is always one less than the number of the term, and

$$a_n = a_1 + (n-1)d. \tag{1}$$

EXAMPLE 2 Find the nineteenth term and a formula for the nth term of the following arithmetic progressions.

(a) $-1, 2, 5, 8, \ldots$
(b) $3, 1, -1, -3, \ldots$

SOLUTION

(a) The nineteenth term of the sequence is $a_{19} = a_1 + 18d$. To find the common difference, we need only subtract one of the terms from the succeeding term. Using the first two terms, we get

$$d = 2 - (-1) = 3.$$

Since $a_1 = -1$, we have

$$a_{19} = -1 + (18)(3) = 53,$$

and

$$a_n = -1 + (n-1)(3) = 3n - 4.$$

(b) The first term here is $a_1 = 3$, and the common difference is $d = 1 - 3 = -2$. Thus

$$a_{19} = a_1 + 18d$$
$$= 3 + (18)(-2)$$
$$= -33,$$

and

$$a_n = a_1 + (n-1)d$$
$$= 3 + (n-1)(-2)$$
$$= 5 - 2n.$$

EXAMPLE 3 A certain arithmetic progression has $a_5 = 5$ and $a_{13} = 21$. Find the first term a_1, the common difference d, and a_{25}.

SOLUTION Using the formula $a_n = a_1 + (n-1)d$ with $a_5 = 5$ and $a_{13} = 21$, we have

$$a_1 + 4d = 5,$$
$$a_1 + 12d = 21.$$

Subtracting the top equation from the bottom one, we get

$$8d = 16,$$
$$d = 2.$$

Using $d = 2$ in $a_1 + 4d = 5$, we have

$$a_1 + 8 = 5,$$
$$a_1 = -3.$$

Then

$$a_{25} = -3 + (24)(2)$$
$$= 45,$$

and this completes the problem.

The series formed by adding the first n terms of a general arithmetic progression is denoted by S_n:

$$S_n = a_1 + a_2 + a_3 + \cdots + a_n.$$

A formula for S_n in terms of a_1 and a_n may be obtained as follows. First we write the sum S_n in the form

$$S_n = a_1 + (a_1 + d) + (a_1 + 2d) + \cdots + [a_1 + (n-1)d]. \qquad (2)$$

We then write the same sum in reverse order, subtracting d from each term to get the next one.

$$S_n = a_n + (a_n - d) + (a_n - 2d) + \cdots + [a_n - (n-1)d] \qquad (3)$$

Adding equations (2) and (3), we have

$$2S_n = (a_1 + a_n) + (a_1 + a_n) + (a_1 + a_n) + \cdots + (a_1 + a_n). \qquad (4)$$

The right member of equation (4) has n of the terms $a_1 + a_n$, so

$$2S_n = n(a_1 + a_n)$$

and

$$S_n = \frac{n}{2}(a_1 + a_n). \qquad (5)$$

We can obtain another formula for S_n in terms of a_1 and d by substituting $a_n = a_1 + (n-1)d$ in this formula. This substitution gives

$$S_n = \frac{n}{2}[a_1 + a_1 + (n-1)d],$$

or

$$S_n = \frac{n}{2}[2a_1 + (n-1)d]. \qquad (6)$$

For convenient reference, we record the formulas relating to arithmetic progressions in the following theorem.

THEOREM 11-4   In an arithmetic progression with first term a_1 and common difference d,

(a) the nth term is given by

$$a_n = a_1 + (n-1)d,$$

and

(b) the sum of the first n terms is given by

$$S_n = \frac{n}{2}(a_1 + a_n),$$

or

$$S_n = \frac{n}{2}[2a_1 + (n-1)d].$$

EXAMPLE 4   Use the information given for each arithmetic progression to find the twentieth term and the sum of the first 20 terms.

(a) $-3, -6, -9, -12, \ldots$
(b) $a_1 = 5, a_{19} = 59$
(c) $S_{10} = 145, a_7 = 19$

SOLUTION   (a) It is clear that $a_1 = -3$ and $d = -3$. By the formula in Theorem 11-4(a), we have

$$a_{20} = a_1 + 19d$$
$$= -3 + (19)(-3)$$
$$= -60.$$

By the second formula in Theorem 11-4(b),

$$S_{20} = \tfrac{20}{2}(2a_1 + 19d)$$
$$= 10(-6 - 57)$$
$$= -630.$$

(b) Substituting $a_1 = 5$ and $a_{19} = 59$ in the formula from Theorem 11-4(a), we have

$$59 = 5 + 18d,$$

and

$$d = 3.$$

Thus

$$a_{20} = a_{19} + d = 59 + 3 = 62,$$

and

$$S_{20} = \tfrac{20}{2}(a_1 + a_{20})$$
$$= 10(5 + 62)$$
$$= 670.$$

(c) Using $S_{10} = 145$ in the second formula from Theorem 11-4(b), we have

$$145 = \tfrac{10}{2}(2a_1 + 9d),$$

and this gives

$$145 = 10a_1 + 45d. \tag{7}$$

Using $a_7 = 19$ in Theorem 11-4(a), we have

$$19 = a_1 + 6d. \tag{8}$$

To find a_1 and d, we solve equations (7) and (8) simultaneously. Multiplying both sides of equation (8) by 10 and rewriting equation (7), we have

$$190 = 10a_1 + 60d,$$
$$145 = 10a_1 + 45d.$$

Subtraction and division yield

$$d = 3.$$

Substituting $d = 3$ in equation (8) yields

$$a_1 = 1.$$

Thus

$$a_{20} = a_1 + 19d$$
$$= 1 + (19)(3)$$
$$= 58,$$

and

$$S_{20} = \tfrac{20}{2}(2a_1 + 19d)$$
$$= 10[2 + (19)(3)]$$
$$= 590.$$

Note that this computation of S_{20} is a lot shorter than actually adding $1 + 4 + 7 + 10 + \cdots + 58$, and is less prone to errors in arithmetic.

EXERCISES 11-2

Use the information given to find the first six terms of an arithmetic progression which satisfies the stated conditions.

1. $a_1 = 1, d = 1$ 2. $a_1 = -3, d = -2$

3. $a_4 = 7, d = -2$ 4. $a_4 = -7, d = -5$

5. $a_3 = -5, a_6 = -17$ 6. $a_3 = 6, a_4 = 5.9$

7. $a_3 - a_2 = 4, a_4 = 14$ 8. $a_5 = -7, d = 2.$

Determine whether or not each sequence is an arithmetic progression. If it is, find d and a_n.

9. $-17, -12, -7, -2, \ldots$ 10. $5, 2, -1, -4, \ldots$

11. $-6, -9, -12, -15, \ldots$ 12. $\frac{5}{2}, 2, \frac{3}{2}, 1, \ldots$

13. $8, 4, 2, 1, \ldots$ 14. $2, 4, 8, 16, \ldots$

15. $7, -7, -21, -35, \ldots$ 16. $3, 3.04, 3.08, 3.12, \ldots$

17. $x, x + y, x + 2y, x + 3y, \ldots$

18. $-k, -k + x, -k + 2x, -k + 3x, \ldots$

19. $x, x^2, x^3, x^4, \ldots$ 20. $x, x/2, x/4, x/8, \ldots$

21. $x, 2x, 3x, 4x, \ldots$ 22. $2x, 4x, 6x, 8x, \ldots$

Use the information given to find a_{11} and S_{13} for an arithmetic progression which satisfies the stated conditions.

23. $a_1 = -5, d = -2$ 24. $a_1 = 7, a_5 = 15$

25. $a_7 = -5, d = 3$ 26. $a_5 = -3, d = -3$

27. $a_5 = 5, a_9 = -7$ 28. $a_3 = 10, a_7 = -2$

29. $a_1 = m, d = -x$ 30. $a_4 = m, d = -x$

31. $a_5 = x + 2k, a_7 = x - 2k$ 32. $a_5 = 3 - 4p, d = p$

Use the information given to find the value of the requested quantities in an arithmetic progression.

33. $a_1 = -7, d = 3$; find a_{21}.   34. $a_1 = 17, d = -5$; find a_{17}.

35. $a_3 = -7, d = 16$; find S_4.   36. $a_{10} = -\frac{5}{2}, d = \frac{1}{2}$; find S_{17}.

37. $-4, -1, 2, 5, \ldots$; find a_{10}.   38. $\frac{5}{2}, \frac{3}{2}, \frac{1}{2}, -\frac{1}{2}, \ldots$; find a_{15}.

39. $a_1 = 16, a_{17} = -3$; find S_{57}.   40. $a_1 = -3, a_{11} = 16$; find S_{200}

41. $S_5 = -30, a_7 = -4$; find a_1 and d.

42. $S_5 = \frac{5}{2}, a_7 = -\frac{7}{2}$; find a_1 and d.

43. $S_7 = 14, a_{10} = -28$; find a_1 and d.

44. $S_8 = 52, a_4 = 5$; find a_1 and d.

Find the sum of each of the following series, using the fact that each is formed from an arithmetic progression. (See Exercises 53 and 54 below.)

45. $\sum_{i=5}^{11} (2i - 3)$ 46. $\sum_{i=3}^{16} (3i - 2)$

47. $\sum_{k=0}^{10} \left(5 - \frac{k}{3}\right)$ 48. $\sum_{k=0}^{20} \left(-3 - \frac{k}{2}\right)$

49. $\sum_{n=1}^{50} 2n$ 50. $\sum_{n=1}^{100} n$

51. $\sum_{n=1}^{13} (5 - 2n)$ 52. $\sum_{n=1}^{19} (2n - \frac{11}{2})$

*53. Prove that if $a_1, a_2, a_3, \ldots$ is a sequence which has a defining function f that is linear [i.e., $a_n = f(n) = an + b$], then the sequence is an arithmetic progression.

*54. Prove that every arithmetic progression has a defining function that is linear (i.e., $a_n = an + b$ for some fixed a and b).

55. A pile of logs is in the shape of a trapezoid with 50 logs on the bottom, 49 logs next to the bottom, 48 logs on the next level, and so on, with 21 logs on the top. Find the total number of logs in the pile.

56. In constructing a candy Christmas tree on a flat plate, Leslie places one candy at the top, Patti places 2 candies just below, then Leslie places 3 candies below these two, and they continue, increasing by one candy at each level. If the bottom row has 27 candies in it, how many candies were used to construct the tree?

57. Insert numbers in the blanks below so that the sequence forms an arithmetic progression.

$$-15, \underline{\quad}, \underline{\quad}, \underline{\quad}, \underline{\quad}, -40$$

58. Insert seven numbers between -5 and -26 so that the sequence thus formed is an arithmetic progression.

59. Find the sum of the multiples of 5 between 7 and 213.

60. Find the sum of the integers from -13 to -71, inclusive.

61. Find the number n of positive integers between 17 and 199 which are multiples of 5.

62. Find the number n of positive integers between 17 and 199 which are multiples of 3.

11-3
Geometric
Progressions

In this section, we consider another special type of sequence, called geometric progressions.

DEFINITION 11-5   A *geometric progression* is a sequence in which each term after the first is obtained by multiplying the preceding term by a fixed nonzero number r. The constant r is called the *common ratio*.

A geometric progression with first term a_1 and common ratio r is given by

$$a_1, \quad a_1 r, \quad a_1 r^2, \quad a_1 r^3, \quad \ldots$$

EXAMPLE 1   Find the first four terms of the geometric progression which has the first term and common ratio as specified below.

(a) $a_1 = -3, r = -2$

(b) $a_1 = 16, r = \frac{1}{4}$

SOLUTION
(a) The first four terms of the geometric progression are computed as follows.

$$a_1 = -3,$$
$$a_2 = a_1 r = (-3)(-2) = 6,$$
$$a_3 = a_1 r^2 = (-3)(-2)^2 = -12,$$
$$a_4 = a_1 r^3 = (-3)(-2)^3 = 24.$$

Thus the first four terms of the geometric progression are $-3, 6, -12, 24$.

(b) To illustrate another method of computation, this time we obtain the terms after the first one by multiplying by r.

$$a_1 = 16,$$
$$a_2 = a_1 r = (16)(\tfrac{1}{4}) = 4,$$
$$a_3 = a_2 r = (4)(\tfrac{1}{4}) = 1,$$
$$a_4 = a_3 r = (1)(\tfrac{1}{4}) = \tfrac{1}{4}$$

Thus the first four terms are given by $16, 4, 1, \tfrac{1}{4}$.

From the pattern in the terms

$$a_1, \quad a_1 r, \quad a_1 r^2, \quad a_1 r^3, \quad a_1 r^4, \quad \dots$$

of a geometric progression, it is easy to see that the nth term in a geometric progression is given by the formula

$$a_n = a_1 r^{n-1}. \tag{1}$$

EXAMPLE 2 Find the sixth term and a formula for the nth term of the following geometric progressions.

(a) $1, 2, 4, 8, \dots$
(b) $a_1 = 4, r = -\tfrac{1}{2}$

SOLUTION
(a) To find r, we need only choose one of the terms, other than the first, and divide it by the preceding term. Using the third and second terms, we obtain $r = \tfrac{4}{2} = 2$. Since $a_1 = 1$, we have

$$a_6 = a_1 r^5 = (1)(2)^5 = 32,$$

and

$$a_n = a_1 r^{n-1} = (1)(2)^{n-1} = 2^{n-1}.$$

(b) Substituting $a_1 = 4$ and $r = -\tfrac{1}{2}$ in equation (1), we have

$$a_6 = (4)(-\tfrac{1}{2})^5 = -\tfrac{1}{8},$$

and

$$a_n = (4)(-\tfrac{1}{2})^{n-1} = \frac{4}{(-2)^{n-1}}.$$

EXAMPLE 3 A geometric progression has $a_4 = -3$ and $a_7 = -24$. Find the first term a_1, and the common ratio r.

SOLUTION Using the formula $a_n = a_1 r^{n-1}$ with $a_4 = -3$ and $a_7 = -24$, we have

$$-3 = a_1 r^3,$$
$$-24 = a_1 r^6.$$

We can solve for r by forming quotients of corresponding sides in these two equations:

$$\frac{a_1 r^6}{a_1 r^3} = \frac{-24}{-3}.$$

This gives

$$r^3 = 8,$$

and

$$r = 2.$$

Substituting $r = 2$ and $a_4 = -3$ in $a_4 = a_1 r^3$, we have

$$-3 = a_1(2)^3$$

and

$$a_1 = -\tfrac{3}{8}.$$

As was the case with arithmetic progressions, we can find a formula for the series formed by adding the first n terms of a general geometric progression. To obtain the formula, let

$$S_n = a_1 + a_1 r + a_1 r^2 + \cdots + a_1 r^{n-3} + a_1 r^{n-2} + a_1 r^{n-1}. \qquad (2)$$

When both sides of this equation are multiplied by r, we have

$$r S_n = a_1 r + a_1 r^2 + a_1 r^3 + \cdots + a_1 r^{n-2} + a_1 r^{n-1} + a_1 r^n. \qquad (3)$$

Subtracting equation (3) from equation (2) yields

$$S_n - r S_n = a_1 - a_1 r^n,$$

or

$$(1 - r)S_n = a_1(1 - r^n).$$

If $r \neq 1$, both sides of this equation can be divided by $1 - r$ to obtain

$$S_n = a_1 \frac{1 - r^n}{1 - r} \qquad \text{if } r \neq 1. \qquad (4)$$

This formula cannot be used if $r = 1$, but equation (2) easily gives $S_n = na_1$ for $r = 1$.

For easy reference, the formulas in equations (1) and (4) are stated in the following theorem.

THEOREM 11-6   In a geometric progression with first term a_1 and common ratio r,

(a) the nth term is

$$a_n = a_1 r^{n-1},$$

and

(b) the sum of the first n terms is

$$S_n = a_1 \frac{1 - r^n}{1 - r} \qquad \text{if } r \neq 1.$$

EXAMPLE 4   Find the sum of the first six terms in the geometric progression

$$-\tfrac{1}{3}, -\tfrac{1}{9}, -\tfrac{1}{27}, -\tfrac{1}{81}, \ldots.$$

SOLUTION   Using $a_1 = -\tfrac{1}{3}$ and $r = \tfrac{1}{3}$ in the formula from Theorem 11-6(b), we have

$$S_6 = -\frac{1}{3} \cdot \frac{1 - (\tfrac{1}{3})^6}{1 - \tfrac{1}{3}}$$

$$= -\frac{1}{3} \cdot \frac{1 - \tfrac{1}{729}}{1 - \tfrac{1}{3}}$$

$$= -\frac{1}{3} \cdot \frac{\tfrac{728}{729}}{\tfrac{2}{3}}$$

$$= -\frac{364}{729}.$$

EXAMPLE 5   Use Theorem 11-6 to find the value of

$$\sum_{i=1}^{5} 3 \cdot 2^i.$$

SOLUTION   Expanding the sum, we have

$$\sum_{i=1}^{5} 3 \cdot 2^i = 3 \cdot 2 + 3 \cdot 2^2 + 3 \cdot 2^3 + 3 \cdot 2^4 + 3 \cdot 2^5.$$

From this expansion, it is clear that the terms in the sum form a geometric progression with $a_1 = 3 \cdot 2 = 6$ and $r = 2$. By Theorem 11-6(b),

$$\sum_{i=1}^{5} 3 \cdot 2^i = 6 \cdot \frac{1 - 2^5}{1 - 2}$$

$$= 6 \cdot \frac{-31}{-1}$$

$$= 186.$$

This use of Theorem 11-6 is a lot shorter than doing the addition.

For the given value of n, write out the first n terms of the geometric progression which satisfies the stated conditions.

1. $a_1 = -1, r = -2, n = 5$
2. $a_1 = 3, r = \frac{1}{2}, n = 4$
3. $a_1 = 4, r = \frac{1}{4}, n = 4$
4. $a_1 = -6, r = 2, n = 5$
5. $a_1 = \frac{3}{4}, r = 4, n = 4$
6. $a_1 = -\frac{1}{8}, r = -2, n = 5$
7. $a_2 = -\frac{1}{2}, a_3 = 1, n = 5$
8. $a_3 = \frac{2}{3}, a_4 = 1, n = 4$
9. $a_1 = -3, a_3 = -12, n = 3$
10. $a_1 = 4, a_3 = \frac{1}{16}, n = 4$

Find the fifth term, the nth term, and the sum of the first five terms of a geometric progression which satisfies the stated conditions.

11. $\frac{1}{4}, \frac{1}{2}, 1, 2, \ldots$
12. $-1, -\sqrt{3}, -3, -3\sqrt{3}, \ldots$
13. $a_1 = \frac{1}{3}, r = -3$
14. $a_1 = 4, r = \frac{1}{4}$
15. $a_3 = -2, a_4 = 4$
16. $a_3 = -3, a_4 = \frac{3}{2}$

Determine whether or not each of the following sequences is a geometric progression. If it is, find r and a_n.

17. $7, \frac{7}{2}, \frac{7}{4}, \frac{7}{8}, \ldots$
18. $\sqrt{3}, 3, 3\sqrt{3}, 9, \ldots$
19. $4, 2\sqrt{2}, 2, \sqrt{2}, \ldots$
20. $\frac{5}{6}, \frac{5}{3}, \frac{10}{3}, \frac{20}{3}, \ldots$
21. $2, -4, 6, -8, \ldots$
22. $1, 3, 7, 15, \ldots$
23. $1, 4, 8, 12, \ldots$
24. $3, 6, 18, 108, \ldots$
25. $-4, 2, -1, \frac{1}{2}, \ldots$
26. $-5, \frac{5}{3}, -\frac{5}{9}, \frac{5}{27}, \ldots$
27. $343, -49, 7, -1, \ldots$
28. $\frac{1}{9}, -\frac{1}{3}, 1, -3, \ldots$

Find the sums of each of the following series, using the fact that each is a geometric progression.

29. $\sum_{i=1}^{5} 2^{i-1}$
30. $\sum_{n=0}^{5} 4^n$
31. $\sum_{i=1}^{5} (\frac{3}{5})^i$
32. $\sum_{n=4}^{9} \frac{1}{3^n}$
33. $\sum_{n=0}^{4} 5(-\frac{2}{3})^n$
34. $\sum_{j=1}^{6} 128(-\frac{3}{2})^j$
35. $\sum_{j=0}^{4} 64(\frac{5}{4})^j$
36. $\sum_{n=2}^{7} 9(\frac{5}{3})^{n-2}$
37. $\sum_{j=1}^{6} 8 \cdot 2^{1-j}$
38. $\sum_{n=2}^{6} 32(\frac{1}{2})^{n-1}$

Find r and a_1 for the geometric progression which satisfies the given conditions.

39. $a_7 = -9, a_{11} = -81$
40. $a_4 = -\frac{2}{3}, a_7 = -\frac{9}{4}$

41. $a_6 = 4(1.01)^4$, $a_8 = 4(1.01)^6$ 42. $a_4 = 2$, $a_8 = \frac{2}{81}$

43. Leslie was determined to save some money for a vacation. So beginning Nov. 1, she saved 1¢, and on each succeeding day she saved twice as much as the day before. How much did Leslie save on November 15? What was the total amount saved if she stopped saving on November 15?

44. It is known that the population of a country in 1920 was 10,000,000 people. If the population doubles every 20 years, what will the population be in 2000?

45. At a certain rate of inflation, the cost of living doubles every 6 years. If a person earns $12,000 now, how much must she earn 36 years later to keep her salary in pace with inflation?

46. The number of bacteria in a culture is observed to triple every hour. If initially there are 1000 bacteria, approximately how long will it take for 1,000,000 bacteria to be present?

11-4
Infinite Geometric
Progressions

Up to this point, we have considered only series which are formed from a finite sequence. In this section, we examine some cases where the sum of an infinite series has meaning.

Consider an infinite geometric progression with first term a_1 and common ratio r:

$$a_1, \quad a_1 r, \quad a_1 r^2, \quad a_1 r^3, \quad \ldots .$$

We have seen in Section 11-3 that the sum S_n of the first n terms of this progression is given by

$$S_n = a_1 + a_1 r + a_1 r^2 + \cdots + a_1 r^{n-1}$$
$$= a_1 \frac{1 - r^n}{1 - r}, \quad \text{if } r \neq 1.$$

If $|r| < 1$, that is, if $-1 < r < 1$, the term r^n steadily decreases in absolute value as n increases, getting closer and closer to 0. The fact that r^n gets nearer and nearer to 0 as n takes on larger and larger values suggests that the sums S_n should themselves be getting closer and closer to some certain value S. This is actually what happens, and we write

$$\lim_{n \to \infty} S_n = S$$

to indicate that the sums S_n get closer and closer to S as n increases without bound.[1] The symbol "$\lim_{n \to \infty} S_n$" is read "limit of S_n as n approaches infinity."

To illustrate how the sums S_n behave when $|r| < 1$, consider the particular case where $a_1 = 1$ and $r = \frac{1}{2}$. The geometric progression is

---

[1]The concept of a limit is the fundamental concept of the calculus. Limits are treated there with a great deal of care and rigor.

given by

$$1, \tfrac{1}{2}, \tfrac{1}{4}, \tfrac{1}{8}, \tfrac{1}{16}, \ldots,$$

and the sum of the first n terms is

$$S_n = 1 + \frac{1}{2} + \frac{1}{4} + \cdots + \frac{1}{2^{n-1}}$$

$$= 1 \cdot \frac{1 - (\tfrac{1}{2})^n}{1 - \tfrac{1}{2}}$$

$$= 2[1 - (\tfrac{1}{2})^n].$$

As n takes on the values 1, 2, 3, 4, 5, 6, ... in succession, the corresponding values of S_n are given by

$$S_1 = 1 = 2(1 - \tfrac{1}{2}) = 2 - 1,$$
$$S_2 = 1 + \tfrac{1}{2} = 2(1 - \tfrac{1}{4}) = 2 - \tfrac{1}{2},$$
$$S_3 = 1 + \tfrac{1}{2} + \tfrac{1}{4} = 2(1 - \tfrac{1}{8}) = 2 - \tfrac{1}{4},$$
$$S_4 = 1 + \tfrac{1}{2} + \tfrac{1}{4} + \tfrac{1}{8} = 2(1 - \tfrac{1}{16}) = 2 - \tfrac{1}{8},$$
$$S_5 = 1 + \tfrac{1}{2} + \tfrac{1}{4} + \tfrac{1}{8} + \tfrac{1}{16} = 2(1 - \tfrac{1}{32}) = 2 - \tfrac{1}{16},$$
$$S_6 = 1 + \tfrac{1}{2} + \tfrac{1}{4} + \tfrac{1}{8} + \tfrac{1}{16} + \tfrac{1}{32} = 2(1 - \tfrac{1}{64}) = 2 - \tfrac{1}{32}.$$

From these computations it can be seen that, as n increases without bound, the sums S_n get closer and closer to the value 2. Thus

$$\lim_{n \to \infty} S_n = 2$$

for this geometric progression. We say that 2 is the *sum* of the infinite geometric progression, and we write

$$1 + \tfrac{1}{2} + \tfrac{1}{4} + \tfrac{1}{8} + \cdots = 2,$$

even though it is impossible to find the sum by actually performing the indicated additions. Another notation which is commonly used is

$$\sum_{n=1}^{\infty} \frac{1}{2^{n-1}} = 2.$$

The development in the particular case above can be carried out for any geometric progression with $|r| < 1$. Since r^n gets closer and closer to 0 as n increases without bound, we write

$$\lim_{n \to \infty} r^n = 0.$$

With this fact in mind, it is clear that

$$\lim_{n \to \infty} S_n = \lim_{n \to \infty} a_1 \frac{1 - r^n}{1 - r}.$$

$$= \frac{a_1}{1 - r}.$$

If $|r| \geq 1$ and $a_1 \neq 0$, the terms $a_1 r^n$ increase in absolute value without bound as n increases without bound, and the sums S_n do also. This discussion is summarized in the following theorem.

THEOREM 11-7   For an infinite geometric progression $a_1, a_1 r$, $a_1 r^2, \ldots$, let

$$S_n = a_1 + a_1 r + a_1 r^2 + \cdots + a_1 r^{n-1}.$$

(a) If $|r| < 1$, the sum $a_1 + a_1 r + a_1 r^2 + \cdots$ of the infinite geometric progression is given by

$$S = \lim_{n \to \infty} S_n = \frac{a_1}{1 - r}.$$

(b) If $|r| \geq 1$ and $a_1 \neq 0$, the sum of the series $\sum\limits_{n=1}^{\infty} a_1 r^{n-1}$ does not exist.

In the case where $|r| < 1$, we also write

$$a_1 + a_1 r + a_1 r^2 + \cdots = \frac{a_1}{1 - r},$$

or

$$\sum_{n=1}^{\infty} a_1 r^{n-1} = \frac{a_1}{1 - r}.$$

EXAMPLE 1   For each geometric progression below, determine whether or not the sum exists. If the sum exists, find its value.

(a) $\frac{3}{5}, \frac{1}{5}, \frac{1}{15}, \frac{1}{45}, \ldots$
(b) $\frac{2}{3}, 1, \frac{3}{2}, \frac{9}{4}, \ldots$
(c) $9, -6, 4, -\frac{8}{3}, \ldots$

SOLUTION   (a) The common ratio is given by

$$r = \frac{\frac{1}{5}}{\frac{3}{5}} = \frac{1}{3}.$$

Since $|r| < 1$, the sum exists, by Theorem 11-7(a), and

$$\frac{3}{5} + \frac{1}{5} + \frac{1}{15} + \frac{1}{45} + \cdots = \frac{a_1}{1-r}$$

$$= \frac{\frac{3}{5}}{1-\frac{1}{3}}$$

$$= \frac{9}{10}.$$

(b) It is easy to see that $r = \frac{3}{2}$. Since $|r| > 1$, the sum does not exist, by Theorem 11-7(b).

(c) The common ratio is

$$r = \frac{-6}{9} = -\frac{2}{3},$$

so the sum exists, and

$$S = \frac{a_1}{1-r}$$

$$= \frac{9}{1-(-\frac{2}{3})}$$

$$= \frac{27}{5}.$$

The formula in Theorem 11-7(a) can be used to write a repeating decimal as a quotient of integers.

EXAMPLE 2 Write the rational number

$$0.131313 \cdots$$

as a quotient of integers.

SOLUTION The given number can be written as the sum of an infinite geometric progression in the following manner.

$$0.131313 \cdots = 0.13 + 0.0013 + 0.000013 + \cdots$$

The terms in the sum on the right are from the geometric progression with $a_1 = 0.13$ and $r = 0.01$. By Theorem 11-7(a),

$$0.131313 \cdots = \frac{0.13}{1 - 0.01}$$

$$= \frac{0.13}{0.99}$$

$$= \frac{13}{99}.$$

EXAMPLE 3 Write the rational number

$$3.4173173173 \cdots$$

as a quotient of integers.

SOLUTION   We first write the number as follows:

$$3.4173173173 \cdots = 3.4 + 0.0173 + 0.0000173 + \cdots.$$

The terms on the right after 3.4 are from the geometric progression with $a_1 = 0.0173$ and $r = 0.001$. Thus

$$0.0173 + 0.0000173 + \cdots = \frac{0.0173}{1 - 0.001}$$

$$= \frac{0.0173}{0.999}$$

$$= \frac{173}{9990}.$$

This gives

$$3.4173173173 \cdots = 3.4 + \frac{173}{9990}$$

$$= \frac{34}{10} + \frac{173}{9990}$$

$$= \frac{33,966 + 173}{9990}$$

$$= \frac{34,139}{9990}.$$

This may be checked by long division.

EXERCISES 11-4

For each of the following geometric progressions, determine whether or not the sum exists. If the sum exists, find its value.

1. $-3, \frac{3}{2}, -\frac{3}{4}, \frac{3}{8}, \ldots$

2. $-4, \frac{4}{3}, -\frac{4}{9}, \frac{4}{27}, \ldots$

3. $\frac{9}{2}, -\frac{3}{2}, \frac{1}{2}, -\frac{1}{6}, \ldots$

4. $\frac{5}{9}, \frac{1}{9}, \frac{1}{45}, \frac{1}{225}, \ldots$

5. $-\frac{4}{3}, 4, -12, 36, \ldots$

6. $8, -4, 2, -1, \ldots$

7. $1, 1.01, (1.01)^2, (1.01)^3, \ldots$

8. $1, 1.2, 1.44, 1.728, \ldots$

9. $2, \sqrt{2}, 1, 1/\sqrt{2}, \ldots$

10. $-0.9, 0.81, -0.729, 0.6561, \ldots$

11. $5, 0.5, 0.05, 0.005, \ldots$

12. $5, 2.5, 1.25, 0.625, \ldots$

13. $\frac{5}{3}, \frac{10}{9}, \frac{20}{27}, \frac{40}{81}, \ldots$

14. $\frac{4}{9}, \frac{2}{3}, 1, \frac{3}{2}, \ldots$

Find the value of each of the following sums, if it exists. If it does not exist, give a reason.

15. $15 + \frac{15}{2} + \frac{15}{4} + \frac{15}{8} + \cdots$

16. $52 + 4 + \frac{4}{13} + \frac{4}{169} + \cdots$

17. $4 - \frac{4}{3} + \frac{4}{9} - \frac{4}{27} + \cdots$

18. $-17 + \frac{17}{3} - \frac{17}{9} + \frac{17}{27} - \cdots$

19. $1 - 1.02 + (1.02)^2 - (1.02)^3 + \cdots$

20. $5 + 10 + 20 + 40 + \cdots$

21. $3 - \dfrac{3}{0.99} + \dfrac{3}{(0.99)^2} - \dfrac{3}{(0.99)^3} + \cdots$

22. $0.25 + 0.50 + 1 + 2 + \cdots$

23. $512 + 64 + 8 + 1 + \cdots$

24. $\frac{4}{3} + \frac{1}{3} + \frac{1}{12} + \frac{1}{48} + \cdots$

25. $\frac{1}{36} + \frac{1}{6} + 1 + 6 + \cdots$

26. $\frac{1}{16} + \frac{1}{4} + 1 + 4 + \cdots$

27. $\sum_{i=0}^{\infty} 5(\tfrac{1}{2})^i$

28. $\sum_{i=2}^{\infty} -15(\tfrac{2}{3})^i$

29. $\sum_{i=3}^{\infty} 17(-3)^i$

30. $\sum_{n=4}^{\infty} 7(2)^n$

31. $\sum_{i=1}^{\infty} \dfrac{13}{4^i}$

32. $\sum_{n=1}^{\infty} \dfrac{8}{5^n}$

33. $\sum_{n=5}^{\infty} 8(-\tfrac{1}{5})^{n-4}$

34. $\sum_{n=3}^{\infty} -7(-\tfrac{4}{3})^{n-2}$

35. $\sum_{n=2}^{\infty} 6(\tfrac{1}{3})^n$

36. $\sum_{n=2}^{\infty} 4(\tfrac{3}{4})^n$

Express each rational number as a quotient of integers.

37. $0.9999 \cdots$

38. $0.1111 \cdots$

39. $0.010101 \cdots$

40. $0.313131 \cdots$

41. $0.013101310131 \cdots$

42. $0.171171171 \cdots$

43. $3.111111 \cdots$

44. $3.8787878 \cdots$

45. $-2.2917917 \cdots$

46. $-9.01727272 \cdots$

47. $10.1343434 \cdots$

48. $0.79191919 \cdots$

49. Matthew had a ball which, when dropped from any height h to a concrete floor, bounced back to $\frac{2}{5}$ of the height h. Approximately how far would the ball travel before coming to rest if he dropped it from a height of 2 meters?

50. Work problem 49 for a ball which bounces back each time to $\frac{1}{3}$ of the height from which it falls.

11-5
Mathematical
Induction

Mathematical induction is a method of proof which is used mainly to prove theorems which assert that a certain statement holds true for all positive integers.

The method of proof by mathematical induction is based on the following property of the positive integers: if T is a set such that

(i) 1 is in T,

(ii) $k \in T$ always implies $k + 1 \in T$,

then T contains all the positive integers.

To get some intuitive feeling for the method, assume that T is a set which satisfies the two conditions above, and let us apply the conditions a few times. We have

$$1 \in T, \text{ by condition (i)};$$
$$1 \in T \text{ implies } 1 + 1 = 2 \in T, \text{ by condition (ii)};$$
$$2 \in T \text{ implies } 2 + 1 = 3 \in T \text{ by condition (ii)};$$
$$3 \in T \text{ implies } 3 + 1 = 4 \in T, \text{ by condition (ii)};$$
$$4 \in T \text{ implies } 4 + 1 = 5 \in T, \text{ by condition (ii)};$$

and condition (ii) can be applied repeatedly, as long as we wish. By applying condition (ii) enough times, we can arrive at the statement that any given positive integer is in T. Thus T must contain all the positive integers.

A proof by mathematical induction is often compared to climbing an endless ladder. To know that we can climb to any desired step, we need only know that (i) we can climb onto the first step, and that (ii) we can climb from any step to the next higher step.

Mathematical induction is usually employed in connection with a certain statement $S(n)$ about the positive integer n. As an illustration, $S(n)$ might be the statement that

$$1 + 2 + 3 + \cdots + n = \frac{n(n+1)}{2}$$

for an arbitrary positive integer n. As another example, $S(n)$ might be the statement that

$$1 + 2n \le 3^n.$$

Such statements as these can be proved using the Principle of Mathematical Induction, stated in the following theorem.

THEOREM 11-8 (PRINCIPLE OF MATHEMATICAL INDUCTION) For each positive integer n, suppose that $S(n)$ represents a statement about n which is either true or false. If

(i) $S(1)$ is true,

and if

(ii) the truth of $S(k)$ always implies the truth of $S(k + 1)$, then $S(n)$ is true for all positive integers n.

A proof based on the Principle of Mathematical Induction consists of two parts:

1. the statement is verified for $n = 1$;
2. (a) the statement is *assumed*[2] true for $n = k$, and with this assumption made,
   (b) the statement is then proved to be true for $n = k + 1$.

It then follows that the statement is true for all positive integers n.
Two examples of this type of proof are given below.

EXAMPLE 1 Use mathematical induction to prove that

$$1 + 2 + 3 + \cdots + n = \frac{n(n+1)}{2} \tag{1}$$

for any positive integer n.

SOLUTION Let $S(n)$ be the statement that equation (1) holds true.

1. We first verify that $S(1)$ is true. (In a formula such as this, it is understood that, when $n = 1$, there is only one term on the left side, and no addition is performed in this case.) The value of the left side is 1 when $n = 1$, and the value of the right side is

$$\frac{1(1+1)}{2} = \frac{(1)(2)}{2} = 1.$$

Thus $S(1)$ is true.

2. (a) Assume that $S(k)$ is true. That is, assume that

$$1 + 2 + 3 + \cdots + k = \frac{k(k+1)}{2}. \tag{2}$$

(b) We must now prove that $S(k + 1)$ is true. By adding $k + 1$ to both sides of equation (2), we obtain

$$1 + 2 + 3 + \cdots + k + (k+1) = \frac{k(k+1)}{2} + k + 1$$

$$= \frac{k(k+1) + 2(k+1)}{2}$$

$$= \frac{(k+1)(k+2)}{2}$$

$$= \frac{(k+1)[(k+1)+1]}{2}.$$

[2]This assumption is frequently referred to as "the induction hypothesis."

This last expression is exactly the right member of equation (1) with n replaced by $k + 1$. We have shown that the truth of $S(k)$ implies the truth of $S(k + 1)$. Therefore, the statement $S(n)$ is true for all positive integers n.

EXAMPLE 2 Use mathematical induction to prove that

$$\frac{1}{1 \cdot 2} + \frac{1}{2 \cdot 3} + \frac{1}{3 \cdot 4} + \cdots + \frac{1}{n(n + 1)} = \frac{n}{n + 1} \qquad (3)$$

for any positive integer n.

SOLUTION 1. For $n = 1$, the left member of equation (3) is

$$\frac{1}{1 \cdot 2} = \frac{1}{2},$$

and the right member is

$$\frac{1}{1 + 1} = \frac{1}{2}.$$

Thus the statement is true for $n = 1$.

2. (a) Assume that

$$\frac{1}{1 \cdot 2} + \frac{1}{2 \cdot 3} + \frac{1}{3 \cdot 4} + \cdots + \frac{1}{k(k + 1)} = \frac{k}{k + 1}. \qquad (4)$$

(b) To change the left member of equation (4) to the left member of equation (3) when $n = k + 1$, we need to add

$$\frac{1}{(k + 1)[(k + 1) + 1]} = \frac{1}{(k + 1)(k + 2)}$$

to both sides of equation (4). Doing this, we have

$$\frac{1}{1 \cdot 2} + \frac{1}{2 \cdot 3} + \frac{1}{3 \cdot 4} + \cdots + \frac{1}{k(k + 1)} + \frac{1}{(k + 1)(k + 2)}$$

$$= \frac{k}{k + 1} + \frac{1}{(k + 1)(k + 2)}$$

$$= \frac{k(k + 2) + 1}{(k + 1)(k + 2)}$$

$$= \frac{k^2 + 2k + 1}{(k + 1)(k + 2)}$$

$$= \frac{(k + 1)^2}{(k + 1)(k + 2)}$$

$$= \frac{k + 1}{k + 2}$$

$$= \frac{k + 1}{(k + 1) + 1}.$$

The last expression matches the right member of equation (3) with n replaced by $k + 1$. Thus the truth of the statement for $n = k$ implies the truth of the statement for $n = k + 1$. By the Principle of Mathematical Induction, equation (3) holds for all positive integers n.

EXERCISES 11-5

Use mathematical induction to prove that each of the following statements is true for all positive integers n.

1. $1 + 3 + 5 + \cdots + (2n - 1) = n^2$

2. $2 + 4 + 6 + \cdots + (2n) = n(n + 1)$

3. $1^2 + 2^2 + 3^2 + \cdots + n^2 = \dfrac{n(n + 1)(2n + 1)}{6}$

4. $1 \cdot 2 + 2 \cdot 3 + 3 \cdot 4 + \cdots + n(n + 1) = \dfrac{n(n + 1)(n + 2)}{3}$

5. $2 + 4 + 8 + \cdots + 2^n = 2^{n+1} - 2$

6. $4 + 4^2 + 4^3 + \cdots + 4^n = \dfrac{4(4^n - 1)}{3}$

7. $4 + 8 + 12 + 16 + \cdots + 4n = 2n(n + 1)$

8. $3 + 6 + 9 + 12 + \cdots + 3n = \dfrac{3n(n + 1)}{2}$

9. $\dfrac{1}{1 \cdot 4} + \dfrac{1}{4 \cdot 7} + \dfrac{1}{7 \cdot 10} + \cdots + \dfrac{1}{(3n - 2)(3n + 1)} = \dfrac{n}{3n + 1}$

10. $\dfrac{1}{1 \cdot 2 \cdot 3} + \dfrac{1}{2 \cdot 3 \cdot 4} + \dfrac{1}{3 \cdot 4 \cdot 5} + \cdots + \dfrac{1}{n(n + 1)(n + 2)}$
    $= \dfrac{n(n + 3)}{4(n + 1)(n + 2)}$

11. $1^3 + 2^3 + 3^3 + \cdots + n^3 = \dfrac{n^2(n + 1)^2}{4}$

12. $\dfrac{1}{3} + \dfrac{1}{3^2} + \dfrac{1}{3^3} + \cdots + \dfrac{1}{3^n} = \dfrac{1}{2}\left[1 - \left(\dfrac{1}{3}\right)^n\right]$

13. $1 + 2n \leq 3^n$

14. Use mathematical induction to prove that $\overline{z^n} = (\bar{z})^n$ for any complex number z and any positive integer n.

15. Show that if the statement
    $$1 + 2 + 3 + \cdots + n = \dfrac{n(n + 1)}{2} + 2$$
    is assumed to be true for $n = k$, the same equation can be proved to hold for $n = k + 1$. Is the statement true for all positive integers?

16. Show that $n^2 - n + 5$ is a prime integer when $n = 1, 2, 3, 4$, but that it is not true that $n^2 - n + 5$ is always a prime integer. Investigate the same sort of statements for the polynomial $n^2 - n + 11$.

*17. Use mathematical induction to prove that $a - b$ is a factor of $a^n - b^n$ for every positive integer n. [*Hint:* $a^{k+1} - b^{k+1} = a^k(a - b) + (a^k - b^k)b$.]

*18. Use mathematical induction to prove that $a + b$ is a factor of $a^{2n} - b^{2n}$ for every positive integer n.

PRACTICE TEST for Chapter 11

1. Write the first five terms of the sequence which has the general term
$$a_n = \frac{n-1}{n+1}.$$

2. Write the series $\sum_{n=2}^{5} (-2)^n$ in expanded form, and find the value of the sum.

In problems 3 and 4, use the information given to find the value of the requested quantities in an arithmetic progression.

3. $a_4 = 6$, $a_9 = -4$; find a_{11} and S_{11}.

4. $a_6 = 8$, $S_8 = 40$; find a_1 and d.

5. Find a_5 and S_5 for a geometric progression which has $a_3 = -2$, $a_4 = 3$.

6. Determine whether or not each sequence is a geometric progression. If it is, find r and a_n.
   (a) $24, -36, 54, -81, \ldots$ (b) $3, 4, 12, 20, \ldots$

7. Find r and a_1 for the geometric progression which has $a_3 = 81$ and $a_6 = -24$.

8. Find the value of $\sum_{n=1}^{\infty} 25(-\frac{2}{3})^n$, if it exists. If it does not exist, give a reason.

9. Express $3.212121 \cdots$ as a quotient of integers.

10. Use mathematical induction to prove that
$$\frac{1}{1 \cdot 3} + \frac{1}{3 \cdot 5} + \frac{1}{5 \cdot 7} + \cdots + \frac{1}{(2n-1)(2n+1)} = \frac{n}{2n+1}.$$

Further Topics

12-1
The Counting
Principle

In the study of certain branches of mathematics, including probability, we are often faced with the problem of determining the number of ways a certain act can be performed, or a certain selection can be made. As a simple example, suppose that a man has a wardrobe which contains a white, a tan, and a brown shirt, and a green and a gold tie. From these, he wishes to make a selection of a shirt and a tie. Since he can choose any of the 3 shirts and either of the 2 ties, there are $(3) \cdot (2)$, or 6, different ways he can make the selection of a shirt and tie. This is an example of the use of the Counting Principle given in Theorem 12-1. These possible selections are presented in Figure 12.1. A diagram of this form is called a *tree*. The number of smaller "branches" is the number of different selections that can be made, and each possible selection is apparent in the tree. For example, the third from the top of the smaller branches indicates a selection of a tan shirt and a green tie.

THEOREM 12-1 (THE COUNTING PRINCIPLE) Suppose that one selection can be made in m ways, and after that selection is made, a second

339

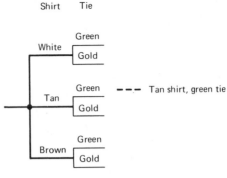

FIGURE 12.1

selection can be made in n ways, where the result of the first selection does not influence the result of the second selection. Then the two selections can be made in that order in $m \cdot n$ ways.

The Counting Principle can be extended for any finite number of selections. We make use of this in the following examples.

EXAMPLE 1 Suppose that identification tags are to be made using two letters from among the letters a, b, c, d, e, and one of the numbers 1, 2, 3, 4, 5, 6, 7, 8, 9. How many distinct tags can be made if (a) any letter can be repeated, and (b) no letter can be repeated?

SOLUTION The tag has the form given by Figure 12.2.

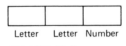

Letter   Letter   Number

FIGURE 12.2

(a) For the first position to be filled by one of the 5 letters, there are 5 choices. There are also 5 choices of letters to be put in the second position. For the last position, there are 9 choices of numbers. Using the Counting Principle, there are

$$5 \cdot 5 \cdot 9 = 225$$

different tags of the form described.

(b) Any one of the 5 letters can be used in the first position. But no letter can be repeated, so once a letter is chosen for the first position, the same letter cannot be used in the second position. Hence there are only 4 choices of a letter for the second position. Any one of the 9 numbers can be used for the last position. Using the Counting Principle, we see that there are

$$5 \cdot 4 \cdot 9 = 180$$

different tags where no letter is repeated.

EXAMPLE 2 Suppose that a contest consists of two parts: first, an ordinary cubical die is rolled, and next, a coin is tossed. Find the total number of possible results in the contest.

SOLUTION Since a cubical die has 6 distinct faces, and any one may face up when it is rolled, there are 6 possible results for the first part of the contest. Similarly, there are 2 possible results when the coin is tossed in the second part of the contest. Thus there is a total of

$$6 \cdot 2 = 12$$

possible results in the contest.

EXERCISES 12-1

1. A boy's clothes consist of 4 pairs of slacks, 5 shirts, and 3 jackets. How many different ways can he choose a pair of slacks, a shirt, and a jacket?

2. In choosing a name for their daughter, a couple settles on 4 first names and 6 middle names as possibilities. How many different names can they select from these possibilities?

3. To travel from town A to town B a traveler can take any of 4 routes. To travel from town B to town C, he can take any of 3 routes. In how many ways can he travel from town A to town C if he must pass through town B?

4. A college offers a freshman 4 history courses, 3 math courses, 5 English courses, and 2 geography courses. If a student takes one course in each area, how many different schedules are possible?

5. A decorator has a choice of 4 carpets, 6 wallpapers, and 5 paints. How many different selections of 1 carpet, 1 wallpaper, and 2 different paints can be made in decorating a room?

6. A cafeteria serves 2 meats, 4 vegetables, 3 salads, and 2 desserts. A meal consists of 1 meat, 2 vegetables, 1 salad, and 1 dessert. How many different meals are possible?

7. A test consists of 10 true-false questions. How many different answer sheets are possible, if all questions are answered?

8. A test consists of 10 multiple-choice questions, each with 5 possible answers. How many different answer sheets are possible if all questions are answered?

9. If 4 coins are tossed on the floor and the sides facing up are observed, how many different possible results can occur?

10. If two cubical dice are tossed and the sides facing up are observed, how many different possible results can occur?

11. If 6 runners enter a race, how many ways can the first 3 winning positions be awarded?

12. A class of 10 boys and 15 girls elects a president and a secretary. In how many ways can a president and a secretary be selected if no student can hold both offices and if
    (a) there are no other restrictions?
    (b) the president must be a girl?
    (c) the secretary must be a girl?

13. A United States social security number consists of 3 digits followed by 2 digits, then followed by a final 4 digits. If any digit may fill any position, how many social security numbers are possible?

14. License tags in a certain state consist of 3 letters and 3 digits. How many tags are possible if
    (a) repeats are not allowed on the letters?
    (b) the final digit may not be zero?
    (c) repeats are not allowed on the letters or digits?
    (d) there are no restrictions?

15. How many numbers greater than 2000 and consisting of 4 digits can be formed?

16. In problem 15, suppose that only the digits 0, 1, 2, 3, 4, and 5 can be used.

17. How many three-digit odd numbers can be made from the digits 1, 2, 3, 4, 5, and 6?

18. How many three-digit even numbers can be made using the digits 1, 2, 3, 4, 5, and 6?

12-2
Permutations

A permutation is an arrangement of objects. For example,

x	xy	yz	xyz	yzx
y	xz	zx	xzy	zxy
z	yx	zy	yxz	zyx

are arrangements, and hence permutations of the three letters x, y, and z. The three arrangements

$$x \quad y \quad z$$

are the permutations of the three letters taken one at a time; the six arrangements

xy	yx	zx
xz	yz	zy

are the permutations of the three letters taken two at a time; and the six arrangements

$$xyz \qquad yxz \qquad zxy$$

$$xzy \qquad yzx \qquad zyx$$

are the permutations of the three letters taken three at a time. In general, a *permutation* of *n* objects taken *r* at a time is an arrangement using *r* of the *n* objects.

In this section, our interest focuses on the determination of the *number* of permutations of *n* objects taken *r* at a time. The *factorial* notation described in the next definition will be used in counting permutations.

DEFINITION 12-2   If *n* is a positive integer, then *n factorial*, denoted by *n*!, is the product of the *n* integers *n*, *n* − 1, *n* − 2, . . . , 2, 1; that is,

$$n! = n \cdot (n-1) \cdot (n-2) \cdots 2 \cdot 1.$$

Also, 0! = 1, by definition.

The reason for defining 0! = 1 will be clear after Theorem 12-3.

EXAMPLE 1   Some illustrations of the use of the factorial notation are given below.

(a) $3! = 3 \cdot 2 \cdot 1 = 6$
(b) $7! = 7 \cdot 6 \cdot 5 \cdot 4 \cdot 3 \cdot 2 \cdot 1 = 5040$
(c) $1! = 1$
(d) $(n-1)! = (n-1)(n-2)(n-3) \cdots (2)(1)$
(e) $n(n-1)! = n(n-1)(n-2)(n-3) \cdots (2)(1) = n!$

We shall denote the number of permutations of *n* objects taken *r* at a time by the symbol

$$P(n, r),$$

where $0 \leq r \leq n$. To determine a formula for $P(n, r)$, first consider $P(n, n)$. We use the Counting Principle to determine the number of distinct arrangements of *n* objects taken *n* at a time. In other words, we have *n* positions to fill with a choice of *n* objects where repetition of any object is not allowed. For the first position there are *n* choices, for the second position *n* − 1 choices, for the third position *n* − 2 choices, and so on, as indicated in Figure 12.3. Thus we have

$$P(n, n) = n \cdot (n-1) \cdot (n-2) \cdots 2 \cdot 1,$$

or

$$P(n, n) = n! \tag{1}$$

Position		1	2	3	$\cdots$	$n-1$	n
Number of choices		n	$n-1$	$n-2$	$\cdots$	2	1

FIGURE 12.3

In a similar manner, suppose that we have r positions to fill with a choice of n objects $(0 \leq r \leq n)$, where repetition of any object is not allowed. For the first position there are n choices, for the second position $n-1$ choices, for the third position $n-2$ choices, and so on. Finally, for the rth position there are $n-(r-1)$, or $n-r+1$, choices. The tabulation appears in Figure 12.4. The Counting Principle gives

Position		1	2	3	$\cdots$	r
Number of choices		n	$n-1$	$n-2$	$\cdots$	$n-r+1$

FIGURE 12.4

$$P(n, r) = n(n-1)(n-2) \cdots (n-r+1). \tag{2}$$

To express this result using the factorial notation, we note the following. Multiplying both sides of equation (2) by $(n-r)!$ yields

$$P(n, r)(n-r)! = n(n-1) \cdots (n-r+1)(n-r)!$$
$$= n(n-1) \cdots (n-r+1)(n-r)(n-r-1) \cdots 2 \cdot 1$$
$$= n!.$$

Solving for $P(n, r)$, we have

$$P(n, r) = \frac{n!}{(n-r)!}.$$

This result is stated in the following theorem.

THEOREM 12-3   If $0 \leq r \leq n$, then the number of distinct permutations of n objects taken r at a time is given by

$$P(n, r) = \frac{n!}{(n-r)!}.$$

Notice that if $r = n$, we have

$$P(n, n) = \frac{n!}{(n-n)!} = \frac{n!}{0!} = \frac{n!}{1} = n!,$$

and this agrees with equation (1).

EXAMPLE 2 Some evaluations of the formula in Theorem 12-3 are given below.

(a) $P(5, 3) = \dfrac{5!}{(5 - 3)!} = \dfrac{5!}{2!} = \dfrac{5 \cdot 4 \cdot 3 \cdot 2!}{2!} = 60$

(b) $P(6, 6) = \dfrac{6!}{0!} = \dfrac{6 \cdot 5 \cdot 4 \cdot 3 \cdot 2 \cdot 1}{1} = 720$

(c) $P(10, 3) = \dfrac{10!}{(10 - 3)!} = \dfrac{10!}{7!} = \dfrac{10 \cdot 9 \cdot 8 \cdot 7!}{7!} = 10 \cdot 9 \cdot 8 = 720$

(d) $P(4, 0) = \dfrac{4!}{(4 - 0)!} = \dfrac{4!}{4!} = 1$

EXAMPLE 3 From a group of 10 people, a president, vice-president, secretary, and treasurer are to be chosen. How many different sets of officers can be chosen if no person can hold more than one office?

SOLUTION Using the Counting Principle, we have 10 choices for president. Once a president has been chosen, there are 9 choices for vice-president, then 8 choices for secretary, and finally, 7 choices for treasurer. Thus there are

$$10 \cdot 9 \cdot 8 \cdot 7 = 5040$$

different sets of officers.

Alternatively, permutations can be used to solve the problem. Each set of officers is a permutation of 10 people taken 4 at a time. Thus the number of different sets of officers is the number of distinct permutations of 10 things taken 4 at a time, or

$$P(10, 4) = \frac{10!}{6!} = 10 \cdot 9 \cdot 8 \cdot 7 = 5040.$$

EXAMPLE 4 In how many ways can 5 boys and 3 girls be seated in a row of 8 chairs if the girls must sit together?

SOLUTION Since the girls must sit together, we must choose seats for 5 boys and the group of girls. This can be done in $P(6, 6) = 6!$ ways. The number of different ways the girls can be arranged within the group is $P(3, 3) = 3!$. The Counting Principle gives

$$6! \cdot 3! = 4320$$

as the number of different seating arrangements of the desired type.

Suppose that we consider next permutations of n objects which are not all distinct. For example, consider the permutations of the letters in the word "BETTER." The "word"

BETRET

is one such permutation. If we interchange the two letters "E" in the arrangement, we obtain the same permutation. To count the number of distinguishable permutations, we use the Counting Principle. Let D represent the number of distinct permutations of the letters in the word "BETTER." There are $P(2, 2) = 2!$ arrangements of the two E's which will not change a permutation, and $P(2, 2) = 2!$ arrangements of the two T's which will not change a permutation. If all letters were distinguishable, there would be

$$D \cdot 2! \cdot 2!$$

arrangements of the six letters taken six at a time. But $P(6, 6) = 6!$ is the number of permutations of six distinguishable letters taken six at a time. Thus

$$D \cdot 2! \cdot 2! = 6!,$$

and

$$D = \frac{6!}{2! \, 2!} = 180.$$

The general situation is described in the next theorem.

THEOREM 12-4   Let D represent the number of distinct permutations of n objects taken n at a time, where there are r $(0 < r \leq n)$ distinguishable types: n_1 are of one type, n_2 are of a second type, $\ldots$, n_r are of the rth type. Then

$$D = \frac{n!}{n_1! n_2! \cdots n_r!}.$$

EXAMPLE 5   How many distinct permutations are possible for the letters in the word "TENNESSEE"?

SOLUTION   The word "TENNESSEE" has nine letters, consisting of one T, four E's, two N's, and two S's. Thus there are

$$D = \frac{9!}{1! \, 4! \, 2! \, 2!} = 3780$$

distinct permutations of the letters in the word "TENNESSEE."

EXERCISES 12-2

1. List all the permutations of 0 and 1.

2. List all the permutations of a, b, c, and d.

3. List all the permutations of 1, 2, 3, 4, and 5, taken 2 at a time.

4. List all the permutations of 1, 2, 3, 4, and 5, taken 3 at a time.

Evaluate each of the following.

5. $3!$

6. $8!$

7. $\dfrac{9!}{2!\,3!\,4!}$

8. $\dfrac{8!}{1!\,3!\,5!}$

9. $\dfrac{6!}{(3!)^2}$

10. $\dfrac{10!}{(5!)^2}$

11. $P(4, 4)$

12. $P(6, 6)$

13. $P(15, 2)$

14. $P(8, 6)$

15. $P(15, 0)$

16. $P(20, 0)$

17. $P(n, 2)$

18. $P(n - 1, 3)$

19. $P(n, n - 2)$

20. $P(n - 1, n - 3)$

Simplify each of the following.

21. $\dfrac{n!}{(n - 2)!}$

22. $\dfrac{(n - 1)!}{(n - 2)!}$

23. $\dfrac{n!}{(n + 2)!}$

24. $\dfrac{(n + 3)!}{(n - 3)!}$

25. In how many ways can 6 first graders be seated in a row of (a) 6 chairs; (b) 7 chairs?

26. In how many ways can 5 Girl Scouts be seated in a row of 10 chairs?

27. In how many ways can a set of 12 different books be arranged on a shelf?

28. In how many ways can 5 different pictures be arranged in a row on a wall?

29. In how many ways can 6 different books be arranged on a shelf if 3 of the 6 are math books, and they must be together?

30. In how many ways can 8 different books be arranged on a shelf if 2 are trigonometry books which must stay together and 3 are algebra books which must stay together?

31. In how many ways can 7 high school students be seated in a row of 7 seats if two of the students must be seated side by side?

32. In how many ways can a group of 4 college freshmen and 5 college sophomores be seated in a row of 9 seats if the 4 freshmen are seated together?

33. How many "words" can be made using the letters in the word COUNT?

34. How many "words" can be made using the letters in the word MATH?

35. How many "words" can be made using the letters in the word (a) ALGEBRA; (b) CALCULUS; (c) ADD; (d) DIVIDE?

36. How many "words" can be made using the letters in the word (a) MISSISSIPPI; (b) ALABAMA; (c) LOUISIANA; (d) ILLINOIS?

37. In how many ways can 6 apples, 4 oranges, and 5 bananas be distributed among 15 children, if each child is to receive one piece of fruit?

38. If 5 identical television sets and 4 identical ovens are to be given to 9 prize winners, how many ways can the prizes be awarded?

39. If signals are to be made using 2 green flags, 4 black flags, and 4 white flags, how many distinct signals can be made by lining up the flags?

40. If signals are to be made using 3 red flags, 5 white flags, and 2 blue flags, how many distinct signals can be made by lining up the flags?

41. In how many ways can 4 girls and 3 boys be seated in 7 seats if the girls and boys must alternate?

42. In how many ways can 4 girls and 5 boys be seated in 9 seats if the girls and boys must alternate?

*43. In how many relative orders can 5 people be seated in 5 chairs at a round table? (*Hint:* Fix the position of one person at the table, and then arrange the remaining people in the remaining positions.)

*44. In how many relative orders can 7 people be seated in 7 chairs at a round table?

*45. At a round table, 4 men and 4 women are to be seated in 8 chairs. How many distinct relative orders are possible if the men and women must alternate? If the men are to sit together?

*46. How many distinct relative orders of seating 5 women and 5 men in 10 chairs are there around a circular table if the women and men must alternate? If the women must sit together?

12-3
Combinations

A combination is a grouping of objects. For example,

$$\{x\}, \quad \{y\}, \quad \{z\}, \quad \{x, y\} \quad \{x, z\}, \quad \{y, z\}, \quad \{x, y, z\}$$

are groupings and hence combinations of the three letters $x, y,$ and z. The three groupings

$$\{x\}, \quad \{y\}, \quad \{z\},$$

are the combinations of the three letters taken one at a time; the three groupings

$$\{x, y\}, \quad \{x, z\}, \quad \{y, z\},$$

are the combinations of the three letters taken two at a time; and the one grouping

$$\{x, y, z\}$$

is the only combination of the three letters taken three at a time. In general, a *combination* of n objects taken r at a time is a group of r of the n objects, where the order in which the objects appear is of no consequence.

Notice that the distinct permuations

$$xy \quad \text{and} \quad yx$$

are in fact the same combination. This important difference between combinations and permutations cannot be overemphasized. A reordering of the objects changes a permutation, but a reordering of the objects does not change a combination.

Let $C(n, r)$ denote the number of combinations of n objects taken r at a time. A formula for evaluating $C(n, r)$ can be obtained by observing the following. We can obtain $P(n, r)$, the number of permutations of n objects taken r at a time by

(a) first choosing r of the n objects, which can be done in $C(n, r)$ ways;

(b) then forming all possible distinct arrangements of the r chosen objects, which can be done in $P(r, r) = r!$.

Thus we have

$$P(n, r) = C(n, r)r!,$$

and

$$C(n, r) = \frac{P(n, r)}{r!}.$$

But since $P(n, r) = n!/(n - r)!$, this can be rewritten as

$$C(n, r) = \frac{n!}{(n - r)!\, r!}.$$

We state our result in the next theorem.

THEOREM 12-5   If $0 \le r \le n$, then the number of combinations of n objects taken r at a time is given by

$$C(n, r) = \frac{n!}{(n - r)!\, r!}.$$

Combinations are used extensively in statistics and probability, and $C(n, r)$ must be calculated often.[1] The symbol $C(n, r)$ which we have used here is not the only accepted notation for the number of combinations of n things taken r at a time. Other symbols that enjoy widespread use are $_nC_r$ and $\binom{n}{r}$, with $\binom{n}{r}$ probably being the most popular. We shall use $C(n, r)$ and $\binom{n}{r}$ interchangeably throughout the rest of this chapter. Similarly, the symbol $_nP_r$ is often used to represent the same number as $P(n, r)$.

The use of Theorem 12-5 is demonstrated in the following examples.

[1] There are tables that give the values of $C(n, r)$ for small values of n.

EXAMPLE 1   How many committees of 4 people can be chosen from a group of 7 people?

SOLUTION   Since the order in which the committee members are chosen is of no consequence, the number of committees consisting of 4 of the 7 people is given by

$$\binom{7}{4} = \frac{7!}{4!\,3!} = 35.$$

EXAMPLE 2   A basketball team consists of 2 guards, 2 forwards, and 1 center. If a coach has 4 guards, 6 forwards, and 3 centers to choose from, how many ways can he choose a team?

SOLUTION   Theorem 12-5 is used three times: once to determine the number of ways he can choose the guards, again for the forwards, and again for the center. From the 4 available guards he can choose 2 in

$$C(4, 2) = \frac{4!}{2!\,2!} = 6$$

ways. From the 6 forwards he can choose 2 in

$$C(6, 2) = \frac{6!}{4!\,2!} = 15$$

ways. From the 3 centers he can choose 1 in

$$C(3, 1) = \frac{3!}{2!\,1!} = 3$$

ways. By the Counting Principle, there are

$$6 \cdot 15 \cdot 3 = 270$$

ways the coach can choose a team.

EXAMPLE 3   A poker hand consists of 5 cards dealt from an ordinary deck of 52 cards. How many different poker hands are possible?

SOLUTION   Since the order in which the cards appear in the hand is of no significance, there are

$$\binom{52}{5} = \frac{52!}{47!\,5!} = 2,598,960$$

different poker hands.

EXERCISES 12-3

1. List all the combinations of 0 and 1.

2. List all the combinations of a, b, c, and d.

3. List all the combinations of 1, 2, 3, 4, and 5 taken 2 at a time.

4. List all the combinations of 1, 2, 3, 4, and 5 taken 3 at a time.

Evaluate each of the following.

5. $C(6, 5)$

6. $C(8, 4)$

7. $C(10, 10)$

8. $C(4, 4)$

9. $\dbinom{5}{0}$

10. $\dbinom{9}{0}$

11. $\dbinom{6}{4}$

12. $\dbinom{6}{2}$

Simplify each of the following.

13. $\dbinom{n}{n-1}$

14. $\dbinom{n}{n-2}$

15. $C(n, n)$

16. $C(n, 0)$

17. A decorator chooses 4 colors in decorating a house. If she has 12 colors to choose from, how many different combinations are possible?

18. From a list of 9 players, 8 are to be chosen. In how many ways can this be done?

19. From a group of 15 people, how many committees of 6 can be chosen? How many committees of 10 can be chosen?

20. From a group of 10 students, 5 are to be chosen to serve as ushers at graduation. In how many ways can this be done?

21. From a club of 8 members, an entertainment committee of 4 is to be selected. In how many ways can this be done if one must serve as chairperson?

22. A grievance committee is to be selected out of an algebra class of 20 students. How many different committees can be selected if there are to be 4 people on the committee, with 1 of those designated as spokesperson?

23. How many committees of 2 men and 3 women can be chosen from a group of 6 men and 12 women?

24. From a group of 8 freshmen and 9 sophomores, a committee consisting of 3 freshmen and 5 sophomores is to be selected. How many different committees can be chosen?

25. In how many ways can a student choose 2 math books from among 6 math books, 4 novels from among 7 novels, and 3 psychology books from 9 psychology books?

26. A baseball team consists of 3 outfielders, 4 infielders, 1 pitcher, and 1 catcher. From a group of 7 outfielders, 9 infielders, 6 pitchers, and 3 catchers, how many different groups of people can be chosen to form a team?

27. An orchestra can play 10 marches, 8 overtures, and 15 pop songs. How many different concerts consisting of 3 marches, 2 overtures, and 5 pop songs can it give if the order of the pieces is of no consequence?

28. A restaurant can prepare 9 main dishes, 7 vegetables, 6 salads, 4 breads, and 12 desserts. If it plans a buffet consisting of 2 main dishes, 3 vegetables, 4 salads, 2 breads, and 5 desserts, how many possible menus are there?

29. A bag contains 7 green, 5 white, and 2 purple balls. Three balls are to be chosen. In how many ways can the following selections be made? (a) 1 green and 2 white; (b) all white balls; (c) each of different color; (d) all purple balls?

30. From a bag containing 6 blue, 4 green, and 5 red balls, four are chosen. Find the number of ways each of the following selections can be made: (a) all green; (b) each of different color; (c) none green; (d) all the same color; (e) 2 blue, 1 red, and 1 green.

31. How many distinct straight lines can be drawn through 9 points, where no three are on the same line? (*Hint:* Two points determine a straight line.)

32. Determine the number of quadrilaterals which can be drawn using the points in problem 31 as vertices.

33. A bridge hand consists of 13 cards dealt from an ordinary deck of 52. How many hands are there consisting of the following? (a) 5 spades, 3 hearts, 2 clubs, and 3 diamonds; (b) 12 spades and 1 heart; (c) 7 clubs and 6 spades.

34. If a poker hand of 5 cards is dealt, determine the number of ways each of the following hands can occur? (a) a royal flush: ace, king, queen, jack, and 10, all of the same suit; (b) a straight: five cards in consecutive numerical order, not all of the same suit (consider a jack as 11, queen as 12, king as 13, and ace as either 1 or 14); (c) four of a kind (e.g., four 2's or four aces).

35. Show that $C(n, r) = C(n, n - r)$ for all $r \leq n$.

*36. Show that $C(n, r - 1) + C(n, r) = C(n + 1, r)$.

12-4
The Binomial
Theorem

One of the most important applications of combinations is in the calculation of the *binomial coefficients*, that is, the coefficients in the expansion of $(a + b)^n$, where n is a positive integer, and a and b are any real or complex numbers. Recall that $(a + b)^n$ represents a product of n factors when n is a positive integer, each factor being $(a + b)$. By direct multiplication, we have the following:

$$(a + b)^1 = a + b,$$
$$(a + b)^2 = a^2 + 2ab + b^2,$$
$$(a + b)^3 = a^3 + 3a^2b + 3ab^2 + b^3,$$
$$(a + b)^4 = a^4 + 4a^3b + 6a^2b^2 + 4ab^3 + b^4,$$
$$(a + b)^5 = a^5 + 5a^4b + 10a^3b^2 + 10a^2b^3 + 5ab^4 + b^5.$$

By examination of these expansions, we can make the following observations about $(a + b)^n$.

---

1. There are $n + 1$ terms.

2. The highest power of a and b is n.

3. In each term, the sum of the powers of a and b is n.

4. As the power of a decreases in each successive term, the power of b increases.

The expansions above illustrate the fact that the terms in the expansion of $(a + b)^n$ are terms involving products of the form

$$a^n, \quad a^{n-1}b, \quad a^{n-2}b^2, \quad \ldots, \quad a^2b^{n-2}, \quad ab^{n-1}, \quad b^n.$$

Now we wish to determine the coefficient of each term. Consider $(a + b)^3$. Applying the distributive, associative, and commutative properties, we have

$$(a + b)(a + b)(a + b) = (a^2 + ba + ab + b^2)(a + b)$$
$$= a^3 + ba^2 + aba + b^2a + a^2b + bab + ab^2 + b^3$$
$$= a^3 + 3a^2b + 3ab^2 + b^3.$$

Each term in the expansion comes from a product consisting of three factors, each factor being one of either a or b. The term a^3 comes from using three a's, one from each factor $a + b$. The term $3a^2b$, involving a^2, comes from using two a's at a time from the three factors $(a + b)$. But there are $\binom{3}{2}$ ways of choosing from which of the three factors $(a + b)$ to take the two a's. The term $3ab^2$, involving a^1, comes from using one a at a time from the three factors $(a + b)$. But there are $\binom{3}{1}$ ways of choosing one of the three factors $(a + b)$ to furnish the one a. The last term b^3, involving a^0, comes from using no a's from the three factors $(a + b)$, or from using three b's, one from each of the three factors $(a + b)$. Thus $(a + b)^3$ can be written as

$$(a + b)^3 = a^3 + \binom{3}{2}a^2b^1 + \binom{3}{1}a^1b^2 + b^3.$$

In general, the coefficient of $a^{n-r}b^r$ in the expansion of $(a + b)^n$ is given by $\binom{n}{n-r}$, and we call $\binom{n}{n-r}$, $r = 0, 1, 2, \ldots, n$, the *binomial coefficients*. These results are recorded in the following theorem.

THEOREM 12-6 (THE BINOMIAL THEOREM) Let n be a positive integer, and let a and b be real or complex numbers. Then $(a + b)^n$ is a sum of $n + 1$ terms of the form

$$\binom{n}{n-r}a^{n-r}b^r; \qquad r = 0, 1, 2, \ldots, n.$$

More specifically,

$$(a + b)^n = a^n + \binom{n}{n-1}a^{n-1}b + \binom{n}{n-2}a^{n-2}b^2 + \cdots$$
$$+ \binom{n}{n-r}a^{n-r}b^r + \cdots + \binom{n}{2}a^2b^{n-2}$$
$$+ \binom{n}{1}ab^{n-1} + b^n.$$

EXAMPLE 1 Expand $(x^2 - 2)^6$.

SOLUTION Using the Binomial Theorem, we first write $(x^2 - 2)^6$ as $[x^2 + (-2)]^6$ and

$$[x^2 + (-2)]^6 = (x^2)^6 + \binom{6}{5}(x^2)^5(-2) + \binom{6}{4}(x^2)^4(-2)^2$$
$$+ \binom{6}{3}(x^2)^3(-2)^3 + \binom{6}{2}(x^2)^2(-2)^4$$
$$+ \binom{6}{1}(x^2)(-2)^5 + (-2)^6$$
$$= x^{12} + 6 \cdot x^{10} \cdot (-2) + 15 \cdot x^8 \cdot (4) + 20 \cdot x^6 \cdot (-8)$$
$$+ 15 \cdot x^4 \cdot (16) + 6 \cdot x^2 \cdot (-32) + 64.$$

Thus

$$(x^2 - 2)^6 = x^{12} - 12x^{10} + 60x^8 - 160x^6 + 240x^4 - 192x^2 + 64.$$

The Binomial Theorem can also be used to determine the coefficient of a particular term in the expansion of $(a + b)^n$.

EXAMPLE 2 Find the coefficient of t^8s^9 in the expansion of $(t^2 - s)^{13}$.

SOLUTION The terms in the expansion of $(t^2 - s)^{13}$ are of the form

$$\binom{n}{n-r}(t^2)^{n-r}(-s)^r.$$

The coefficient required is in the term with $r = 9$ and $n - r = 13 - 9 = 4$. Therefore, the term in the expansion is

$$\binom{13}{4}(t^2)^4(-s)^9,$$

and the required coefficient is

$$-\binom{13}{4} = -\frac{13!}{4!\,9!} = -715.$$

Notice in the binomial expansion that the power of b is always one less than the term number. For example, b^1 occurs in term number 2, b^2 occurs in term number 3, . . . , b^n occurs in term number $n + 1$. Thus b^{r-1} occurs in the rth term. Since the sum of the powers of a and b must be equal to n, then the corresponding power of a is $n - (r - 1)$. Knowing the power of a leads to the correct selection of the binomial coefficient, $\binom{n}{n - r + 1}$. Thus we have the following theorem.

THEOREM 12-7   The rth term $(r \le n)$ in the expansion of $(a + b)^n$ is

$$\binom{n}{n - r + 1}a^{n-r+1}b^{r-1}.$$

EXAMPLE 3   Find the 12th term in the expansion of $(3x^2 - y^3)^{15}$.

SOLUTION   The value of n is 15, and $r - 1$ is 11 in the 12th term. Thus the 12th term is given by

$$\binom{15}{4}(3x^2)^4(-y^3)^{11} = -\frac{15!}{4!\,11!}3^4x^8y^{33}$$

$$= -\frac{15 \cdot \overset{7}{\cancel{14}} \cdot 13 \cdot \cancel{12} \cdot \cancel{11!}}{\cancel{4} \cdot \cancel{3} \cdot \cancel{2} \cdot 1 \cdot \cancel{11!}} \cdot 81x^8y^{33}$$

$$= -110{,}565x^8y^{33}.$$

It is shown in more advanced mathematics courses that the Binomial Theorem is valid under certain conditions when n is any positive or negative rational number. Before we formulate these conditions, we make the following observation. The quantity $\binom{n}{n - r}$ can be rewritten as

$$\binom{n}{n - r} = \frac{n!}{(n - r)!\,r!}$$

$$= \frac{n(n - 1)(n - 2) \cdots (n - r + 1)\cancel{(n - r)!}}{\cancel{(n - r)!}\,r!}$$

$$= \frac{n(n - 1)(n - 2) \cdots (n - r + 1)}{r!}.$$

The binomial expansion can now be restated as

$$(a + b)^n = a^n + na^{n-1}b + \frac{n(n-1)}{2!}a^{n-2}b^2 + \frac{n(n-1)(n-2)}{3!}a^{n-3}b^3$$

$$+ \cdots + \frac{n(n-1)(n-2)\cdots(n-r+1)}{r!}a^{n-r}b^r + \cdots.$$

If n is not a positive integer, the binomial expansion never terminates, but continues indefinitely, since $n - r$ is never zero. In the calculus, it is shown that if $a = 1$ and $b = x$, then the binomial expansion for

$$(a + b)^n = (1 + x)^n$$

is valid for all x such that $|x| < 1$.

EXAMPLE 4 Find the first five terms of the expansion of $(1 + x)^{1/2}$.

SOLUTION The expansion is valid for all x such that $|x| < 1$. We have

$$(1 + x)^{1/2} = 1 + \frac{1}{2} \cdot x + \frac{(\frac{1}{2})(\frac{1}{2} - 1)}{2!}x^2 + \frac{\frac{1}{2}(\frac{1}{2} - 1)(\frac{1}{2} - 2)}{3!}x^3$$

$$+ \frac{(\frac{1}{2})(\frac{1}{2} - 1)(\frac{1}{2} - 2)(\frac{1}{2} - 3)}{4!}x^4 + \cdots$$

$$= 1 + \frac{1}{2}x - \frac{1}{8}x^2 + \frac{1}{16}x^3 - \frac{5}{128}x^4 + \cdots.$$

EXERCISES 12-4

Expand by using the Binomial Theorem.

1. $(x + y)^7$
2. $(a + b)^8$
3. $(2x + y)^4$
4. $(x + 3y)^6$
5. $(x^2 + 2y)^5$
6. $(3x + y^2)^3$
7. $(x^2 - y^2)^4$
8. $(a^3 - b^2)^5$
9. $\left(2x - \frac{1}{y}\right)^4$
10. $\left(x + \frac{1}{x}\right)^3$
11. $\left(3x + \frac{y}{3}\right)^4$
12. $\left(\frac{x}{4} - 2y\right)^6$
13. $\left(x^3 - \frac{x^2}{2}\right)^6$
14. $(3x^2 - x)^4$
15. $(\frac{1}{2}a - 3b^2)^5$
16. $(\frac{1}{3}a^2 + 4b^3)^4$

Find the coefficient of the indicated term in the given expansion.

17. x^8y^2 in $(x - y)^{10}$
18. a^3b^6 in $(a - b)^9$
19. x^4y^9 in $(x^2 + 2y^3)^5$
20. a^2b^{10} in $(3a^2 + b^2)^6$

Find the indicated term in the given expansion.

21. 10th term of $(2x + y)^{12}$

22. 9th term of $\left(\dfrac{x}{2} + y\right)^{13}$

23. 4th term of $(x - 4y^2)^{15}$

24. 5th term of $(x^2 - 2)^9$

25. 9th term of $(p^2 + q^3)^{14}$

26. 8th term of $(x^3 + z)^{10}$

27. 14th term of $(x + 2y)^{13}$

28. 21st term of $\left(\dfrac{x}{3} - y^2\right)^{20}$

Determine the first five terms in the expansions of the following. Also determine the values of the variable for which each expansion is valid.

29. $(1 + x)^{1/4}$

30. $(1 + x)^{1/3}$

31. $(1 - y)^{2/3}$

32. $(1 - y)^{3/4}$

33. $(1 + 2x)^{-1/2}$

34. $(1 + 3x)^{-1/4}$

35. $(1 + x^2)^{-2}$

36. $(1 + x^2)^{-3}$

37. $\left(1 - \dfrac{a}{2}\right)^{-2}$

38. $(1 - 3a)^{-3}$

Expand by using the Binomial Theorem.

*39. $(x + y + 1)^4$

*40. $(x - 1 - z)^3$

*41. $(x - y + z - w)^3$

*42. $(x + y + z + w)^4$

Approximate the following by using the first four terms in the binomial expansion.

43. $(1.01)^6$

44. $(1.03)^4$

45. $(0.99)^8$

46. $(2.99)^3$

47. $(3.01)^3$

48. $(1.03)^8$

49. $\sqrt{1.02}$

50. $\sqrt[3]{1.1}$

12-5
Partial Fractions   In Section 2-5, we studied the procedure for combining (adding or subtracting) rational integral expressions (fractions). Often, in the calculus, it is necessary to perform the reverse process: that of separating a fraction into a sum or difference of simpler, or *partial fractions*. Suppose that $P(x)/Q(x)$ is a rational integral expression, reduced to lowest terms, where the degree of $P(x)$ is less than the degree of $Q(x)$. We first factor $Q(x)$ into its linear and irreducible quadratic factors, and consider four cases.

Case I   The irreducible factors of $Q(x)$ are all of the first degree, and none are repeated. That is,

$$Q(x) = (a_1x + b_1)(a_2x + b_2) \cdots (a_nx + b_n),$$

where no one of the factors $a_i x + b_i$ is equal to a constant multiple of another factor $a_j x + b_j$, $i \neq j$. Then $P(x)/Q(x)$ can be expressed as a sum of fractions:

$$\frac{P(x)}{Q(x)} = \frac{A_1}{a_1 x + b_1} + \frac{A_2}{a_2 x + b_2} + \cdots + \frac{A_n}{a_n x + b_n},$$

where $A_1, A_2, \ldots, A_n$ are constants.

EXAMPLE 1   Resolve

$$\frac{10x - 2}{x^3 - x}$$

into partial fractions.

SOLUTION   The irreducible factors of the denominator $Q(x)$ are x, $x - 1$, and $x + 1$ since $Q(x) = x^3 - x = x(x - 1)(x + 1)$. Thus we can write

$$\frac{10x - 2}{x^3 - x} = \frac{A}{x} + \frac{B}{x - 1} + \frac{C}{x + 1}. \tag{1}$$

We must determine constants A, B, and C so that the sum of the three fractions on the right will be the fraction on the left. Multiplying both sides of equation (1) by the least common denominator, $x(x - 1)(x + 1)$, yields

$$10x - 2 \equiv A(x - 1)(x + 1) + B(x)(x + 1) + C(x)(x - 1). \tag{2}$$

The notation "$\equiv$" (read "identically equals") means that the equation is true for all values of x. There are two methods for determining the constants A, B, and C.

Method 1   Since equation (2) is an identity, it must be true for all values of x. Each time a value is assigned to x, there results an equation in A, B, and C. The best choices for values of x are those which make one of the linear factors zero, because these give simple equations in A, B, and C. In equation (2), the best choices are $x = 0$, $x = 1$, and $x = -1$.

If $x = 0$, then $-2 = A(-1)(1)$, and $A = 2$.

If $x = 1$, then $8 = B(1)(2)$, and $B = 4$.

If $x = -1$, then $-12 = C(-1)(-2)$, and $C = -6$.

Therefore, the partial fraction decomposition is

$$\frac{10x - 2}{x^3 - x} = \frac{2}{x} + \frac{4}{x - 1} - \frac{6}{x + 1}.$$

This may be checked by adding the three fractions on the right.

Method 2   Combining like terms in equation (2) yields

$$10x - 2 \equiv (A + B + C)x^2 + (B - C)x + (-A).$$

Since this equation is an identity, the coefficients of like powers of x on each side of the equation must be equal. Equating coefficients of like powers of x yields a system of three linear equations.

$$A + B + C = 0$$
$$B - C = 10$$
$$-A = -2$$

The solution to this system is $A = 2$, $B = 4$, $C = -6$, and this gives the partial fraction decomposition that was obtained by Method 1.

Case II The irreducible factors of $Q(x)$ are all of the first degree, and some are repeated. That is, at least one of the irreducible factors of $Q(x)$ is of the form $(ax + b)^n$, where $n > 1$. In the partial fraction decomposition, there will correspond to this factor a sum of n partial fractions of the form

$$\frac{A_1}{ax + b} + \frac{A_2}{(ax + b)^2} + \frac{A_3}{(ax + b)^3} + \cdots + \frac{A_n}{(ax + b)^n},$$

where $A_1, A_2, \ldots, A_n$ are constants. All of the linear factors of $Q(x)$ that are not repeated are handled as described in Case I.

EXAMPLE 2 Decompose

$$\frac{x^3 + 1}{x^2(x - 1)^2}$$

into partial fractions.

SOLUTION Since both linear factors of the denominator are repeated, the partial fraction decomposition will have the form

$$\frac{x^3 + 1}{x^2(x - 1)^2} = \frac{A}{x} + \frac{B}{x^2} + \frac{C}{x - 1} + \frac{D}{(x - 1)^2}.$$

Multiplying by $x^2(x - 1)^2$ gives

$$x^3 + 1 \equiv Ax(x - 1)^2 + B(x - 1)^2 + Cx^2(x - 1) + Dx^2. \qquad (3)$$

We shall illustrate both methods for determining the constants A, B, C, and D.

Method 1 Since equation (3) is true for all values of x, we can assign values to x and obtain equations in the unknown constants.

If $x = 0$, then $1 = B(-1)^2$, and $B = 1$.

If $x = 1$, then $2 = D(1)^2$, and $D = 2$.

If $x = 2$, then $9 = A(2)(1)^2 + B(1)^2 + C(2)^2(1) + D(2)^2$,

$$\text{and } 0 = A + 2C.$$

If $x = -1$, then $0 = A(-1)(-2)^2 + B(-2)^2 + C(-1)^2(-2) + D(-1)^2$,

and $3 = 2A + C$.

Solving the system

$$A + 2C = 0$$
$$2A + C = 3,$$

we obtain $A = 2$ and $C = -1$. Therefore, the partial fraction decomposition is

$$\frac{x^3 + 1}{x^2(x - 1)^2} = \frac{2}{x} + \frac{1}{x^2} - \frac{1}{x - 1} + \frac{2}{(x - 1)^2}.$$

Method 2 Collecting like terms in equation (3) gives

$$x^3 + 1 \equiv (A + C)x^3 + (-2A + B - C + D)x^2 + (A - 2B)x + B.$$

When we equate coefficients of like powers of x, we obtain the following system of four linear equations:

$$A \qquad\quad + C \qquad\quad = 1$$
$$-2A + \quad B - C + D = 0$$
$$A - 2B \qquad\qquad = 0$$
$$B \qquad\qquad = 1.$$

The solution to this system is

$$A = 2, \qquad B = 1, \qquad C = -1, \qquad D = 2,$$

and these constants give the same partial fraction decomposition as obtained by Method 1.

Case III The denominator, $Q(x)$, contains irreducible factors of the second degree, and none are repeated. For each irreducible quadratic factor of the form $ax^2 + bx + c$, there corresponds a partial fraction of the form

$$\frac{Ax + B}{ax^2 + bx + c},$$

where A and B are constants.

EXAMPLE 3 Write

$$\frac{2x^2 + x + 1}{x^3 + x}$$

as a sum of partial fractions.

SOLUTION The irreducible factors of the denominator are x and $x^2 + 1$. Thus, the partial fraction decomposition is of the form

$$\frac{2x^2 + x + 1}{x^3 + x} = \frac{A}{x} + \frac{Bx + C}{x^2 + 1}.$$

Multiplying by $x(x^2 + 1)$ gives the identity

$$2x^2 + x + 1 \equiv A(x^2 + 1) + (Bx + C)x. \tag{4}$$

It is possible to use either method to determine the constants A, B, and C, but we shall use Method 2. Collecting like terms in equation (4) yields

$$2x^2 + x + 1 \equiv (A + B)x^2 + Cx + A.$$

Equating coefficients, we have the system

$$A + B = 2$$
$$C = 1$$
$$A = 1,$$

and the solution is easily seen to be $A = 1$, $B = 1$, $C = 1$. Thus,

$$\frac{2x^2 + x + 1}{x^3 + x} = \frac{1}{x} + \frac{x + 1}{x^2 + 1}.$$

Case IV The denominator, $Q(x)$, contains irreducible factors of the second degree which are repeated. For each irreducible quadratic factor of the form $(ax^2 + bx + c)^n$, $n > 1$, there corresponds a sum of n partial fractions of the form

$$\frac{A_1 x + B_1}{ax^2 + bx + c} + \frac{A_2 x + B_2}{(ax^2 + bx + c)^2} + \cdots + \frac{A_n x + B_n}{(ax^2 + bx + c)^n},$$

where all A_i and B_i are constants.

EXAMPLE 4 Resolve

$$\frac{5x^2 - 2x}{(x^2 + 1)^2}$$

into partial fractions.

SOLUTION Since the irreducible quadratic factor is repeated, the partial fraction decomposition has the form

$$\frac{5x^2 - 2x}{(x^2 + 1)^2} = \frac{Ax + B}{x^2 + 1} + \frac{Cx + D}{(x^2 + 1)^2}.$$

Multiplying by $(x^2 + 1)^2$ yields

$$5x^2 - 2x \equiv (Ax + B)(x^2 + 1) + Cx + D.$$

Using Method 2 to determine the constants A, B, C, and D, we collect like terms and obtain

$$5x^2 - 2x \equiv Ax^3 + Bx^2 + (A + C)x + (B + D).$$

Equating coefficients of like powers of x yields the system

$$
\begin{aligned}
A & & = 0 \\
B & & = 5 \\
A & + C & = -2 \\
B & + D & = 0.
\end{aligned}
$$

The solution to this system is

$$A = 0, \qquad B = 5, \qquad C = -2, \qquad D = -5,$$

and the partial fraction decomposition is

$$\frac{5x^2 - 2x}{(x^2 + 1)^2} = \frac{5}{x^2 + 1} - \frac{2x + 5}{(x^2 + 1)^2}.$$

There are two important points that are worth remembering in decomposing rational integral expressions into partial fractions.

---

1. The number of constants to be found is always equal to the degree of the denominator of the original fraction.

2. A decomposition can be checked by adding the partial fractions to see if the sum is equal to the original fraction.

EXERCISES 12-5

Write each of the following as a sum of partial fractions.

1. $\dfrac{3x + 2}{x^2 - 5x + 6}$

2. $\dfrac{5x - 1}{x^2 - 1}$

3. $\dfrac{3x - 2}{x^2 + 2x}$

4. $\dfrac{2x}{x^2 - 4}$

5. $\dfrac{3x - 4}{x^3 - 3x^2 + 2x}$

6. $\dfrac{3x^2 - 4x - 5}{2x^3 - x^2 - x}$

7. $\dfrac{x - 1}{x^3 - x^2 - 2x}$

8. $\dfrac{1 - 4x^2}{(x - 2)(x - 1)(x + 3)}$

9. $\dfrac{4x^2 + 10x + 8}{(3x + 2)(x + 1)(x + 2)}$

10. $\dfrac{1 + 3x - 8x^2}{(1 - 4x)(1 - x^2)}$

11. $\dfrac{x - 3}{(x - 2)^2}$

12. $\dfrac{x}{(x + 3)^2}$

13. $\dfrac{x^2 + 1}{(x - 1)^3}$

14. $\dfrac{x^2 + x + 1}{(x + 2)^4}$

15. $\dfrac{x^2 + 1}{(x + 1)(x - 1)(x + 1)}$

16. $\dfrac{3x^2 - x + 6}{(x + 2)(1 - 2x)(x + 2)}$

17. $\dfrac{1}{x^4 - 4x^2}$

18. $\dfrac{x - 1}{x^4 + x^3}$

19. $\dfrac{3x - 2}{x^2(x - 1)^2}$

20. $\dfrac{12x^3 - 13x^2 + 7x - 1}{x^2(1 - 2x)^2}$

21. $\dfrac{8x^2 + 5x + 3}{(x - 2)^2(3x - 1)}$

22. $\dfrac{3x^2 + 1}{x(x - 1)^3}$

23. $\dfrac{-4}{(x - 1)(x^2 + x + 2)}$

24. $\dfrac{3x - 2}{x^3 + 2x^2 + 2x}$

25. $\dfrac{5x^2 - 1}{(x - 3)(x^2 + x - 1)}$

26. $\dfrac{3x^3 - 7x^2 + 7x - 1}{(x - 1)^2(x^2 + 1)}$

27. $\dfrac{5x^3 + 3x^2 - 3x + 3}{(x^2 - 1)(x^2 + 1)}$

28. $\dfrac{4x^3 - 17x^2 - 60}{(x^2 - 4)(x^2 + 4)}$

29. $\dfrac{x^3 - 1}{(1 + x^2)^2}$

30. $\dfrac{3x^2 + 1}{(1 + x + x^2)^2}$

31. $\dfrac{4x^3 + x}{(2x^2 - x + 1)^2}$

32. $\dfrac{6x^2 + 4x - 5}{(3x^2 + x - 5)^2}$

*33. $\dfrac{x^3 - 2x^2 + 4x}{x^2 - 2x + 3}$

*34. $\dfrac{10 + 5x - 17x^2 - 6x^3}{1 + x - 6x^2}$

*35. $\dfrac{2x^4 - 5x^3 + x^2 + 6x - 3}{x(x - 1)^2}$

*36. $\dfrac{x^5 - 2x^4 + 6x^2 - 12x + 10}{(x - 2)(x^2 + 1)}$

12-6

Graphs of Rational Functions

In Section 2-5, a rational integral expression was defined as a quotient of polynomials. Similarly, a *rational function* is a function which is defined by an equation of the form

$$f(x) = \frac{P(x)}{Q(x)}.$$

where $P(x)$ and $Q(x)$ are polynomials in the variable x, and $Q(x)$ is not the zero polynomial. In this section, we are considering only those rational functions which are quotients of polynomials with real coefficients, and our goal is to learn to sketch the graph of such a function with some degree of efficiency. If such a rational function f is defined by

$$y = f(x) = \frac{P(x)}{Q(x)},$$

it is understood that the domain of f is the set of all real numbers x such that $Q(x) \neq 0$.

In very simple cases, the graph can be sketched satisfactorily by plotting several points. Consider the following example.

EXAMPLE 1 Sketch the graph of the function defined by

$$y = \frac{2}{x - 1}.$$

The domain of the function is the set of all real numbers x such that $x \neq 1$. This means that there is no point on the graph where $x = 1$, and the graph is separated into two parts: those points (x, y) with $x > 1$, and those with $x < 1$. We concentrate first on those points with $x > 1$. Using the values $x = 2, 5, 10$ in succession, we obtain the first three sets of coordinates in the table below. Then using $x = \frac{3}{2}, 1.1, 1.01$, we obtain the next three sets.

x	2	5	10	$\frac{3}{2}$	1.1	1.01
y	2	$\frac{1}{2}$	$\frac{2}{9}$	4	20	200

After locating the points

$$A(2, 2), \quad B(5, \tfrac{1}{2}), \quad C(10, \tfrac{2}{9}),$$

it is clear that, as x continues to increase, the corresponding y values are positive, but steadily decreasing toward 0. Also, after locating the points

$$D(\tfrac{3}{2}, 4), \quad E(1.1, 20), \quad F(1.01, 200),$$

it is clear that, as x moves closer to 1 from the right, the corresponding y values increase steadily without bound. Thus the points in the table lead us to draw the right-hand part of the curve as shown in Figure 12.5.

In similar fashion, we obtain the following table for some values of $x < 1$:

x	0	-2	-10	$\frac{1}{2}$	0.9	0.99
y	-2	$-\frac{2}{3}$	$-\frac{2}{11}$	-4	-20	-200

After plotting these points, we draw the left-hand part of the curve as shown in Figure 12.5. The dashed line $x = 1$ is used as a guide in drawing the curve, and is *not* a part of the graph of the function.

The line $x = 1$ is useful in drawing the graph, because the points on the graph are closer to the line when the chosen values of x are closer to 1. A line such as this is called a *vertical asymptote* of the graph. From Figure 12.5, it is clear that the graph has the same sort of relationship to the x-axis. As $|x|$ increases without bound, the points on the graph are closer and closer to the line $y = 0$. A line such as this is called a *horizontal asymptote* of the curve.

It is not an accident that the vertical asymptote in Example 1 occurred at a zero of the denominator of the function. In the general

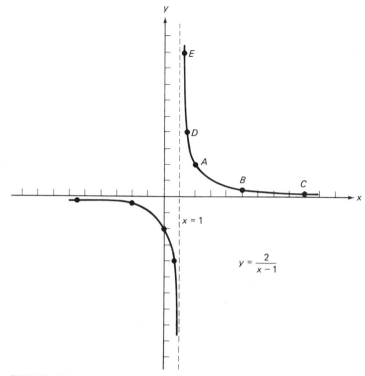

FIGURE 12.5

case, suppose that

$$y = \frac{P(x)}{Q(x)}$$

and $x = a$ is a value of x for which $Q(a) = 0$ and $P(a) \neq 0$. As x takes on values very close to a, $Q(x)$ is very close to 0, and $P(x)$ is very close to $P(a)$. This means that the quotient $P(x)/Q(x)$ grows larger numerically (i.e., in absolute value) as x gets closer to a, and $|y|$ increases without bound. A rigorous discussion of the situation here calls for the concept of a limit, which is the fundamental concept of the calculus. However, our discussion should make the following theorem plausible.

THEOREM 12-8   Let f be a rational function defined by

$$f(x) = \frac{P(x)}{Q(x)} = \frac{a_n x^n + a_{n-1}x^{n-1} + \cdots + a_1 x + a_0}{b_m x^m + b_{m-1}x^{m-1} + \cdots + b_1 x + b_0},$$

where $P(x)$ and $Q(x)$ are polynomials with real coefficients, and $Q(x)$ is not the zero polynomial. The horizontal and vertical asymptotes of the graph of f may be found by the following rules.

Vertical Asymptotes:

If a is a real number such that $Q(a) = 0$ and $P(a) \neq 0$, then the the line $x = a$ is a vertical asymptote of the graph of f.

Horizontal Asymptotes:

(a) If $n < m$, then $y = 0$ is a horizontal asymptote.

(b) If $n = m$, then $y = a_n/b_n$ is a horizontal asymptote.

(c) If $n > m$, there are no horizontal asymptotes.

EXAMPLE 2 Sketch the graph of the rational function defined by

$$y = \frac{2x^2}{(x-1)^2}.$$

Locate all vertical or horizontal asymptotes.

SOLUTION We note that the domain is the set of all real numbers $x \neq 1$, and that y is never negative, since

$$y = 2\left(\frac{x}{x-1}\right)^2.$$

Since $Q(x) = (x-1)^2$ is zero at $x = 1$ and $P(x) = 2x^2$ is not zero at $x = 1$, the line $x = 1$ is a vertical asymptote. Since

$$y = \frac{2x^2}{x^2 - 2x + 1},$$

$y = 2$ is a horizontal asymptote, by Theorem 12-8(b). We note that the x-intercept is 0, and the y-intercept is 0. Plotting a few points on either side of $x = 1$, we obtain the following table. The graph is shown in Figure 12.6.

x	2	3	5	10	0.5	0	−1	−2	−9
y	8	4.5	3.1	2.5	2	0	0.5	0.9	1.6

EXAMPLE 3 Sketch the graph of

$$y = \frac{2x^2 - 1}{x^2 - 3x},$$

locating all vertical or horizontal asymptotes.

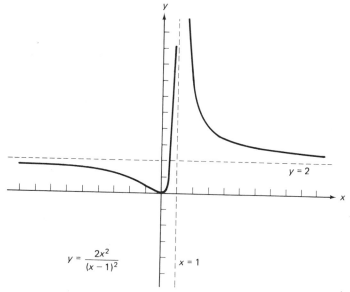

$$y = \frac{2x^2}{(x-1)^2}$$

$x = 1$

$y = 2$

FIGURE 12.6

SOLUTION To find vertical asymptotes, we set the denominator equal to 0:

$$x^2 - 3x = 0,$$

or

$$x(x - 3) = 0.$$

Thus $x = 0$ and $x = 3$ are vertical asymptotes. The numerator and denominator have the same degree, so $y = 2$ is a horizontal asymptote, by Theorem 12-8(b). Since 0 is not in the domain, there is no y-intercept. To find the x-intercepts, we set $y = 0$. That is,

$$\frac{2x^2 - 1}{x^2 - 3x} = 0.$$

This gives

$$2x^2 - 1 = 0,$$

and

$$x = \pm \frac{1}{\sqrt{2}}.$$

Thus the x-intercepts are given approximately by $x \approx \pm 0.7$. A table of values is given below, where the function values are rounded to the nearest tenth.

x	−0.7	−1	−2	−3	0.7	1	2	4	5	8
y	0	0.2	0.7	1	0	−0.5	−3.5	7.8	4.9	3.2

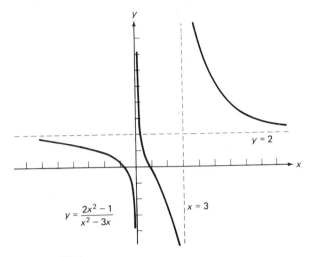

$$y = \frac{2x^2 - 1}{x^2 - 3x}$$

$y = 2$

$x = 3$

FIGURE 12.7

The graph is shown in Figure 12.7.

Under certain conditions, the graph of a rational function may have an asymptote which is of a type different from those described up to this point. If the degree of $A(x)$ is greater than or equal to the degree of $B(x)$, the equation

$$y = \frac{A(x)}{B(x)} = Q(x) + \frac{R(x)}{B(x)},$$

where degree $R(x) <$ degree $B(x)$, means that $y = Q(x)$ is an asymptote since $R(x)/B(x)$ approaches 0 as $|x|$ increases without bound. Thus

$$y = \frac{x^4 + 2x}{x^2 + 1} = x^2 - 1 + \frac{2x + 1}{x^2 + 1}$$

is asymptotic to the parabola $y = x^2 - 1$ when $|x|$ is large. In particular, if the degree of $A(x)$ is exactly 1 more than the degree of $B(x)$, then

$$\boxed{y = \frac{A(x)}{B(x)} = ax + b + \frac{R(x)}{B(x)}}$$

has the line $y = ax + b$ as an *oblique asymptote* when $|x|$ is large.

EXAMPLE 4 Sketch the graph of

$$y = \frac{x^2}{x - 1},$$

and locate all asymptotes of the graph.

SOLUTION According to Theorem 12-8, the graph has a vertical asymptote at $x = 1$, and it has no horizontal asymptote. The only point where the graph crosses a coordinate axis is at the origin $(0, 0)$. To find the oblique asymptote, we divide x^2 by $x - 1$ and obtain

$$y = x + 1 + \frac{1}{x - 1}.$$

From this equation, it can be seen that for $|x|$ very large, the value of $1/(x - 1)$ is near zero, and points (x, y) on the graph are close to the line

$$y = x + 1.$$

That is, $y = x + 1$ is an oblique asymptote. A table of values, rounded to the nearest tenth, is given below.

x	−1	−2	−3	−4	0	0.5	2	2.5	3	4
y	−0.5	−1.3	−2.2	−3.2	0	−0.5	4	4.2	4.5	5.3

The graph is shown in Figure 12.8.

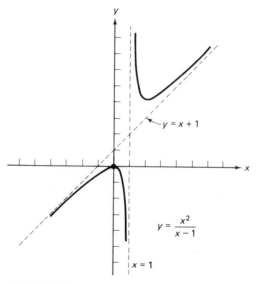

FIGURE 12.8

A systematic routine for using the techniques illustrated in this section is given below.

> *Procedure for Graphing a Rational Function*
>
> 1. Locate all asymptotes of the graph. (See Theorem 12-8 and Example 4.)
> 2. Locate the x-intercepts and y-intercepts, if there are any.
> 3. Plot a few points on either side of each vertical asymptote.
> 4. Sketch the graph, using the points plotted and the asymptotes as guides. The graph will be a smooth curve, except for breaks at the vertical asymptotes.

EXERCISES 12-6

Sketch the graphs of the rational functions defined by the following equations. Locate all vertical, horizontal, and oblique asymptotes.

1. $y = \dfrac{2}{x + 4}$

2. $y = -\dfrac{3}{2 - x}$

3. $y = \dfrac{2x + 1}{x - 3}$

4. $y = \dfrac{3x - 4}{2x + 1}$

5. $y = \dfrac{5 - 4x}{3x + 5}$

6. $y = \dfrac{3x + 5}{5 - 4x}$

7. $y = \dfrac{1}{x^2 - 3x}$

8. $y = \dfrac{1}{x^2 - x}$

9. $y = \dfrac{1}{(x + 1)^2}$

10. $y = \dfrac{1}{x^2}$

11. $y = \dfrac{1}{x^2 + 4}$

12. $y = \dfrac{1}{x^2 + 1}$

13. $y = \dfrac{2x}{(x + 2)^2}$

14. $y = \dfrac{2x}{(x - 1)(x + 2)}$

15. $y = \dfrac{1}{x^2(x + 1)}$

16. $y = \dfrac{1}{x^2(x - 2)}$

17. $y = \dfrac{4}{(x + 2)(x - 3)}$

18. $y = \dfrac{2}{(x - 1)(x + 3)}$

19. $y = \dfrac{x^2 - x - 2}{x^2 - x}$

20. $y = \dfrac{2x^2 + 5x - 3}{x^2 + 2x}$

21. $y = \dfrac{x^2 - 4}{2x^2 - x - 3}$

22. $y = \dfrac{x^2 - 9}{2x^2 - 5x - 3}$

23. $y = \dfrac{2x^2}{x + 1}$

24. $y = \dfrac{3x^2 - 1}{x}$

25. $y = \dfrac{3x - 4}{x^2 - 3x + 2}$

26. $y = \dfrac{2x + 1}{x^2 - 2x - 3}$

27. $y = \dfrac{x^2 - 1}{x^2(x + 2)}$

28. $y = \dfrac{x^2 - 4}{x^2(x + 3)}$

29. $y = \dfrac{x^2 + 4}{x - 2}$

30. $y = \dfrac{x^2 + 9}{x - 3}$

31. $y = \dfrac{x^2 - 9}{x^3}$

32. $y = \dfrac{x^2 - 4}{x^3}$

*33. $y = \dfrac{x^3 - 8}{x - 1}$

*34. $y = \dfrac{x^3 + 1}{x + 2}$

*35. $y = \dfrac{x^2 - 4}{x + 2}$

*36. $y = \dfrac{x^2 - 9}{x - 3}$

*37. $y = \dfrac{6x^2 + x - 12}{3x - 4}$

*38. $y = \dfrac{4x^2 - 7x - 15}{x - 3}$

*39. $y = \dfrac{x^3 - 27}{x - 3}$

*40. $y = \dfrac{x^3 + 8}{x + 2}$

1. License tags in a certain state consist of 3 digits, followed by one of the letters A, B, C, D, E, F, G, H, X, then followed by another 3 digits. How many different tags are possible?

2. How many three-digit even numbers can be made using the digits 1, 2, 3, 4, 5 6, and 7?

3. In how many ways can 6 people be seated in a row of 8 chairs?

4. How many "words" can be made using the letters in the word COLLEGE?

5. From a group of 12 people, how many committees of 4 can be chosen?

6. Expand $(x - 2y)^6$ by using the Binomial Theorem.

7. Find the 6th term of $(2x - y^2)^9$.

Write each fraction as a sum of partial fractions.

8. $\dfrac{5x^2 + 5x + 6}{(x + 1)^2(x - 2)}$

9. $\dfrac{2x^2 + 5x + 1}{(x^2 + x + 1)^2}$

10. Sketch the graph of the rational function defined by the following equation. Locate all vertical, horizontal, and oblique asymptotes.

$$y = \frac{2x - 4}{x + 1}$$

Solutions for Practice Tests

CHAPTER 1

1. (a) $A \cap B = \{1, 3, 5, 6, 7, 8\} \cap \{2, 4, 6, 8, 9\} = \{6, 8\}$
(b) $A \cup B = \{1, 3, 5, 6, 7, 8\} \cup \{2, 4, 6, 8, 9\} = \{1, 2, 3, 4, 5, 6, 7, 8, 9\}$
(c) $A' = \{0, 2, 4, 9\}$
(d) $A - B = \{1, 3, 5, 6, 7, 8\} - \{2, 4, 6, 8, 9\} = \{1, 3, 5, 7\}$
(e) $A \cap (B' \cup A) = A$, si ce $A \subseteq B' \cup A$

2. (a) $-4 \in I, -4 \in Q, -4 \in \mathcal{R}$ (b) $-\frac{3}{2} \in Q, -\frac{3}{2} \in \mathcal{R}$
(c) $\sqrt{3} \in \mathcal{R}, \sqrt{3} \in I_r$ (d) $-0.7666 \cdots \in Q, -0.7666 \cdots \in \mathcal{R}$
(e) $\pi \in \mathcal{R}, \pi \in I_r$

3. (a) The set I of integers is *not closed* under division. For example, $3 \in I$ and $2 \in I$, but $3 \div 2 = 1.5 \notin I$.
(b) The set of even integers is *closed* under multiplication.

4. If $x = 2$ and $y = 3$, then
$$[2x(3 - y)] - (7 - y)(2x - 3y) = [4(3 - 3)] - (7 - 3)(4 - 9)$$
$$= [4(0)] - (4)(-5)$$
$$= 0 - (-20)$$
$$= 20.$$

5. *Proof*: $(a)(-b) = (-b)(a)$ (Commutative property, $\cdot$)
$= -(ba)$ (Given fact)
$= -(ab)$ (Commutative property, $\cdot$)

6. (a)
$$-7x + 4 < 2x - 1$$
$$-7x + 4 - 2x - 4 < 2x - 1 - 2x - 4$$
$$-9x < -5$$
$$(-\tfrac{1}{9})(-9x) > (-\tfrac{1}{9})(-5)$$
$$x > \tfrac{5}{9}$$

(b) $-3 < 2x - 7 \le -4$
$$4 < 2x \le 3$$
$$2 < x \le \tfrac{3}{2}$$

7. $-2 < x \le 1$ means that either $x = 1$, or x is between -2 and 1.
$x > 3$ means that x lies to the right of 3 on the number line.
Putting these together, the representation of $-2 < x \le 1$ or $x > 3$ is
given by

8. (a) $|x - 3| < 4$ means that the distance between x and 3 is less than 4,
so x lies between $3 - 4 = -1$ and $3 + 4 = 7$.

(b) Since $|x + 2| = |x - (-2)|$, $|x + 2| \ge 3$ means that the distance
between x and -2 is equal to or greater than 3. Thus x is either at
or to the right of $-2 + 3 = 1$, or x is at or to the left of $-2 - 3$
$= -5$.

9. (a) $|6 - \sqrt{37}| = -(6 - \sqrt{37})$, since $6 < \sqrt{37}$
$$= \sqrt{37} - 6$$

(b) If $x < 0$, then $|x| = -x$ and $\left|\dfrac{x}{3}\right| = \dfrac{|x|}{|3|} = \dfrac{-x}{3} = -\dfrac{x}{3}$.

10. (a) If $y > -3$, then $y + 3 > 0$ and $|y + 3| = y + 3$.

(b) If $x < 1$ then
$$3x < 3$$
$$3x - 5 < 3 - 5,$$
$$3x - 5 < -2.$$
Therefore, $|3x - 5| = -(3x - 5) = 5 - 3x$.

CHAPTER 2

1. (a) $\left(\dfrac{-2x^2}{y}\right)^3 = \dfrac{(-2x^2)^3}{y^3} = \dfrac{(-2)^3(x^2)^3}{y^3} = \dfrac{-8x^6}{y^3}$

(b) $(x^2y)^2(2xy^3)^2 = (x^2)^2y^2 2^2 x^2(y^3)^2 = x^4y^2 \cdot 4 \cdot x^2y^6 = 4x^6y^8$

2. (a) $(a + b)^0 = \begin{cases} 1 & \text{if } a \ne -b \\ \text{undefined} & \text{if } a = -b \end{cases}$

(b) $\dfrac{3a^{-3}y^{-2}}{6ay^{-4}} = \dfrac{y^4}{2a^3ay^2} = \dfrac{y^2}{2a^4}$

3. (a)
$$\begin{array}{r} 5x^2 - 3x + 2 \\ -7x^3 - 3x^2 + x - 1 \\ \hline -7x^3 + 2x^2 - 2x + 1 \end{array}$$

(b)
$$\begin{array}{r} 9x^4 \quad\quad - 3x^2 + 7x \\ - 5x^3 - 4x^2 + 3x + 2 \\ \hline 9x^4 - 5x^3 - 7x^2 + 10x + 2 \end{array}$$

4. (a)
$$
\begin{array}{r}
6x^2 - 3x + 1 \\
5x^3 - 3x \\
\hline
30x^5 - 15x^4 + 5x^3 \\
- 18x^3 + 9x^2 - 3x \\
\hline
30x^5 - 15x^4 - 13x^3 + 9x^2 - 3x
\end{array}
$$

(b) $(3x + y)(9x^2 - 3xy + y^2) = [3x + y][(3x)^2 - (3x)(y) + y^2]$
$$= (3x)^3 + y^3 = 27x^3 + y^3$$

5. (a) $36x^2 - 12x + 1 = (6x - 1)^2$

(b) $8x^3 - 27 = (2x)^3 - (3)^3 = (2x - 3)[(2x)^2 + (2x)(3) + (3)^2]$
$$= (2x - 3)(4x^2 + 6x + 9)$$

6. (a) $6x^3 - 14x^2 - 12x = 2x(3x^2 - 7x - 6) = 2x(3x + 2)(x - 3)$

(b) $x^3 - 4x^2 - 9x + 36 = x^2(x - 4) - 9(x - 4)$
$$= (x^2 - 9)(x - 4)$$
$$= (x - 3)(x + 3)(x - 4)$$

7. (a) $\dfrac{4x^5 - 20x^3 + 14x^2 + 3x - 1}{2x^2} = 2x^3 - 10x + 7 + \dfrac{3x - 1}{2x^2}$

(b)
$$
\begin{array}{r}
x^2 - 2x + 1 \\
3x - 1 \overline{\smash{\big)}\ 3x^3 - 7x^2 + 5x - 2} \\
\underline{3x^3 - x^2} \\
- 6x^2 + 5x \\
\underline{- 6x^2 + 2x} \\
3x - 2 \\
\underline{3x - 1} \\
- 1
\end{array}
$$

$$\frac{3x^3 - 7x^2 + 5x - 2}{3x - 1} = x^2 - 2x + 1 - \frac{1}{3x - 1}$$

8.
$$
\begin{array}{r}
1 \,\overline{\smash{|}}\ 1 \quad 0 \quad -7 \quad 3 \quad 5 \\
\ 1 \quad 1 \quad -6 \quad -3 \\
\hline
1 \quad 1 \quad -6 \quad -3 \quad 2
\end{array}
$$

$$\frac{x^4 - 7x^2 + 3x + 5}{x - 1} = x^3 + x^2 - 6x - 3 + \frac{2}{x - 1}$$

9. (a) $\dfrac{1}{2x - 2} - \dfrac{x}{x^2 - 4x + 3} = \dfrac{1}{2(x - 1)} - \dfrac{x}{(x - 3)(x - 1)}$

$$= \frac{x - 3}{2(x - 1)(x - 3)} - \frac{2x}{2(x - 3)(x - 1)}$$

$$= \frac{x - 3 - 2x}{2(x - 1)(x - 3)}$$

$$= -\frac{x + 3}{2(x - 1)(x - 3)}$$

(b) $\dfrac{(x - y)^2}{2x^2 + 3xy + y^2} \div \dfrac{x^2 - y^2}{x^2 + 2xy + y^2}$

$$= \frac{(x - y)^2}{2x^2 + 3xy + y^2} \cdot \frac{x^2 + 2xy + y^2}{x^2 - y^2}$$

$$= \frac{(x - y)^2}{(2x + y)(x + y)} \cdot \frac{(x + y)^2}{(x + y)(x - y)}$$

$$= \frac{x - y}{2x + y}$$

$$10. \quad \dfrac{\dfrac{x}{y} - \dfrac{y}{x}}{\dfrac{1}{x} - \dfrac{1}{y}} = \dfrac{\dfrac{x^2}{xy} - \dfrac{y^2}{xy}}{\dfrac{y}{xy} - \dfrac{x}{xy}} = \dfrac{\dfrac{x^2 - y^2}{xy} \cdot xy}{\dfrac{y - x}{xy} \cdot xy}$$

$$= \dfrac{x^2 - y^2}{y - x} = \dfrac{(x-y)(x+y)}{y - x}$$

$$= \dfrac{-(y-x)(x+y)}{y-x} = -(x+y)$$

CHAPTER 3

1. (a) $\sqrt[3]{-\dfrac{125}{8}} = \sqrt[3]{\dfrac{(-5)^3}{2^3}} = \dfrac{\sqrt[3]{(-5)^3}}{\sqrt[3]{2^3}} = -\dfrac{5}{2}$

(b) $\sqrt{\sqrt[3]{64}} = \sqrt{\sqrt[3]{4^3}} = \sqrt{4} = 2$

2. (a) $\sqrt[3]{\dfrac{216x^6}{64y^3}} = \sqrt[3]{\dfrac{6^3(x^2)^3}{4^3 y^3}} = \dfrac{\sqrt[3]{6^3}\,\sqrt[3]{(x^2)^3}}{\sqrt[3]{4^3}\,\sqrt[3]{y^3}} = \dfrac{6x^2}{4y} = \dfrac{3x^2}{2y}$

(b) $\sqrt[4]{\dfrac{16x^8}{81y^{12}}} = \sqrt[4]{\dfrac{2^4(x^2)^4}{3^4(y^3)^4}} = \dfrac{\sqrt[4]{2^4}\,\sqrt[4]{(x^2)^4}}{\sqrt[4]{3^4}\,\sqrt[4]{(y^3)^4}} = \dfrac{2x^2}{3y^3}$

3. (a) $(3x)^{-2/5} = (\sqrt[5]{3x})^{-2} = \dfrac{1}{(\sqrt[5]{3x})^2} = \dfrac{1}{\sqrt[5]{9x^2}}$

(b) $\sqrt[4]{y^3} = (y^3)^{1/4} = y^{3/4}$

4. (a) $(-32)^{3/5} = (\sqrt[5]{-32})^3 = (-2)^3 = -8$

(b) $(\tfrac{1}{125})^{-2/3} = (125)^{2/3} = (\sqrt[3]{125})^2 = 5^2 = 25$

5. (a) $3\sqrt{32} - 5\sqrt{50} = 3\sqrt{(16)(2)} - 5\sqrt{(25)(2)} = 3 \cdot 4\sqrt{2} - 5 \cdot 5\sqrt{2}$

$$= (12 - 25)\sqrt{2} = -13\sqrt{2}$$

(b) $\sqrt[3]{81x^4y} - \sqrt[3]{24x^4y} = \sqrt[3]{27x^3}\,\sqrt[3]{3xy} - \sqrt[3]{8x^3}\,\sqrt[3]{3xy}$

$$= 3x\sqrt[3]{3xy} - 2x\sqrt[3]{3xy}$$

$$= x\sqrt[3]{3xy}$$

6. (a) $\dfrac{\sqrt[3]{40}}{\sqrt[3]{5a^3}} = \sqrt[3]{\dfrac{40}{5a^3}} = \sqrt[3]{\dfrac{8}{a^3}} = \dfrac{2}{a}$

(b) $(3 - 2\sqrt{5})(2 + 4\sqrt{5}) = 6 + 12\sqrt{5} - 4\sqrt{5} - 8(\sqrt{5})^2$

$$= 6 + 8\sqrt{5} - 40$$

$$= -34 + 8\sqrt{5}$$

7. (a) $3a\sqrt{7} = \sqrt{9a^2}\,\sqrt{7} = \sqrt{63a^2}$

(b) $2\sqrt[4]{5x} = \sqrt[4]{2^4}\,\sqrt[4]{5x} = \sqrt[4]{16}\,\sqrt[4]{5x} = \sqrt[4]{80x}$

8. (a) $\sqrt[12]{125} = \sqrt[12]{5^3} = 5^{3/12} = 5^{1/4} = \sqrt[4]{5}$

(b) $\sqrt[9]{125x^{12}} = \sqrt[9]{5^3(x^4)^3} = \sqrt[9]{(5x^4)^3} = (5x^4)^{3/9} = (5x^4)^{1/3} = \sqrt[3]{5x^4}$

$$= x\sqrt[3]{5x}$$

9. (a) $\dfrac{3}{2 - \sqrt{2}} = \dfrac{3(2 + \sqrt{2})}{(2 - \sqrt{2})(2 + \sqrt{2})} = \dfrac{3(2 + \sqrt{2})}{4 - 2} = \dfrac{3(2 + \sqrt{2})}{2}$

(b) $\sqrt[3]{\dfrac{4x}{3y}} = \sqrt[3]{\dfrac{4x(3y)^2}{(3y)(3y)^2}} = \dfrac{\sqrt[3]{4x(3y)^2}}{\sqrt[3]{(3y)^3}} = \dfrac{\sqrt[3]{36xy^2}}{3y}$

10. (a) $\sqrt[3]{9}\sqrt{36} = \sqrt[3]{9 \cdot 6} = \sqrt[3]{3^2 \cdot 3 \cdot 2} = \sqrt[3]{3^3 \cdot 2} = 3\sqrt[3]{2}$

(b) $\sqrt[5]{x^2y^3} \cdot \sqrt{xy} = (x^2y^3)^{1/5}(xy)^{1/2} = (x^2y^3)^{2/10}(xy)^{5/10}$

$$= \sqrt[10]{(x^2y^3)^2}\,\sqrt[10]{(xy)^5} = \sqrt[10]{x^4y^6x^5y^5}$$

$$= \sqrt[10]{x^9y^{11}} = y\sqrt[10]{x^9y}$$

CHAPTER 4

1. (a) $7x + 4(3 - 2x) = 5 + x$ (b) $\dfrac{3}{x + 1} + 5 = \dfrac{4x}{x + 1}$

$\quad\quad 7x + 12 - 8x = 5 + x$

$\quad\quad 7x - 8x - x = 5 - 12$

$\quad\quad\quad\quad\quad -2x = -7$

$\quad\quad\quad\quad\quad\quad x = \frac{7}{2}$

$\quad\quad$ Solution set: $\{\frac{7}{2}\}$

$\quad\quad\quad 3 + 5(x + 1) = 4x \quad (x \neq -1)$

$\quad\quad\quad\quad 3 + 5x + 5 = 4x$

$\quad\quad\quad\quad\quad 5x - 4x = -8$

$\quad\quad\quad\quad\quad\quad\quad x = -8$

$\quad\quad\quad$ Solution set: $\{-8\}$

2. $\dfrac{|3 - 2x|}{|2 - x|} = 2$

$\quad \left|\dfrac{3 - 2x}{2 - x}\right| = 2$

$\quad \dfrac{3 - 2x}{2 - x} = \pm 2$

$\dfrac{3 - 2x}{2 - x} = 2 \quad$ or $\quad \dfrac{3 - 2x}{2 - x} = -2$

$3 - 2x = 2(2 - x) \quad\quad 3 - 2x = -2(2 - x)$

$3 - 2x = 4 - 2x \quad\quad\quad 3 - 2x = -4 + 2x$

$\quad\quad\quad 3 = 4 \quad\quad\quad\quad\quad\quad -4x = -7$

$\quad\quad$ Impossible $\quad\quad\quad\quad\quad\quad x = \frac{7}{4}$

Solution set: $\{\frac{7}{4}\}$

3. Let x be the number of hours it takes Jane to mow the yard alone. Then $1/x$ is the portion of the yard mowed by Jane in 1 hour. Ron mows $\frac{1}{4}$ of the yard in 1 hour. Since it takes them 1 hour to mow the yard together,

$$\frac{1}{x} + \frac{1}{4} = 1$$

$$4 + x = 4x$$

$$4 = 3x$$

$$x = \tfrac{4}{3} \text{ hr} \quad \text{or}$$

$$x = 1 \text{ hr } 20 \text{ min.}$$

4. $\quad\quad |7 - 4x| < 5$

$\quad -5 < 7 - 4x < 5$

$-12 < \quad -4x < -2$

$\quad 3 > x > \frac{1}{2}, \quad$ or $\quad \frac{1}{2} < x < 3$

5. Let $x =$ number of cubic meters of concrete that can be made. There is 1 part cement in 6 parts total volume, so

$$\frac{1}{6} = \frac{4}{x}$$

$$x = (6)(4) = 24 \text{ cubic meters.}$$

6. $\quad z = \dfrac{kyx^2}{w^3} \quad\quad\quad\quad z = \dfrac{32xy^2}{3w^3}$

$\quad 12 = \dfrac{(k)(2)(36)}{64} \quad\quad z = \dfrac{(32)(3)(16)}{(3)(8)}$

$\quad 12 = \dfrac{(k)(9)}{8} \quad\quad\quad\quad z = 64$

$\quad 96 = 9k$

$\quad\quad k = \frac{96}{9} = \frac{32}{3}$

7. $6x^2 + 7x - 20 = 0$

$(3x - 4)(2x + 5) = 0$

$3x - 4 = 0$ or $2x + 5 = 0$

$x = \frac{4}{3}$ or $x = -\frac{5}{2}$

8. Let $z = \sqrt{x}$ in $7 - 15\sqrt{x} + 2x = 0$.

$$7 - 15z + 2z^2 = 0$$

$$(7 - z)(1 - 2z) = 0$$

$7 - z = 0$ or $1 - 2z = 0$

$z = 7$ or $z = \frac{1}{2}$

$\sqrt{x} = 7$ or $\sqrt{x} = \frac{1}{2}$

$x = 49$ or $x = \frac{1}{4}$

Checking $x = 49$: $7 - (15)(7) + (2)(7) = 21 - 105 \neq 0$.

Checking $x = \frac{1}{4}$: $7 - (15)(\frac{1}{2}) + (2)(\frac{1}{4}) = 7 - \frac{15}{2} + \frac{1}{2} = 0$.

Solution set: $\{\frac{1}{4}\}$.

9. $5x^2 - 3x - 1 = 0$

$a = 5, b = -3, c = -1$

$$x = \frac{-b \pm \sqrt{b^2 - 4ac}}{2a} = \frac{3 \pm \sqrt{9 - (4)(5)(-1)}}{(2)(5)} = \frac{3 \pm \sqrt{29}}{10}.$$

10. $3x = 5 - \sqrt{7 - 3x}$

$3x - 5 = -\sqrt{7 - 3x}$

$9x^2 - 30x + 25 = 7 - 3x$

$9x^2 - 27x + 18 = 0$

$x^2 - 3x + 2 = 0$

$(x - 1)(x - 2) = 0$

$x = 1$ or $x = 2$

Checking $x = 1$: RHS $= 5 - \sqrt{7 - 3} = 5 - 2 = 3 =$ LHS.

Checking $x = 2$: RHS $= 5 - \sqrt{7 - 6} = 5 - 1 = 4 \neq$ LHS.

Solution set: $\{1\}$.

CHAPTER 5

1. $y = \sqrt{x^2 - 9} - 4$

Since $x^2 - 9 \geq 0$, then $x^2 \geq 9$. Thus the domain is $\{x \mid x \geq 3$ or $x \leq -3\}$.

Since $\sqrt{x^2 - 9} \geq 0$, the range is $\{y \mid y \geq -4\}$.

2. $f(x) = x^2 - 2x, g(x) = \sqrt{x} + 3$

(a) $f(4) = 4^2 - (2)(4) = 16 - 8 = 8$

(b) $f(g(9)) = f(\sqrt{9} + 3) = f(6) = 6^2 - (2)(6) = 36 - 12 = 24$

(c) $(g \circ f)(4) = g(f(4)) = g(8) = \sqrt{8} + 3 = 2\sqrt{2} + 3$

(d) $f(g(a)) = f(\sqrt{a} + 3)$

$= (\sqrt{a} + 3)^2 - 2(\sqrt{a} + 3)$

$= a + 6\sqrt{a} + 9 - 2\sqrt{a} - 6$

$= a + 4\sqrt{a} + 3$

(e) $f(x + h) - f(x) = (x + h)^2 - 2(x + h) - (x^2 - 2x)$

$= x^2 + 2xh + h^2 - 2x - 2h - x^2 + 2x$

$= 2xh + h^2 - 2h$

$= h(2x + h - 2)$

3. $y = |3x + 2|$

If $3x + 2 \geq 0$, or $x \geq -\frac{2}{3}$, then $y = 3x + 2$.

If $3x + 2 < 0$, or $x < -\frac{2}{3}$, then $y = -3x - 2$.

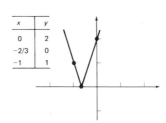

4. (a) Setting $x = 0$ in $3x - 2y = 12$ gives $-2y = 12$, or $y = -6$. Thus -6 is the y-intercept. Setting $y = 0$ yields $3x = 12$, or $x = 4$. Thus 4 is the x-intercept. The graph is as shown.
   (b) The straight line $y = 5$ has no x-intercept, and 5 is the y-intercept. The graph is as shown.

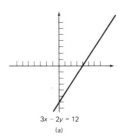

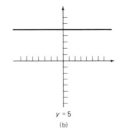

$3x - 2y = 12$
(a)

$y = 5$
(b)

5. The slope of the line through $(3, -1)$ and $(2, 7)$ is $m = \dfrac{7 - (-1)}{2 - 3} = -8$.

   The point-slope form is $y - (-1) = -8(x - 3)$. This gives $y + 1 = -8x + 24$, and the standard form is $8x + y = 23$.

6. $x = \frac{2}{3}(y - 1)^2 + 4$

   Since $\frac{2}{3} > 0$, the parabola opens to the right. The vertex is at $(4, 1)$, and the axis of symmetry has equation $y = 1$. Setting $y = 0$ yields $x = \frac{2}{3}(-1)^2 + 4 = \frac{2}{3} + 4 = \frac{14}{3}$. Also, if $y = 2$, then $x = \frac{2}{3}(1)^2 + 4 = \frac{14}{3}$.

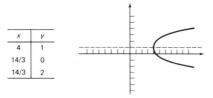

7. $$x = -4 + \sqrt{y - 4}$$
   $$x + 4 = \sqrt{y - 4}$$
   $$(x + 4)^2 = y - 4$$
   $$y = (x + 4)^2 + 4$$

   The graph of $y = (x + 4)^2 + 4$ is a parabola opening upward with vertex at $(-4, 4)$ and axis of symmetry $x = -4$. But the graph of the original equation $x = -4 + \sqrt{y - 4}$ is only the right half of the parabola since $\sqrt{y - 4} \geq 0$ implies that $x \geq -4$.

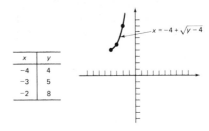

x	y
-4	4
-3	5
-2	8

8. $x^2 + 3x + 2 > 6$

$x^2 + 3x - 4 > 0$

The solution set is the set of all x for which the graph of $y = x^2 + 3x - 4$ is above the x-axis. The graph of $y = x^2 + 3x - 4$ is a parabola opening upward with x-intercepts found by setting $y = 0$:

$x^2 + 3x - 4 = 0,$

$(x + 4)(x - 1) = 0,$

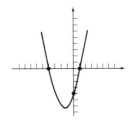

$x = -4, \qquad x = 1.$

The graph is above the x-axis for $x > 1$ or $x < -4$. The solution set is $\{x \mid x > 1 \text{ or } x < -4\}$.

9. $w^2 + 6w + 8 < 0$

$w^2 + 6w + 8 = 0$

$(w + 4)(w + 2) = 0$

Thus $w = -4$ and $w = -2$ are zeros of the function $f(w) = w^2 + 6w + 8$ and separate the real numbers into the three regions $w < -4$, $-4 < w < -2$, $w > -2$.

Testing points $w = -5$, $w = -3$, $w = 0$, we have

$$f(-5) = (-5)^2 + 6(-5) + 8 = 25 - 30 + 8 = 3 \not< 0,$$
$$f(-3) = (-3)^2 + 6(-3) + 8 = 9 - 18 + 8 = -1 < 0,$$
$$f(0) = (0)^2 + 6(0) + 8 = 8 \not< 0.$$

The solution set is given by $\{w \mid -4 < w < -2\}$.

10. $\left| \dfrac{3x - 1}{x + 1} \right| > 2, \quad x \neq -1$

$$\dfrac{(3x - 1)^2}{(x + 1)^2} > 4$$
$$(3x - 1)^2 > 4(x + 1)^2 \qquad (x \neq -1)$$
$$9x^2 - 6x + 1 > 4x^2 + 8x + 4$$
$$5x^2 - 14x - 3 > 0$$
$$(5x + 1)(x - 3) > 0$$

The zeros of $f(x) = (5x + 1)(x - 3)$ are $-\frac{1}{5}$ and 3. They separate the real numbers into $x < -\frac{1}{5}$, $-\frac{1}{5} < x < 3$, $x > 3$. Testing points in each region, we have $f(-1) = (-4)(-4) > 0$, $f(0) = (1)(-3) \not> 0$, $f(4) = (21)(1) > 0$.

The solutions to $(5x + 1)(x - 3) > 0$ are $x < -\frac{1}{5}$ or $x > 4$.

Since $x \neq -1$ in $\left|\dfrac{3x - 1}{x + 1}\right| > 2$, the solution set is $\{x \,|\, x < -1$ or $-1 <$

$x < -\frac{1}{5}$ or $x > 4\}$.

CHAPTER 6

1. Let $(x_1, y_1) = (7, 2)$ and $(x_2, y_2) = (3, -1)$. Then
$$d = \sqrt{(3 - 7)^2 + (-1 - 2)^2} = \sqrt{(-4)^2 + (-3)^2} = \sqrt{16 + 9} = \sqrt{25}$$
$$= 5.$$

2.
$$x^2 - 2x + y^2 = 12y - 38$$
$$(x^2 - 2x + 1) + (y^2 - 12y + 36) = -38 + 1 + 36$$
$$(x - 1)^2 + (y - 6)^2 = -1$$
Since the sum of the squares of two real numbers can never be negative, the equation $x^2 - 2x + y^2 = 12y - 38$ does not represent a circle.

3. An equation of the circle with center $(2, -3)$ and radius 4 is $(x - 2)^2$ $+ [y - (-3)]^2 = 4^2$ or $(x - 2)^2 + (y + 3)^2 = 16$.

4.
$$2y = \sqrt{1 - 4x^2}$$
$$4y^2 = 1 - 4x^2$$

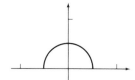

$$4x^2 + 4y^2 = 1$$
$$x^2 + y^2 = \tfrac{1}{4}$$
This is an equation of the circle with center $(0, 0)$ and radius $\frac{1}{2}$. But $2y = \sqrt{1 - 4x^2}$ is that portion of the circle where $y \geq 0$.

5. $25y^2 - 4x^2 = 100$
$$\frac{25y^2}{100} - \frac{4x^2}{100} = \frac{100}{100}$$
$$\frac{y^2}{4} - \frac{x^2}{25} = 1$$

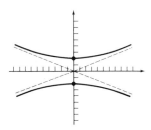

Hyperbola with y-intercepts at ± 2.

Asymptotes: $\dfrac{y^2}{4} - \dfrac{x^2}{25} = 0$
$$\frac{y^2}{4} = \frac{x^2}{25}$$
$$y^2 = \frac{4x^2}{25}$$
$$y = \pm\tfrac{2}{5}x$$

6. $2x = -\sqrt{16 - y^2}$

$$4x^2 = 16 - y^2$$
$$4x^2 + y^2 = 16$$
$$\frac{x^2}{4} + \frac{y^2}{16} = 1$$

Ellipse with x-intercepts ± 2 and y-intercepts ± 4. Since $-\sqrt{16 - y^2} \le 0$, then $x \le 0$, and the graph of $2x = -\sqrt{16 - y^2}$ is the left half of an ellipse.

7.
$$x^2 - 3y = 1$$
$$2x - y = 3$$
$$y = 2x - 3$$
$$x^2 - 3(2x - 3) = 1$$
$$x^2 - 6x + 9 - 1 = 0$$
$$x^2 - 6x + 8 = 0$$
$$(x - 4)(x - 2) = 0$$
$$x = 4, \qquad y = 2(4) - 3 = 5$$
$$x = 2, \qquad y = 2(2) - 3 = 1$$
The solutions are $(4, 5)$ and $(2, 1)$.

8.
$$x^2 + 3xy - 6y^2 = 8$$
$$\underline{x^2 - \;\;xy - 6y^2 = 4}$$
$$4xy \qquad\quad = 4$$
$$xy = 1$$
$$y = \frac{1}{x}$$

Substituting in the first equation yields
$$x^2 + 3x\left(\frac{1}{x}\right) - 6\left(\frac{1}{x^2}\right) = 8$$
$$x^2 + 3 - \frac{6}{x^2} = 8$$
$$x^2 - 5 - \frac{6}{x^2} = 0$$
$$x^4 - 5x^2 - 6 = 0$$
$$(x^2 - 6)(x^2 + 1) = 0$$
$$x^2 - 6 = 0 \qquad\qquad x^2 + 1 = 0$$
$$x^2 = 6 \qquad\qquad\quad x^2 = -1$$
$$x = \pm\sqrt{6} \qquad \text{No solution.}$$
Solutions: $\{(\sqrt{6}, \sqrt{6}/6), (-\sqrt{6}, -\sqrt{6}/6)\}$.

9.
$$x^2 + y^2 > \;\;9$$
$$16x^2 + 9y^2 \le 144$$
The solutions to $x^2 + y^2 > 9$ are those points outside the circle $x^2 + y^2$

== 9. $16x^2 + 9y^2 \leq 144$ is equivalent to $\dfrac{16x^2}{144} + \dfrac{9y^2}{144} \leq 1$, or $\dfrac{x^2}{9} + \dfrac{y^2}{16}$ ≤ 1. The solutions here are the points interior to or on the ellipse $\dfrac{x^2}{3^2} + \dfrac{y^2}{4^2} = 1$. The solutions to the system are those points exterior to the circle and interior or on the ellipse.

Points of intersection:
$$16x^2 + 9y^2 = 144$$

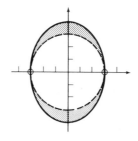

$$\begin{aligned} 9x^2 + 9y^2 &= 81 \\ \overline{7x^2} \quad\quad &= 63 \\ x^2 &= 9 \\ x &= \pm 3 \\ (\pm 3)^2 + y^2 &= 9 \\ 9 + y^2 &= 9 \\ y^2 &= 0 \\ y &= 0 \end{aligned}$$

The points of intersection are $(3, 0)$ and $(-3, 0)$.

10. $y = \sqrt{x-2}$

(a) $f = \{(x, y) \mid y = \sqrt{x-2}\}$
$f^{-1} = \{(x, y) \mid x = \sqrt{y-2}\}$
$x = \sqrt{y-2}$
$x^2 = y - 2$
$y = x^2 + 2$
$f^{-1}(x) = x^2 + 2, \quad x \geq 0$

(b)

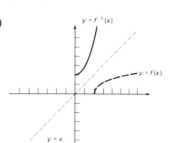

(c) Domain of $f^{-1} = \{x \mid x \geq 0\}$.
Range of $f^{-1} = \{y \mid y \geq 2\}$.

1. $\begin{bmatrix} 2 & -3 \\ -4 & -5 \\ 0 & 2 \end{bmatrix} - 2\begin{bmatrix} -1 & 0 \\ 3 & -11 \\ 1 & 0 \end{bmatrix} = \begin{bmatrix} 2 & -3 \\ -4 & -5 \\ 0 & 2 \end{bmatrix} - \begin{bmatrix} -2 & 0 \\ 6 & -22 \\ 2 & 0 \end{bmatrix}$

$= \begin{bmatrix} 0 & -3 \\ -10 & 17 \\ -2 & 2 \end{bmatrix}$

2. Not possible, since the two matrices have different dimensions.

3. Not possible, since the number of columns in the first matrix does not equal the number of rows in the second matrix.

4. $\begin{bmatrix} 1 & -2 & 0 \\ 0 & 3 & 2 \\ 5 & 0 & 1 \end{bmatrix}\begin{bmatrix} -1 & 0 \\ 3 & 4 \\ 0 & -1 \end{bmatrix} = \begin{bmatrix} -7 & -8 \\ 9 & 10 \\ -5 & -1 \end{bmatrix}$

5. $[A \mid B] = \begin{bmatrix} 1 & -2 & -4 & | & -1 \\ 3 & 0 & -1 & | & 4 \\ 1 & 4 & 7 & | & 2 \end{bmatrix} \xrightarrow{-3R_1+R_2} \begin{bmatrix} 1 & -2 & -4 & | & -1 \\ 0 & 6 & 11 & | & 7 \\ 1 & 4 & 7 & | & 2 \end{bmatrix}$

$\xrightarrow{-R_1+R_2} \begin{bmatrix} 1 & -2 & -4 & | & -1 \\ 0 & 6 & 11 & | & 7 \\ 0 & 6 & 11 & | & 3 \end{bmatrix} \xrightarrow{-R_2+R_3} \begin{bmatrix} 1 & -2 & -4 & | & -1 \\ 0 & 6 & 11 & | & 7 \\ 0 & 0 & 0 & | & -4 \end{bmatrix}$

There is no solution.

6. $[A \mid B] = \begin{bmatrix} -1 & 2 & 1 & | & -1 \\ 3 & 1 & -1 & | & 7 \\ 0 & 1 & 1 & | & -3 \end{bmatrix} \xrightarrow{-R_1} \begin{bmatrix} 1 & -2 & -1 & | & 1 \\ 3 & 1 & -1 & | & 7 \\ 0 & 1 & 1 & | & -3 \end{bmatrix}$

$\xrightarrow{-3R_1+R_2} \begin{bmatrix} 1 & -2 & -1 & | & 1 \\ 0 & 7 & 2 & | & 4 \\ 0 & 1 & 1 & | & -3 \end{bmatrix} \xrightarrow{R_2 \leftrightarrow R_3} \begin{bmatrix} 1 & -2 & -1 & | & 1 \\ 0 & 1 & 1 & | & -3 \\ 0 & 7 & 2 & | & 4 \end{bmatrix}$

$\xrightarrow{2R_2+R_1} \begin{bmatrix} 1 & 0 & 1 & | & -5 \\ 0 & 1 & 1 & | & -3 \\ 0 & 7 & 2 & | & 4 \end{bmatrix} \xrightarrow{-7R_2+R_3} \begin{bmatrix} 1 & 0 & 1 & | & -5 \\ 0 & 1 & 1 & | & -3 \\ 0 & 0 & -5 & | & 25 \end{bmatrix}$

$\xrightarrow{-\frac{1}{5}R_3} \begin{bmatrix} 1 & 0 & 1 & | & -5 \\ 0 & 1 & 1 & | & -3 \\ 0 & 0 & 1 & | & -5 \end{bmatrix} \xrightarrow{-R_3+R_1} \begin{bmatrix} 1 & 0 & 0 & | & 0 \\ 0 & 1 & 1 & | & -3 \\ 0 & 0 & 1 & | & -5 \end{bmatrix}$

$\xrightarrow{-R_3+R_2} \begin{bmatrix} 1 & 0 & 0 & | & 0 \\ 0 & 1 & 0 & | & 2 \\ 0 & 0 & 1 & | & -5 \end{bmatrix} = [I \mid C]$

The solution is $x = 0$, $y = 2$, $z = -5$.

7. $\begin{bmatrix} 2 & -3 \\ -1 & 2 \end{bmatrix}^{-1} = \frac{1}{4-3}\begin{bmatrix} 2 & 3 \\ 1 & 2 \end{bmatrix} = \begin{bmatrix} 2 & 3 \\ 1 & 2 \end{bmatrix}$

8. $[A \mid I] = \begin{bmatrix} 1 & 1 & -3 & | & 1 & 0 & 0 \\ 1 & 0 & 3 & | & 0 & 1 & 0 \\ -2 & 1 & -12 & | & 0 & 0 & 1 \end{bmatrix}$

$\xrightarrow{-R_1+R_2} \begin{bmatrix} 1 & 1 & -3 & | & 1 & 0 & 0 \\ 0 & -1 & 6 & | & -1 & 1 & 0 \\ -2 & 1 & -12 & | & 0 & 0 & 1 \end{bmatrix}$

$\xrightarrow{2R_1+R_3} \begin{bmatrix} 1 & 1 & -3 & | & 1 & 0 & 0 \\ 0 & -1 & 6 & | & -1 & 1 & 0 \\ 0 & 3 & -18 & | & 2 & 0 & 1 \end{bmatrix}$

$$\xrightarrow{3R_2+R_3} \begin{bmatrix} 1 & 1 & -3 & | & 1 & 0 & 0 \\ 0 & -1 & 6 & | & -1 & 1 & 0 \\ 0 & 0 & 0 & | & -1 & 3 & 1 \end{bmatrix}$$

The inverse does not exist.

9. $A^{-1} = \dfrac{1}{(4)(-3)-(-1)(7)}\begin{bmatrix} -3 & 1 \\ -7 & 4 \end{bmatrix} = -\dfrac{1}{5}\begin{bmatrix} -3 & 1 \\ -7 & 4 \end{bmatrix}$

$X = A^{-1}B = -\dfrac{1}{5}\begin{bmatrix} -3 & 1 \\ -7 & 4 \end{bmatrix}\begin{bmatrix} 2 \\ 1 \end{bmatrix} = -\dfrac{1}{5}\begin{bmatrix} -5 \\ -10 \end{bmatrix} = \begin{bmatrix} 1 \\ 2 \end{bmatrix}$

The solution is $x = 1$, $y = 2$.

10. $[A \,|\, I] = \begin{bmatrix} 2 & 1 & -1 & | & 1 & 0 & 0 \\ 0 & 1 & -2 & | & 0 & 1 & 0 \\ 1 & 3 & 0 & | & 0 & 0 & 1 \end{bmatrix} \xrightarrow{R_1 \leftrightarrow R_3} \begin{bmatrix} 1 & 3 & 0 & | & 0 & 0 & 1 \\ 0 & 1 & -2 & | & 0 & 1 & 0 \\ 2 & 1 & -1 & | & 1 & 0 & 0 \end{bmatrix}$

$\xrightarrow{-2R_1+R_3} \begin{bmatrix} 1 & 3 & 0 & | & 0 & 0 & 1 \\ 0 & 1 & -2 & | & 0 & 1 & 0 \\ 0 & -5 & -1 & | & 1 & 0 & -2 \end{bmatrix}$

$\xrightarrow{-3R_2+R_1} \begin{bmatrix} 1 & 0 & 6 & | & 0 & -3 & 1 \\ 0 & 1 & -2 & | & 0 & 1 & 0 \\ 0 & -5 & -1 & | & 1 & 0 & -2 \end{bmatrix}$

$\xrightarrow{5R_2+R_3} \begin{bmatrix} 1 & 0 & 6 & | & 0 & -3 & 1 \\ 0 & 1 & -2 & | & 0 & 1 & 0 \\ 0 & 0 & -11 & | & 1 & 5 & -2 \end{bmatrix}$

$\xrightarrow{-\frac{1}{11}R_3} \begin{bmatrix} 1 & 0 & 6 & | & 0 & -3 & 1 \\ 0 & 1 & -2 & | & 0 & 1 & 0 \\ 0 & 0 & 1 & | & -\frac{1}{11} & -\frac{5}{11} & \frac{2}{11} \end{bmatrix}$

$\xrightarrow{2R_3+R_2} \begin{bmatrix} 1 & 0 & 6 & | & 0 & -3 & 1 \\ 0 & 1 & 0 & | & -\frac{2}{11} & \frac{1}{11} & \frac{4}{11} \\ 0 & 0 & 1 & | & -\frac{1}{11} & -\frac{5}{11} & \frac{2}{11} \end{bmatrix}$

$\xrightarrow{-6R_3+R_1} \begin{bmatrix} 1 & 0 & 0 & | & \frac{6}{11} & -\frac{3}{11} & -\frac{1}{11} \\ 0 & 1 & 0 & | & -\frac{2}{11} & \frac{1}{11} & \frac{4}{11} \\ 0 & 0 & 1 & | & -\frac{1}{11} & -\frac{5}{11} & \frac{2}{11} \end{bmatrix}$

$= [I \,|\, A^{-1}]$

$X = A^{-1}B = \dfrac{1}{11}\begin{bmatrix} 6 & -3 & -1 \\ -2 & 1 & 4 \\ -1 & -5 & 2 \end{bmatrix}\begin{bmatrix} 1 \\ 9 \\ 1 \end{bmatrix} = \dfrac{1}{11}\begin{bmatrix} -22 \\ 11 \\ -44 \end{bmatrix} = \begin{bmatrix} -2 \\ 1 \\ -4 \end{bmatrix}$

The solution is $x = -2$, $y = 1$, $z = -4$.

CHAPTER 8

1. $\begin{vmatrix} 4 & -2 \\ -1 & -3 \end{vmatrix} = (4)(-3) - (-1)(-2) = -12 - 2 = -14$

2. $\begin{vmatrix} 3 & -2 & 1 \\ 4 & 0 & 1 \\ 2 & 5 & -2 \end{vmatrix} = -(-2)\begin{vmatrix} 4 & 1 \\ 2 & -2 \end{vmatrix} - (5)\begin{vmatrix} 3 & 1 \\ 4 & 1 \end{vmatrix}$

$= 2(-8 - 2) - 5(3 - 4)$
$= 2(-10) - 5(-1) = -20 + 5 = -15$

3. $\begin{vmatrix} -2 & x & -2 \\ 3 & 1 & 0 \\ 1 & 2x & 1 \end{vmatrix} = -2\begin{vmatrix} 3 & 1 \\ 1 & 2x \end{vmatrix} + 1\begin{vmatrix} -2 & x \\ 3 & 1 \end{vmatrix}$

$$= -2(6x - 1) + (-2 - 3x)$$
$$= -12x + 2 - 2 - 3x = -15x$$

Therefore, $-15x = 0$, or $x = 0$.

4. $\begin{vmatrix} 2 & 1 & 1 \\ -3 & 5 & -3 \\ 4 & 2 & 1 \end{vmatrix} \overset{-2R_1+R_3}{=} \begin{vmatrix} 2 & 1 & 1 \\ -3 & 5 & -3 \\ 0 & 0 & -1 \end{vmatrix}$. Thus $x = -1$.

5. $\begin{vmatrix} 1 & -2 & -1 \\ 3 & 0 & 2 \\ 2 & 1 & 1 \end{vmatrix} \overset{C_1 \leftrightarrow C_3}{=} (-1)\begin{vmatrix} -1 & -2 & 1 \\ 2 & 0 & 3 \\ 1 & 1 & 2 \end{vmatrix}$. Thus $x = -1$.

6. $\begin{vmatrix} -2 & 4 & 8 \\ -4 & 2 & -6 \\ 2 & 2 & -4 \end{vmatrix} = (2)\begin{vmatrix} -1 & 2 & 4 \\ -4 & 2 & -6 \\ 2 & 2 & -4 \end{vmatrix} = (2)(2)\begin{vmatrix} -1 & 2 & 4 \\ -2 & 1 & -3 \\ 2 & 2 & -4 \end{vmatrix}$

$$= (2)(2)(2)\begin{vmatrix} -1 & 2 & 4 \\ -2 & 1 & -3 \\ 1 & 1 & -2 \end{vmatrix}$$

Thus $x = 8$.

7. $\begin{vmatrix} -1 & 0 & 2 & 0 \\ 2 & 3 & 1 & 2 \\ 1 & 1 & 1 & 0 \\ -1 & 2 & -1 & 3 \end{vmatrix} \overset{2C_1+C_3}{=} \begin{vmatrix} -1 & 0 & 0 & 0 \\ 2 & 3 & 5 & 2 \\ 1 & 1 & 2 & 0 \\ -1 & 2 & -3 & 3 \end{vmatrix} = (-1)\begin{vmatrix} 3 & 5 & 2 \\ 1 & 2 & 0 \\ 2 & -3 & 3 \end{vmatrix}$

$$\overset{-2C_1+C_2}{=} (-1)\begin{vmatrix} 3 & -1 & 2 \\ 1 & 0 & 0 \\ 2 & -7 & 3 \end{vmatrix} = (-1)(-1)\begin{vmatrix} -1 & 2 \\ -7 & 3 \end{vmatrix}$$

$$= -3 - (-14) = 11$$

8. $D = \begin{vmatrix} 3 & -1 \\ 6 & -2 \end{vmatrix} = -6 - (-6) = 0$, $D_x = \begin{vmatrix} -2 & -1 \\ -4 & -2 \end{vmatrix} = 4 - 4 = 0$,

$D_y = \begin{vmatrix} 3 & -2 \\ 6 & -4 \end{vmatrix} = -12 - (-12) = 0$.

The system is dependent.

9. $D = \begin{vmatrix} 5 & 2 \\ -1 & 2 \end{vmatrix} = 10 - (-2) = 12$, $D_x = \begin{vmatrix} 4 & 2 \\ -8 & 2 \end{vmatrix} = 8 - (-16) = 24$,

$D_y = \begin{vmatrix} 5 & 4 \\ -1 & -8 \end{vmatrix} = -40 - (-4) = -36$.

$x = \dfrac{D_x}{D} = \dfrac{24}{12} = 2$, $y = \dfrac{D_y}{D} = \dfrac{-36}{12} = -3$.

10. $D = \begin{vmatrix} 1 & 1 & 2 \\ -1 & 2 & 1 \\ 3 & 0 & 3 \end{vmatrix} \overset{-C_1+C_3}{=} \begin{vmatrix} 1 & 1 & 1 \\ -1 & 2 & 2 \\ 3 & 0 & 0 \end{vmatrix} = (3)\begin{vmatrix} 1 & 1 \\ 2 & 2 \end{vmatrix} = 0$,

$D_x = \begin{vmatrix} 1 & 1 & 2 \\ 5 & 2 & 1 \\ 1 & 0 & 3 \end{vmatrix} \overset{-3C_1+C_3}{=} \begin{vmatrix} 1 & 1 & -1 \\ 5 & 2 & -14 \\ 1 & 0 & 0 \end{vmatrix} = (1)\begin{vmatrix} 1 & -1 \\ 2 & -14 \end{vmatrix} = -12 \neq 0$.

The system is inconsistent.

1. $\dfrac{3-i}{2+5i} = \dfrac{(3-i)(2-5i)}{(2+5i)(2-5i)} = \dfrac{1-17i}{29} = \dfrac{1}{29} - \dfrac{17}{29}i$

2. $i^{67} = i^{(4)(16)+3} = (i^4)^{16} \cdot i^3 = (1)^{16} \cdot i^3 = i^3 = -i$

3. $2x^2 - 3ix + 3 = 0$
$a = 2, \quad b = -3i, \quad c = 3$
$x = \dfrac{3i \pm \sqrt{(-3i)^2 - 4(2)(3)}}{2(2)}$
$= \dfrac{3i \pm \sqrt{-9 - 24}}{4}$
$= \dfrac{3i \pm \sqrt{-33}}{4}$
$= \dfrac{3i \pm \sqrt{33}i}{4}$
$= \dfrac{(3 \pm \sqrt{33})i}{4}$

4. $P(x) = 5x^4 - 2x^3 - 4x + 1$

$$\begin{array}{r|rrrrr} -2 & 5 & -2 & 0 & -4 & 1 \\ & & -10 & 24 & -48 & 104 \\ \hline & 5 & -12 & 24 & -52 & 105 \end{array}$$

5. The polynomial with real coefficients that has $-1, -2$, and $3i$ as zeros is
$P(x) = [x - (-1)][x - (-2)](x - 3i)[x - (-3i)]$
$= (x + 1)(x + 2)(x - 3i)(x + 3i)$
$= (x^2 + 3x + 2)(x^2 + 9)$
$= x^4 + 3x^3 + 11x^2 + 27x + 18.$

6. $P(x) = 2x^4 + x^3 - 4x - 3$

One variation in sign indicates there is one positive zero of $P(x)$.
$P(-x) = 2x^4 - x^3 + 4x - 3$

Three variations in sign indicate that there are three or one negative zero of $P(x)$. The possibilities are: (i) one positive zero, three negative zeros; (ii) one positive zero, one negative zero, two complex zeros.

7. $P(x) = 2x^3 + 4x^2 - 3x - 6$

(a)
$$\begin{array}{r|rrrr} & 2 & 4 & -3 & -6 \\ \hline 1 & 2 & 6 & 3 & -3 \\ 2 & 2 & 8 & 13 & 20 \end{array}$$

Thus 2 is the smallest positive integer which is an upper bound for the zeros of $P(x)$.

(b)
$$\begin{array}{r|rrrr} & 2 & 4 & -3 & -6 \\ \hline -1 & 2 & 2 & -5 & -1 \\ -2 & 2 & 0 & -3 & 0 \\ -3 & 2 & -2 & 3 & -15 \end{array}$$

Thus -3 is the largest negative integer which Theorem 9-16 detects as a lower bound for the zeros of $P(x)$.

8. $P(x) = 2x^3 + 4x^2 - 3x - 6$
From 7(b), we see that $x = -2$ is a zero of $P(x)$, and that
$P(x) = [x - (-2)](2x^2 - 3)$
$= (x + 2)(2)(x^2 - \frac{3}{2})$
$= 2(x + 2)(x - \sqrt{6}/2)(x + \sqrt{6}/2).$
Thus $x = -2$ is the only rational zero of $P(x)$.

9. To determine the value, correct to the nearest tenth, of the negative zero of $P(x) = 2x^4 + x^2 - 4x - 3$, we construct the following synthetic division table:

	2	0	1	−4	−3
0	2	0	1	−4	−3
−1	2	−2	3	−7	4
−0.5	2	−1	1.5	−4.75	−0.635
−0.6	2	−1.2	1.72	−5.032	0.0192
−0.55	2	−1.1	1.605	−4.8828	−0.3145

The negative zero lies between −0.55 and −0.6, and has the value −0.6 to the nearest tenth.

10. $y = (x − 1)^2(x + 2)$

Since $x − 1$ is a factor with even multiplicity, the graph is tangent to the x-axis at $x = 1$. Since $x + 2$ is a factor with odd multiplicity, the graph crosses the x-axis at $x = −2$. The graph and a table of values are as shown.

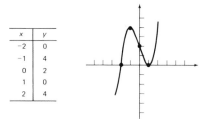

x	y
−2	0
−1	4
0	2
1	0
2	4

CHAPTER 10

1. $f(x) = \left(\dfrac{2}{3}\right)^x$

(a) $f(−2) = \left(\dfrac{2}{3}\right)^{−2} = \left(\dfrac{3}{2}\right)^2 = \dfrac{9}{4}$

(b) $f\left(\dfrac{1}{2}\right) = \left(\dfrac{2}{3}\right)^{1/2} = \sqrt{\dfrac{2}{3}} = \dfrac{\sqrt{2}}{\sqrt{3}} \cdot \dfrac{\sqrt{3}}{\sqrt{3}} = \dfrac{\sqrt{6}}{3}$

2. $(81)^x = \dfrac{1}{27}$

$(3^4)^x = \dfrac{1}{3^3}$

$(3)^{4x} = (3)^{−3}$

$4x = −3$

$x = −\dfrac{3}{4}$

3. $f(x) = 2^{−x}$

x	$f(x)$
−2	4
−1	2
0	1
1	1/2
2	1/4

4. (a) $\log_4 x = 2$ (b) $x = \log_2 \frac{1}{8}$
$$x = (4)^2$$
$$= 16$$
$$= \log_2 \frac{1}{2^3}$$
$$= \log_2(2)^{-3}$$
$$= -3$$

5. $y = \log_3 x$

x	y
1/3	−1
1	0
3	1
9	2

6. (a) $\log_a \frac{x^2 z}{y^3} = \log_a x^2 z - \log_a y^3$
$$= \log_a x^2 + \log_a z - \log_a y^3$$
$$= 2\log_a x + \log_a z - 3\log_a y$$

(b) $2\log_a x^2 yz + 3\log_a 2xyz^2 - \log_a 4x^3 y^2 z^4$
$$= \log_a(x^2 yz)^2 + \log_a(2xyz^2)^3 - \log_a 4x^3 y^2 z^4$$
$$= \log_a x^4 y^2 z^2 + \log_a 8x^3 y^3 z^6 - \log_a 4x^3 y^2 z^4$$
$$= \log_a(x^4 y^2 z^2)(8x^3 y^3 z^6) - \log_a 4x^3 y^2 z^4$$
$$= \log_a 8x^7 y^5 z^8 - \log_a 4x^3 y^2 z^4$$
$$= \log_a \frac{8x^7 y^5 z^8}{4x^3 y^2 z^4}$$
$$= \log_a 2x^4 y^3 z^4$$

7. (a)
$$10\left[6\left[\begin{matrix}\log 421.0 = 2.6243 \\ \log 421.6 = x\end{matrix}\right]d\right]0.0010$$
$$\log 422.0 = 2.6253$$

$$\frac{d}{0.0010} = \frac{6}{10}$$
$$d = \frac{6(0.0010)}{10} = 0.0006$$
$$x = 2.6243 + d$$
$$= 2.6249$$

(b)
$$10\left[d\left[\begin{matrix}\log 0.03270 = 8.5145 - 10 \\ \log x \quad = 8.5155 - 10\end{matrix}\right]0.0010\right]0.0014$$
$$\log 0.03280 = 8.5159 - 10$$

$$\frac{d}{10} = \frac{0.0010}{0.0014}$$
$$d = \frac{0.0100}{0.0014} = 7.3 = 7$$
$$x = 0.03277$$

8. $N = \dfrac{26.4\sqrt{593}}{(7.87)^2}$
$$\log N = \log 26.4 + \log\sqrt{593} - \log(7.87)^2$$
$$= \log 26.4 + \tfrac{1}{2}\log 593 - 2\log 7.87$$
$$= 1.4216 + \tfrac{1}{2}(2.7731) - 2(0.8960)$$
$$= 2.8082 - 1.7920$$
$$= 1.0162$$
$$N = 10.4$$

9. $5^{x+1} = 192$

$(x + 1) \log 5 = \log 192$

$x + 1 = \dfrac{\log 192}{\log 5}$

$x + 1 = \dfrac{2.2833}{0.6990}$

$\log (x + 1) = 0.5140$

$x + 1 = 3.27$

$x = 2.27$

$$10\left[3\begin{bmatrix}\log 2.280 = 0.3579 \\ \log 2.283 = \underline{} \\ \log 2.290 = 0.3598\end{bmatrix}d\right]0.0019$$

$\dfrac{d}{0.0019} = \dfrac{3}{10}$

$d = \dfrac{0.0057}{10} = 0.0006$

$\begin{array}{rl}\log 2.283 = & 10.3585 - 10 \\ -\log 0.6990 = & -9.8445 + 10 \\ \hline \log (x + 1) = & 0.5140\end{array}$

10. $\log x + \log (3x + 1) = 1$

$\log (x)(3x + 1) = 1$

$\log (3x^2 + x) = 1$

$3x^2 + x = 10$

$3x^2 + x - 10 = 0$

$(3x - 5)(x + 2) = 0$

$3x - 5 = 0 \quad \text{or} \quad x + 2 = 0$

$x = \tfrac{5}{3} \quad \text{or} \quad x = -2$

Check $x = \tfrac{5}{3}$:

$\begin{aligned}\log (\tfrac{5}{3}) + \log [(3)(\tfrac{5}{3}) + 1] &= \log (\tfrac{5}{3}) + \log (6) \\ &= \log [(\tfrac{5}{3})(6)] \\ &= \log (10) \\ &= 1\end{aligned}$

Check $x = -2$: $\log (-2)$ does not exist, since logarithms of negative numbers do not exist.

Solution set: $\{\tfrac{5}{3}\} = \{1.67\}$.

CHAPTER 11

1. $a_n = \dfrac{n - 1}{n + 1}$

$a_1 = \dfrac{1 - 1}{1 + 1} = 0, a_2 = \dfrac{2 - 1}{2 + 1} = \dfrac{1}{3}, a_3 = \dfrac{3 - 1}{3 + 1} = \dfrac{1}{2},$

$a_4 = \dfrac{4 - 1}{4 + 1} = \dfrac{3}{5}, a_5 = \dfrac{5 - 1}{5 + 1} = \dfrac{2}{3}$

2. $\displaystyle\sum_{n=2}^{5} (-2)^n = (-2)^2 + (-2)^3 + (-2)^4 + (-2)^5 = 4 - 8 + 16 - 32$

$= -20$

3. $a_4 = 6, a_9 = -4$

$\begin{aligned}a_4 = \quad& 6 = a_1 + 3d \\ a_9 = & -4 = a_1 + 8d \\ \hline 10 = & \qquad -5d \\ d = & -2\end{aligned}$

$-6 = a_1 + 3(-2)$

$-6 = a_1 - 6$

$a_1 = 0$

$a_{11} = a_1 + 10d$

$= 0 + 10(-2)$

$= -20$

$S_{11} = \tfrac{11}{2}(a_1 + a_{11})$

$= \tfrac{11}{2}(0 - 20)$

$= -110$

4. $a_6 = 8, S_8 = 40$

$a_6 = 8 = a_1 + 5d$

$S_8 = 40 = \tfrac{8}{2}(2a_1 + 7d)$

$40 = 4(2a_1 + 7d)$

$10 = 2a_1 + 7d$

$a_1 = 8 - 5d$

$10 = 2(8 - 5d) + 7d$

$10 = 16 - 10d + 7d$

$-6 = -3d$

$d = 2$

$a_1 = 8 - 5(2)$

$= -2$

5. $a_3 = -2, \quad a_4 = 3$

$$r = \frac{a_4}{a_3} = -\frac{3}{2}$$

$$a_5 = ra_4 = -\frac{3}{2}(3) = -\frac{9}{2}$$

$$a_3 = a_1 r^2$$

$$-2 = a_1\left(-\frac{3}{2}\right)^2$$

$$-2 = a_1\left(\frac{9}{4}\right)$$

$$a_1 = -\frac{8}{9}$$

$$S_5 = a_1\frac{1 - r^5}{1 - r}$$

$$= -\frac{8}{9}\frac{1 - (-\frac{3}{2})^5}{1 - (-\frac{3}{2})}$$

$$= -\frac{25}{18}$$

7. $a_3 = 81, \quad a_6 = -24$

$$\frac{a_6}{a_3} = \frac{a_1 r^5}{a_1 r^2} = r^3$$

$$\frac{-24}{81} = r^3$$

$$-\frac{8}{27} = r^3$$

$$r = -\frac{2}{3}$$

$$a_3 = a_1 r^2$$

$$81 = a_1\left(-\frac{2}{3}\right)^2$$

$$81 = a_1\left(\frac{4}{9}\right)$$

$$a_1 = \frac{729}{4}$$

6. (a) $\dfrac{a_2}{a_1} = \dfrac{-36}{24} = -\dfrac{3}{2}$

$$\frac{a_3}{a_2} = \frac{54}{-36} = -\frac{3}{2}$$

$$\frac{a_4}{a_3} = \frac{-81}{54} = -\frac{3}{2}$$

Geometric progression, $r = -\frac{3}{2}$.

$$a_n = a_1 r^{n-1}$$
$$= 24(-\tfrac{3}{2})^{n-1}$$
$$= 24(-\tfrac{3}{2})^n(-\tfrac{3}{2})^{-1}$$
$$= 24(-\tfrac{3}{2})^n(-\tfrac{2}{3})$$
$$= -16(-\tfrac{3}{2})^n$$

(b) $\dfrac{a_2}{a_1} = \dfrac{4}{3}$

$$\frac{a_3}{a_2} = \frac{12}{4} = 3$$

Not a geometric progression.

8. $\displaystyle\sum_{n=1}^{\infty} 25(-\tfrac{2}{3})^n$ is a geometric series with $a_1 = -\frac{50}{3}$ and $r = -\frac{2}{3}$. Since $|r| < 1$, the sum exists and is given by

$$\sum_{n=1}^{\infty} 25(-\tfrac{2}{3})^n = \frac{-\frac{50}{3}}{1 - (-\frac{2}{3})}$$
$$= -10.$$

9. $3.212121\cdots = 3 + 0.21 + 0.0021 + 0.000021 + \cdots$

$$= 3 + \frac{0.21}{1 - 0.01}$$

$$= 3 + \frac{0.21}{0.99}$$

$$= 3 + \frac{7}{11}$$

$$= \frac{40}{11}$$

10. Prove: $\dfrac{1}{1\cdot 3} + \dfrac{1}{3\cdot 5} + \dfrac{1}{5\cdot 7} + \cdots + \dfrac{1}{(2n-1)(2n+1)} = \dfrac{n}{2n+1}$

1. For $n = 1$, the left member is $\dfrac{1}{1\cdot 3} = \dfrac{1}{3}$, and the right

member is $\dfrac{1}{2+1} = \dfrac{1}{3}$. Thus the statement is true for $n = 1$.

2. (a) Assume that

$$\frac{1}{1\cdot 3} + \frac{1}{3\cdot 5} + \frac{1}{5\cdot 7} + \cdots + \frac{1}{(2k-1)(2k+1)} = \frac{k}{2k+1}.$$

(b) Adding

$$\frac{1}{[2(k+1)-1][2(k+1)+1]} = \frac{1}{(2k+1)(2k+3)}$$

to both sides, we obtain

$$\frac{1}{1\cdot 3} + \frac{1}{3\cdot 5} + \frac{1}{5\cdot 7} + \cdots + \frac{1}{(2k-1)(2k+1)} + \frac{1}{(2k+1)(2k+3)}$$

$$= \frac{k}{2k+1} + \frac{1}{(2k+1)(2k+3)}$$

$$= \frac{k(2k+3)+1}{(2k+1)(2k+3)}$$

$$= \frac{2k^2+3k+1}{(2k+1)(2k+3)}$$

$$= \frac{(2k+1)(k+1)}{(2k+1)(2k+3)}$$

$$= \frac{k+1}{2k+3}$$

$$= \frac{k+1}{2(k+1)+1}.$$

The last expression matches the right member of the equation to be proven when n is replaced by $k+1$. Thus the equation is true for all positive integers n, by the Principle of Mathematical Induction.

CHAPTER 12

1.

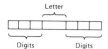

Letter

Digits   Digits

The Counting Principle gives $10\cdot 10\cdot 10\cdot 9\cdot 10\cdot 10\cdot 10 = 9\cdot 10^6$ $= 9{,}000{,}000$ different tags possible.

2. The problem is to find the number of three-digit even numbers that can be made using the digits 1, 2, 3, 4, 5, 6, 7. Any one of the 7 digits can be used as the hundreds and tens digits, whereas only the digits 2, 4, or 6 can be used as the units digit. By the Counting Principle, the number is $7\cdot 7\cdot 3 = 147$.

3. The number of ways that 6 people can be seated in a row of 8 chairs is
$$P(8,6) = \frac{8!}{2!} = \frac{8\cdot 7\cdot 6\cdot 5\cdot 4\cdot 3\cdot 2\cdot 1}{2\cdot 1} = 20{,}160.$$

4. The number of distinct "words" that can be made using the letters in the word COLLEGE is
$$\frac{7!}{2!\,2!} = \frac{7\cdot 6\cdot 5\cdot 4\cdot 3\cdot 2\cdot 1}{(2\cdot 1)(2\cdot 1)} = 1260.$$

5. The number of committees of 4 out of 12 people is
$$C(12,4) = \frac{12!}{8!\,4!} = \frac{12\cdot 11\cdot 10\cdot 9\cdot 8!}{8!\,4\cdot 3\cdot 2\cdot 1} = 495.$$

6. $(x-2y)^6 = x^6 + \binom{6}{5}x^5(-2y) + \binom{6}{4}x^4(-2y)^2 + \binom{6}{3}x^3(-2y)^3$
$$+ \binom{6}{2}x^2(-2y)^4 + \binom{6}{1}x(-2y)^5 + (-2y)^6$$
$$= x^6 + 6(-2)x^5y + 15(4)x^4y^2 + 20(-8)x^3y^3 + 15(16)x^2y^4$$
$$+ 6(-32)xy^5 + 64y^6$$
$$= x^6 - 12x^5y + 60x^4y^2 - 160x^3y^3 + 240x^2y^4 - 192xy^5$$
$$+ 64y^6$$

7. The 6th term of $(2x - y^2)^9$ is

$$\binom{9}{4}(2x)^4(-y^2)^5 = -126(64)x^4y^{10} = -8064x^4y^{10}.$$

8. $\dfrac{5x^2 + 5x + 6}{(x + 1)^2(x - 2)} = \dfrac{A}{x + 1} + \dfrac{B}{(x + 1)^2} + \dfrac{C}{x - 2}$

$$= \dfrac{A(x + 1)(x - 2) + B(x - 2) + C(x + 1)^2}{(x + 1)^2(x - 2)}$$

$5x^2 + 5x + 6 \equiv A(x + 1)(x - 2) + B(x - 2) + C(x + 1)^2$

Set $x = -1$: $6 = -3B$ or $B = -2$.

Set $x = 2$: $36 = 9C$ or $C = 4$.

Set $x = 0$: $6 = -2A - 2B + C$,

$$6 = -2A + 4 + 4,$$

$$A = 1.$$

Thus

$$\dfrac{5x^2 + 5x + 6}{(x + 1)^2(x - 2)} = \dfrac{1}{x + 1} - \dfrac{2}{(x + 1)^2} + \dfrac{4}{x - 2}.$$

9. $\dfrac{2x^2 + 5x + 1}{(x^2 + x + 1)^2} = \dfrac{Ax + B}{x^2 + x + 1} + \dfrac{Cx + D}{(x^2 + x + 1)^2}$

$$= \dfrac{(Ax + B)(x^2 + x + 1) + Cx + D}{(x^2 + x + 1)^2}$$

$2x^2 + 5x + 1 \equiv Ax^3 + (A + B)x^2 + (A + B + C)x + (B + D)$

Equating coefficients, we have

$$0 = A$$
$$2 = A + B$$
$$5 = A + B + C$$
$$1 = \quad B + \quad D,$$

and $A = 0$, $B = 2$, $C = 3$, $D = -1$.

Thus

$$\dfrac{2x^2 + 5x + 1}{(x^2 + x + 1)^2} = \dfrac{2}{x^2 + x + 1} + \dfrac{3x - 1}{(x^2 + x + 1)^2}.$$

10. $y = \dfrac{2x - 4}{x + 1}$

Vertical asymptote: $x = -1$. Horizontal asymptote: $y = 2$.

x	-3	-2	-0.5	0	1	2	3	4
y	5	8	-10	-4	-1	0	0.5	0.8

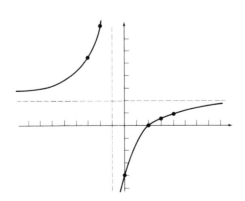

Answers to Selected Exercises

1. $B = \{0, 2, 4, 6, 8\}$ 3. $A = \{x^2 \,|\, x$ is a natural number less than 6$\}$
5. (a) $\{3, 9\}$ (b) $\{1, 5, 7\}$ (c) $\{0, 2, 3, 4, 6\}$ (d) A (e) $\{0, 2, 3, 4, 5, 6\}$
   (f) $\{3, 4, 6\}$ (g) $\{3, 4, 6\}$ (h) $\{4, 6\}$ (i) $\{3, 4, 6\}$ (j) $\{3, 4, 6\}$
   (k) $\{3, 9\}$ (l) $\{0, 2, 9\}$ (m) $\{8, 10\}$ (n) $\{5\}$ (o) $\{3, 4, 6, 9\}$ (p) $\{9\}$
   (q) $\{9\}$ (r) $\{0, 2, 3, 9\}$ (s) $\varnothing$ (t) $\{0, 2, 3, 8, 9, 10\}$
7. (a) True   (b) false   (c) false   (d) false   (e) true   (f) false
9. $\{a, b\}, \{a\}, \{b\}, \varnothing$
11. (a) $A \subseteq B$ (b) $A = B = \varnothing$ (c) $A = B = U$ (d) $B \subseteq A$
    (e) $A \subseteq B$
13. (a) N, I, Q, R (b) Q, R (c) R, I_r (d) Q, R (e) R, I_r (f) I, Q, R
    (g) Q, R (h) Q, R (i) Q, R (j) none of them

1. Additive inverse   3. Distributive property and commutative property, ·
5. Multiplicative identity   7. Additive identity
9. Multiplicative inverse   11. Closed   13. Closed
15. Not closed. 3 and 5 are odd integers, but $3 + 5 = 8$ is not.
17. No. $7 - (5 - 3) = 5$, but $(7 - 5) - 3 = -1$.
19. No. $5 - 7 = -2$, but $7 - 5 = 2$.   21. 5   23. $-\frac{2}{7}$

1. Suppose that $a/b = c/d$. Then

$$a(1/b) = c(1/d) \qquad \text{(Def. of division)}$$
$$[a(1/b)][bd] = [c(1/d)][bd] \qquad \text{(Substitution prop., =)}$$
$$[a(1/b)][bd] = [c(1/d)][db] \qquad \text{(Comm. prop., ·)}$$
$$\{[a(1/b)]b\}d = \{[c(1/d)]d\}b \qquad \text{(Assoc. prop., ·)}$$
$$\{a[(1/b)b]\}d = \{c[(1/d)d]\}b \qquad \text{(Assoc. prop., ·)}$$
$$\{a · 1\}d = \{c · 1\}b \qquad \text{(Mult. inverses)}$$
$$ad = cb \qquad \text{(Mult. identity)}$$
$$ad = bc \qquad \text{(Comm. prop., ·)}$$

Conversely, suppose that $ad = bc$. Then

$$ad = cb \qquad \text{(Comm. prop., ·)}$$
$$\{a · 1\}d = \{c · 1\}b \qquad \text{(Mult. identity)}$$
$$\{a[(1/b)b]\}d = \{c[(1/d)d]\}b \qquad \text{(Mult. inverses)}$$
$$\{[a(1/b)]b\}d = \{[c(1/d)]d\}b \qquad \text{(Assoc. prop., ·)}$$
$$[a(1/b)][bd] = [c(1/d)][db] \qquad \text{(Assoc. prop., ·)}$$
$$[a(1/b)][bd] = [c(1/d)][bd] \qquad \text{(Comm. prop., ·)}$$
$$a(1/b) = c(1/d) \qquad \text{(Canc. law, ·)}$$
$$a/b = c/d \qquad \text{(Def. of division)}$$

3.
$$\begin{aligned}
(a/b) · (c/d) &= [a(1/b)][c(1/d)] & \text{(Def. of division)} \\
&= \{[a · (1/b)]c\}\{1/d\} & \text{(Assoc. prop., ·)} \\
&= \{a[(1/b) · c]\}\{1/d\} & \text{(Assoc. prop., ·)} \\
&= \{a[c · (1/b)]\}\{1/d\} & \text{(Comm. prop., ·)} \\
&= \{[ac][1/b]\}\{1/d\} & \text{(Assoc. prop., ·)} \\
&= [ac][(1/b) · (1/d)] & \text{(Assoc. prop., ·)} \\
&= [ac][1/bd] & \text{(Theorem 1-5(c))} \\
&= ac/bd & \text{(Def. of division)}
\end{aligned}$$

5.
$$\begin{aligned}
a + (b - c) &= a + [b + (-c)] & \text{(Def. of subtraction)} \\
&= [a + b] + (-c) & \text{(Assoc. prop., +)} \\
&= (a + b) - c & \text{(Def. of subtraction)}
\end{aligned}$$

7.
$$\begin{aligned}
(-a)(-b) + [-(ab)] &= (-a)(-b) + (-a)b & \text{[Theorem 1-6(a)]} \\
&= (-a)[(-b) + b] & \text{(Distributive property)} \\
&= (-a) · (0) & \text{(Additive identity)} \\
&= 0 & \text{(Theorem 1-1)} \\
&= ab + [-(ab)] & \text{(Additive inverse)}
\end{aligned}$$

Therefore, $(-a)(-b) = ab$ by the cancellation law for addition.

9.
$$\begin{aligned}
a - (-b) &= a + [-(-b)] & \text{(Def. of subtraction)} \\
&= a + b & \text{(Problem 8)}
\end{aligned}$$

11. Suppose that $b \neq 0$. Then $-b \neq 0$. By problem 7, $(-a)(-b) = ab$.
Therefore, $\dfrac{-a}{b} = \dfrac{a}{-b}$, by problem 1. To prove the other equality, consider

$$\begin{aligned}
\frac{a}{b} + \frac{-a}{b} &= \frac{ab + b(-a)}{b · b} & \text{(Problem 4)} \\
&= \frac{ab + (-a)b}{b · b} & \text{(Comm. prop., ·)} \\
&= \frac{ab + (-(ab))}{b · b} & \text{[Theorem 1-6(a)]} \\
&= \frac{0}{b · b} & \text{(Additive inverse)} \\
&= 0 · \left(\frac{1}{b · b}\right) & \text{(Def. of division)} \\
&= 0 & \text{(Theorem 1-1)}
\end{aligned}$$

Since $\frac{a}{b} + \left(-\frac{a}{b}\right) = 0$, we have

$$\frac{a}{b} + \frac{-a}{b} = \frac{a}{b} + \left(-\frac{a}{b}\right).$$

Therefore

$$\frac{-a}{b} + \frac{a}{b} = \left(-\frac{a}{b}\right) + \frac{a}{b} \qquad \text{(Comm. prop., +)}$$

and $\frac{-a}{b} = -\frac{a}{b}$ by the cancellation law for addition.

13. Suppose that $b \neq 0$. Then

$$\begin{aligned}
\frac{a}{b} - \frac{c}{b} &= a\left(\frac{1}{b}\right) + (-c)\left(\frac{1}{b}\right) & &\begin{array}{l}\text{(Def. of subtraction and}\\ \text{division)}\end{array}\\[2mm]
&= \left(\frac{1}{b}\right)a + \left(\frac{1}{b}\right)(-c) & &\text{(Comm. prop., $\cdot$)}\\[2mm]
&= \left(\frac{1}{b}\right)(a - c) & &\begin{array}{l}\text{(Dist. prop. and def. of}\\ \text{subtraction)}\end{array}\\[2mm]
&= (a - c)\left(\frac{1}{b}\right) & &\text{(Comm. prop., $\cdot$)}\\[2mm]
&= \frac{a - c}{b} & &\text{(Def. of division)}
\end{aligned}$$

15. $\begin{aligned}[t]
a(b - c) &= a[b + (-c)] & &\text{(Def. of subtraction)}\\
&= ab + [a(-c)] & &\text{(Dist. prop.)}\\
&= ab + [(-c)(a)] & &\text{(Comm. prop.)}\\
&= ab + [-(ca)] & &\text{[Theorem 1-6(a)]}\\
&= ab + [-(ac)] & &\text{(Comm. prop., $\cdot$)}\\
&= ab - ac & &\text{(Def. of subtraction)}
\end{aligned}$

17. $\begin{aligned}[t]
(a - b) + (b - c) &= [a + (-b)] + [b + (-c)] & &\text{(Def. of subtraction)}\\
&= \{[a + (-b)] + b\} + (-c) & &\text{(Assoc. prop., +)}\\
&= \{a + [(-b) + b]\} + (-c) & &\text{(Assoc. prop., +)}\\
&= \{a + 0\} + (-c) & &\text{(Additive inverse)}\\
&= a + (-c) & &\text{(Additive identity)}\\
&= a - c & &\text{(Def. of subtraction)}
\end{aligned}$

19. Suppose that $a \neq 0$ and $b \neq 0$. Since $(b)(1/b) \neq 0$, $1/b \neq 0$ by Theorem 1-1. Therefore, $a/b = a(1/b) \neq 0$ by Theorem 1-2, and $1/(a/b)$ exists.

$$\begin{aligned}
\frac{a}{b} \cdot \frac{b}{a} &= \frac{ab}{ba} & &\text{(Problem 3)}\\[2mm]
&= \frac{ab}{ab} & &\text{(Comm. prop., $\cdot$)}\\[2mm]
&= (ab) \cdot \left(\frac{1}{ab}\right) & &\text{(Def. of division)}\\[2mm]
&= 1 & &\text{(Mult. inverse)}\\[2mm]
&= \left(\frac{a}{b}\right) \cdot \frac{1}{a/b} & &\text{(Mult. inverse)}
\end{aligned}$$

Therefore, $b/a = 1/(a/b)$ by the cancellation law for multiplication.

EXERCISES 1-4,
page 23

1. Theorem 1-10(a)   **3.** Theorem 1-8(a)   **5.** Theorem 1-8(c)
7. Theorems 1-8(b) or 1-10(b)   **9.** Theorem 1-10(a)   **11.** 7   **13.** -1
15. 18   **17.** -3 **19.** $1 \leq x < 6$ **21.** $3 > x \geq -1$

23. $\underset{-7 \quad -2}{\longleftrightarrow}$ **25.** $\underset{-4 \quad 0}{\longleftrightarrow}$ **27.** $\underset{1 \quad 3}{\longleftrightarrow}$ **29.** $\underset{-1 \quad 1}{\longleftrightarrow}$

31. all real numbers

33.

35.

37.

39. Suppose that $a > 0$ and $b > 0$. Then a is positive and b is positive, by Definition 1-7. Since the sum of two positive numbers is positive, this means that $a + b$ is positive, and therefore $a + b > 0$ by Definition 1-7. The product of two positive numbers is positive, so ab is also positive, and therefore $ab > 0$ by Definition 1-7.

41. Suppose that $a > b$ and $c > 0$. Then $a - b$ is positive and c is positive, by Definition 1-7. Since the product of two positive numbers is positive, this means that $(a - b)c$ is positive. That is, $ac - bc$ is positive, and $ac > bc$ by Definition 1-7.

43. Suppose that $a < b$. Then $a + a < a + b$ and $a + b < b + b$, by Theorem 1-10(a). That is, $2a < a + b < 2b$. Since $\frac{1}{2} > 0$,
$$\tfrac{1}{2}(2a) < \tfrac{1}{2}(a + b) < \tfrac{1}{2}(2b),$$
by Theorem 1-10(b). That is, $a < \dfrac{a + b}{2} < b$.

EXERCISES 1-5,
page 27

1. $-\pi, -3, -2, |-2|, |-3|, \pi$ **3.** $-|\frac{22}{7}|, -\pi, 0, |-3|, \pi, |-\frac{22}{7}|$
5. $-8 - |3|, -|8 - 3|, 0, |8 - 3|, |8| + 3$ **7.** 7   **9.** 4   **11.** 10
13. 2   **15.** 16   **17.** -2 **19.** $y - 4$ **21.** $4 - y$ **23.** $9 - \sqrt{80}$
25. $10 - 2x$ **27.** $b - a$ **29.** $b - 2a$ **31.** $-x/5$ **33.** 1   **35.** a^2
37. $-5 - 2a$ **39.** $3a - 14$ **41.** Theorem 1-13(a)   **43.** Theorem 1-13(a)
45. Theorem 1-13(c)   **47.** Theorem 1-12(c)   **49.** Theorem 1-12(a)

51.   **53.**   **55.**   **57.**

59.   **61.**   **63.**   **65.**

67.   **69.**   **71.**   **73.**

EXERCISES 2-1,
page 34

1. 8   **3.** $\frac{256}{625}$ **5.** $9y^2/25$ **7.** $27x^6y^3$ **9.** 64   **11.** $9x^{2n}/a^{2h}$ **13.** $\frac{1}{9}$
15. $\frac{81}{16}$ **17.** x^3/y^4 **19.** $2x/y^4$ **21.** $6x^3$ **23.** y^6/x^6 **25.** 1
27. $y^2/25x^{10}$ **29.** x^{n-2} **31.** $27y^2/x^2$ **33.** $16x/9$ **35.** $x/8y^2$ **37.** $\frac{25}{8}$
39. $12/a^3y^5$ **41.** y/x^3 **43.** $y^2 + x^2$ **45.** $3x^{-3}$ **47.** $3 \cdot 5^{-1}x^{-2}y^2$
49. $5y^{-4}z^2$

EXERCISES 2-2,
page 38

1. $3x^3 - 7x^2 - 3x + 13; 3$ **3.** $4x^2 - 12x + 9; 2$
5. $15x^5 - 35x^4 - 31x^3 + 61x^2 - 14x - 8; 5$
7. $x^5 + 2x^4 - 3x^3 - 7x^2 + 3; 5$
9. $4x^6 - 2x^5 - 4x^4 + 13x^3 - 4x^2 - 3x + 6; 6$
11. $-x^4 + 7x^3 - 5x^2 + 14x + 4; 4$ **13.** $6m^4 - 2m^3 - 7m^2 - 4m; 4$
15. $-2x + 5; 1$ **17.** $c^2 - k^2$ **19.** $x^2 - 4$ **21.** $9x^2 - 24xy + 16y^2$
23. $8x^2 + 2x - 15$ **25.** $4x^3 - 12x$ **27.** $6x^2 - 8x - 14$
29. $21y^2 + 29y - 10$ **31.** $x^4 + 4x^2 + 4$ **33.** $x^2 - 4xyz^2 + 4y^2z^4$
35. $a^2/4 - b^2/9$ **37.** $a^3 - b^3$ **39.** $27x^3 - 1$
41. $x^3 - 6x^2y + 12xy^2 - 8y^3$ **43.** $2a^4 - a^2b^2 - 15b^4$

45. $15u^4 - 31u^2v^2 + 14v^4$ **47.** $8x^3 - 36x^2y + 54xy^2 - 27y^3$
49. $27x^3 + 64$ **51.** $100x^2 + 120xy + 36y^2$

EXERCISES 2-3,
page 41

1. $4(3x - 1)$ **3.** $x(b + 1 + c^2)$ **5.** $(x - 1)(y - 1)$ **7.** $(5y - 1)^2$
9. $(4y - 1)(4y + 1)$ **11.** $a(x - 3y^2)(x + 3y^2)$ **13.** $(u - 4)^2$
15. $(x + 6)(x - 4)$ **17.** $(x + 3)(x + 7)$ **19.** $(4 + x)(1 - x)$
21. $(3x - 2a)(x + 3a)$ **23.** $(4w^3 + 3)(2w^3 - 3)$ **25.** $4(x + 2y)$
27. $(3c + d)(a + b)$ **29.** $(x^2 + d^2)(a + b)$
31. $(4x - 3y - 5)(4x - 3y + 5)$ **33.** $(x - z)(x^2 + xz + z^2)$
35. $(2 + u)(4 - 2u + u^2)$ **37.** $8(3x - b)(9x^2 + 3xb + b^2)$
39. $(2 - 5x^3)(1 + 2x)$ **41.** $(4a - b - 2x + z)(4a - b + 2x - z)$

EXERCISES 2-4,
page 46

1. $Q = 2x^2 - 4x + 8$, $R = 0$ **3.** $Q = 8x^2 - 3x + 5$, $R = 10x - 1$
5. $Q = -14y^3 - 21y$, $R = -7y + 1$
7. $\dfrac{2x^3 - 11x^2 + 19x - 10}{2x - 5} = x^2 - 3x + 2 + \dfrac{0}{2x - 5}$
9. $\dfrac{x^3 - 4x^2 + 6x - 3}{x - 1} = x^2 - 3x + 3 + \dfrac{0}{x - 1}$
11. $\dfrac{4x^4 + 5x^2 - 7x + 3}{2x^2 + 3x - 2} = 2x^2 - 3x + 9 + \dfrac{-40x + 21}{2x^2 + 3x - 2}$
13. $\dfrac{4m^3 - 3m - 2}{m + 1} = 4m^2 - 4m + 1 - \dfrac{3}{m + 1}$
15. $\dfrac{3x^5 + 3x^4 - 10x^3 - 10x^2 - 8x - 8}{x^2 + 3x + 2}$
$= 3x^3 - 6x^2 + 2x - 4 + \dfrac{0}{x^2 + 3x + 2}$
17. $\dfrac{4x^4 - 13ax^3 + 12a^2x^2 - 5a^3x + 2a^4}{4x^2 - ax + a^2}$
$= x^2 - 3ax + 2a^2 + \dfrac{0}{4x^2 - ax + a^2}$
19. $\dfrac{x^2 - 9x + 20}{x - 4} = x - 5 + \dfrac{0}{x - 4}$
21. $\dfrac{2x^3 - 2x + 7}{x - 2} = 2x^2 + 4x + 6 + \dfrac{19}{x - 2}$
23. $\dfrac{-3x^3 + 2x - 75}{x + 3} = -3x^2 + 9x - 25 + \dfrac{0}{x + 3}$
25. $\dfrac{x^3 - 4x^2 + 7x - 6}{x + 4} = x^2 - 8x + 39 - \dfrac{162}{x + 4}$
27. $\dfrac{x^4 - a^4}{x + a} = x^3 - ax^2 + a^2x - a^3 + \dfrac{0}{x + a}$

EXERCISES 2-5,
page 51

1. $x + y$ **3.** $(x + 3)/(x + 2)$ **5.** $4(a + 2)^2/(a - 1)$
7. $(x^2 + 3xy + 9y^2)(x + 2y)$ **9.** 140 **11.** $72x^5$ **13.** $x^2 - y^2$
15. $2(x - 3)(x + 2)^2$ **17.** $(3y - 4xz)/12x^2$ **19.** $(xy - y^2 - 3)/(x^2 - y^2)$
21. $(x - 8)/(2x - 3)(x + 2)$ **23.** $(-z^2 + 3z - 14)/4(z - 3)(z + 3)$
25. $(2w^2 - 9w - 9)/(2w - 3)(w - 6)(w - 5)$ **27.** $\frac{1}{3}$
29. $(4 - x)/(6 - x)$ **31.** $(5 - y)/4$
33. $a(2w + 5)/(w + 3)(w^2 + 4w + 16)$

35. $(x^2 - xy + y^2)(2x + 3y)/2$ **37.** $(x^2 + 3x + 9)/3(2x^2 + 5)$
39. $1/(w + 5)$ **41.** $(4w + 5)/(4w - 5)$
43. $(a^2 + axb + x^2b^2)(a + 2b)/2xy(3a - 2b)$
45. $(x - 6)/(x + 2)(x + 1)$
47. $(x + y)/(x - y)$ **49.** $2b(b^2 + ba + a^2)/a^2$ **51.** $(y - x)/xy$
53. $(3x + 2y)(y - 2x)$

EXERCISES 3-1,
page 56

1. 5   **3.** $\frac{3}{4}$ **5.** $\frac{2}{9}$ **7.** $\frac{12}{5}$ **9.** 1.1   **11.** 0.06   **13.** 4   **15.** $-\frac{3}{5}$ **17.** $\frac{1}{5}$
19. $\frac{4}{3}$ **21.** -0.4 **23.** 73   **25.** $\frac{4}{3}$ **27.** 10   **29.** 7   **31.** 2   **33.** 3
35. $2b^2$ **37.** y^2w^3 **39.** $0.4x^2$ **41.** $2x/3y^2$ **43.** -1 **45.** $6x/5$
47. $19y^5$ **49.** xy^2 **51.** x^3/z **53.** $z^3/x^2\sqrt{3}$

EXERCISES 3-2,
page 59

1. $\sqrt[5]{a}$ **3.** $3\sqrt[4]{a}$ **5.** $a^{5/4}$ **7.** $x^{10/9}$ **9.** a^2 **11.** $\sqrt[3]{4x^2y^4}$
13. $(bx^2)^{1/7}$ **15.** $\sqrt[3]{(xy)^5}$ **17.** 2   **19.** Does not exist   **21.** $\frac{1}{256}$
23. Does not exist   **25.** $\frac{1}{36}$ **27.** $\frac{1}{16}$ **29.** 27

EXERCISES 3-3,
page 61

1. $10\sqrt{5}$ **3.** $4ab^2\sqrt{2a}$ **5.** $4x^2\sqrt[3]{2x}$ **7.** $x^2\sqrt[5]{x^3}$ **9.** $-4a^3\sqrt[3]{2}$
11. $3\sqrt[4]{2}$ **13.** $\sqrt{2} + \sqrt{3}$ **15.** $9\sqrt{2}$ **17.** $3\sqrt{6}$
19. $(3a + 4b)\sqrt[3]{2}$ **21.** $4xy\sqrt[3]{3y}$ **23.** $5\sqrt[3]{2}$ **25.** $\sqrt{5}$ **27.** $\sqrt{3}$
29. $2xy\sqrt[3]{3xy}$ **31.** $\sqrt{3}$ **33.** $14\sqrt{2} - 9\sqrt{5}$ **35.** $\sqrt{20x}$ **37.** $\sqrt[3]{27x}$
39. $\sqrt{a/6b}$ **41.** $\sqrt{8x/3y}$ **43.** $\sqrt{2/3}$ **45.** $\sqrt[3]{a^2}$ **47.** $\sqrt[3]{5}$ **49.** $\sqrt{0.3x}$
51. $\sqrt[3]{u^2}$

EXERCISES 3-4,
page 66

1. $\sqrt{5}/5$ **3.** $\sqrt{10}/4$ **5.** $5\sqrt{7}/7$ **7.** $\sqrt[3]{2}/2$ **9.** $3\sqrt{35}/14$
11. $\sqrt[3]{3}$ **13.** $(8 + \sqrt{7})/3$ **15.** $(3\sqrt{15} - \sqrt{10} + \sqrt{6} - 9)/25$
17. $\sqrt{3a}/3$ **19.** $\sqrt{10ab}/2b$ **21.** $\sqrt[3]{6cdx}/3x$ **23.** $\sqrt{5u}/5u^3$
25. $y\sqrt[4]{2a^2xy^2z^3}/xz^2$ **27.** $\sqrt{2bx}(4b + x^2)/2bx$ **29.** $2v\sqrt[5]{uv^2}$ **31.** $\sqrt{6x}$
33. $\sqrt[4]{10}/2$ **35.** $b^2\sqrt[6]{a^2b^3}$ **37.** $x\sqrt[3]{6cxu^2}/6u^3$ **39.** $(43 - 12a)\sqrt{2}/2$
41. $y - 6\sqrt{xy} + 9x$ **43.** $\sqrt[3]{25v}/5$ **45.** $\sqrt[3]{3y^2}$ **47.** $x^{12}\sqrt{1024x^5}$
49. $(2 + y^2)\sqrt[3]{4y^2}/y^2$ **51.** $9 - \sqrt[3]{4}$ **53.** $(a + 2b)^3$ **55.** $x^4\sqrt{x}/y$
57. $3(a + b)^2\sqrt{3(a + b)}$ **59.** $3 - 2\sqrt{2}$ **61.** $-2x\sqrt[3]{x}$ **63.** x
65. $1/(\sqrt{x + h} + \sqrt{x})$ **67.** $1/(\sqrt{a} - \sqrt{b})$
69. $1/(\sqrt[3]{(x + h)^2} + \sqrt[3]{x(x + h)} + \sqrt[3]{x^2})$

EXERCISES 4-1,
page 73

1. $\{3\}$ **3.** $\{\frac{8}{3}\}$ **5.** $\{\frac{27}{20}\}$ **7.** $\{-3\}$ **9.** $\{-3\}$ **11.** $\{\frac{4}{3}\}$ **13.** $\{4\}$
15. $\varnothing$ **17.** $x = (c - by)/a$ **19.** $h = 2A/b$ **21.** $t = (2s - an)/n$
23. ± 3 **25.** $\frac{7}{2}$ **27.** $\frac{9}{4}$ **29.** $-\frac{3}{2}, -5$ **31.** 1, 7

EXERCISES 4-2,
page 77

1. 48, 49, 50   **3.** 92   **5.** 47   **7.** 7 dimes, 5 nickels
9. 3 cartons of milk, 5 cartons of eggs
11. length $= 36$ m, width $= 20$ m
13. 4 lb peanuts, 8 lb cashews   **15.** 25 g of 12% silver, 15 g of 20% silver
17. 24 liters   **19.** $4\frac{4}{7}$ hr   **21.** 21 km/hr, 24 km
23. 70 km/hr, 75 km/hr   **25.** $3\frac{1}{13}$ hr   **27.** 45 hr   **29.** 60 hr

1. $\{x \mid x < 4\}$ 3. $\{x \mid x > 5\}$ 5. $\{y \mid y < 6\}$ 7. $\{x \mid x < 2\}$
9. $\{x \mid x > \sqrt{2}\}$ 11. $\{x \mid x \geq -36\}$ 13. $\{x \mid x \leq -3\}$ 15. $\{x \mid x \geq 0\}$
17. $\{x \mid x \leq -13\}$ 19. $\{x \mid x > -7\}$ 21. $\{x \mid 1 \leq x \leq 4\}$ 23. $\varnothing$
25. $\{x \mid -1 < x < 3\}$ 27. $\{x \mid x > -2\}$ 29. $\{x \mid x \in \mathcal{R}\}$
31. $\{x \mid x \leq \frac{8}{5}\}$ 33. $\{x \mid 1 < x < 4\}$ 35. $\varnothing$ 37. $\{2\}$

39. $\{x \mid -\frac{9}{10} < x < \frac{3}{2}\}$ 41. $\{x \mid x \in \mathcal{R}\}$ 43. $\{x \mid x \neq 1\}$

45. $\{x \mid x < \frac{1}{4}$ or $x > \frac{5}{4}\}$ 47. $\{x \mid x \leq \frac{1}{2}$ or $x \geq \frac{7}{2}\}$

1. $\frac{3}{4}$ 3. 8   5. 3 boys/1 girl   7. 30 mi/hr   9. 432 lb/sq ft
11. 5 mi/qt   13. 216 mi   15. $\frac{24}{25}$ pt   17. $15, $25   19. 9   21. $\pm\frac{1}{2}$
23. $\frac{103}{128}$ 25. $25\sqrt{3}/4$ sq m   27. $\frac{1}{2}$ in.   29. 40   31. 196 ft
33. 87.5 lb/sq in.
35. Since y varies inversely as x, $y = k/x$. The constant k is not 0 since y
is not constant. Therefore, $xy = k$, and $x = k/y$. That is, x varies inversely
as y.

1. $2, -5$ 3. -3 5. $\pm\frac{5}{2}$ 7. $-3, 0$ 9. $-\frac{2}{7}, 0, \frac{1}{7}$ 11. $0, 4$
13. $(-1 \pm \sqrt{5})/2$ 15. $(-5 \pm \sqrt{17})/4$ 17. $-3, -1$ 19. $-\frac{3}{2}, -\frac{2}{3}$
21. $\frac{1}{2}, 7$ 23. $-\frac{2}{9}, \frac{1}{3}$ 25. $-1 \pm \sqrt{2}$ 27. $(-a \pm \sqrt{a^2 - 4b})/2$
29. $-\frac{2}{3}, -\frac{1}{2}$ 31. $\pm\sqrt{2}$ 33. $-2, \frac{1}{3}$ 35. $-4, 5$ 37. $-1, -\frac{1}{5}$
39. $-1, \frac{1}{8}$ 41. $9, 25$

1. $-7, 4$ 3. $\frac{1}{2}, \frac{3}{2}$ 5. $-1, 0$ 7. $-6, 2$ 9. $\pm\frac{5}{4}$ 11. $(-7 \pm \sqrt{17})/8$
13. $(-5 \pm \sqrt{5})/2$ 15. $(-4 \pm \sqrt{7})/3$ 17. $(11 \pm \sqrt{73})/12$
19. $-\frac{9}{7}, \frac{5}{4}$ 21. 16; 2 distinct real roots   23. -31; no real roots
25. 37; 2 distinct real roots   27. 0; 2 equal real roots
29. $x^2 + x - 2 = 0$ 31. $x^2 + 3x - 10 = 0$ 33. $x^2 + 4x + 4 = 0$
35. $x^2 - 2x + 1 = 0$ 37. $x^2 + 3x - 4 = 0$ 39. $x^2 - 2x = 0$
41. 21, 22 or $-22, -21$ 43. 30 m by 60 m   45. 14   47. 20 m
49. 55 mi/hr

1. $\{6\}$ 3. $\varnothing$ 5. $\{2\}$ 7. $\{-10, -9\}$ 9. $\{6\}$ 11. $\{28\}$ 13. $\{2\}$
15. $\{-1\}$ 17. $\{0, 3\}$ 19. $\varnothing$ 21. $\{14\}$ 23. $\{5\}$ 25. $\{-5\}$ 27. $\{2\}$
29. $\{15\}$

1. $D = \mathcal{R}, R = \mathcal{R}$ 3. $D = \{x \mid x \geq 4\}, R = \{y \mid y \leq 0\}$
5. $D = \mathcal{R}, R = \{y \mid y \geq -3\}$ 7. $D = \{x \mid x \neq 1\}, R = \{y \mid y \neq 0\}$
9. $D = \mathcal{R}, R = \{y \mid y \geq -3\}$ 11. $D = \mathcal{R}, R = \mathcal{R}$ 13. 10   15. 16

17. $4b - 2$ **19.** 62 **21.** $4b^2 - 2$ **23.** 8 **25.** 12 **27.** 4 **29.** $2x^2 - x$
31. 2^{x+h} **33.** $h(2x + h - 1)$

35.

37.

39.

41.

43.

45. Function

47. Not a function
49. $g^{-1} = \{(3, 0), (2, 1), (2, -1)\}$; g is a function; g^{-1} is not a function.
51. $h^{-1} = \{(x, y) \,|\, y = x\}$; h and h^{-1} are functions.
53. $p^{-1} = \{(x, y) \,|\, x = |y|\}$; p is a function; p^{-1} is not a function.
55. $f^{-1} = \{(x, y) \,|\, y = (x + 2)/3\}$; f and f^{-1} are functions.
57. $g^{-1} = \{(x, y) \,|\, y = 2x^2\}$; g is not a function; g^{-1} is a function.

EXERCISES 5-2,
page 119

1. Linear function **3.** Not a linear function **5.** Not a linear function
7. Not a linear function **9.** Not a linear function
11. Not a linear function **13.** Not a linear function
15. Not a linear function
17. x-intercept $\frac{3}{2}$; **19.** x-intercept $\frac{5}{4}$; **21.** x-intercept 0;
  y-intercept -3 no y-intercept y-intercept 0

23. x-intercept 2; **25.** x-intercept $\frac{7}{2}$; **27.** x-intercept 3;
  y-intercept -8 y-intercept $\frac{7}{5}$ y-intercept 3

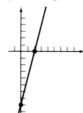

29. $-\frac{4}{3}$ **31.** $\frac{2}{3}$ **33.** Slope is undefined **35.** $\frac{1}{2}$ **37.** $\frac{3}{4}$ **39.** 0
41. $\frac{7}{5}$ **43.** 7 **45.** $-\frac{5}{2}$ **47.** $\frac{1}{2}$

1. $x + y = 0$ 3. $2x - y = -4$ 5. $2x - 3y = -6$ 7. $y = 5$
9. $x - 2y = -6$ 11. $x + y = 7$ 13. $4x - y = 28$ 15. $x - 2y = 2$
17. $4x - y = -3$ 19. $y = 3$ 21. $x = 5$ 23. $3x - y = -9$
25. $6x + 5y = -45$ 27. $x + y = 0$ 29. $5x - 7y = 7$

31. $m_1 = \dfrac{7 - 3}{4 - (-1)} = \dfrac{4}{5}$, $m_2 = \dfrac{10 - 7}{6 - 4} = \dfrac{3}{2}$. No. The slope of the line
between $(-1, 3)$ and $(4, 7)$ is $\frac{4}{5}$, and the slope of the line between $(4, 7)$
and $(6, 10)$ is $\frac{3}{2}$.

1. $V(0, 1)$. $A: x = 0$
   Pts. $(-1, 2)$, $(1, 2)$

3. $V(-1, 3)$. $A: x = -1$
   Pts. $(-2, 4)$, $(0, 4)$

5. $V(-1, -1)$. $A: x = -1$
   Pts. $(-2, -2)$, $(0, -2)$

7. $V(-2, 5)$. $A: x = -2$
   Pts. $(-3, 8)$, $(-1, 8)$

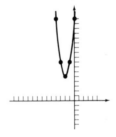

9. $V(2, 4)$. $A: x = 2$
   Pts. $(0, 0)$, $(4, 0)$

11. $V(0, -4)$. $A: x = 0$
    Pts. $(-2, 0)$, $(2, 0)$

13. $V(2, -1)$. $A: x = 2$
    Pts. $(-1, 5)$, $(5, 5)$

15. $V(-4, -3)$. $A: x = -4$
    Pts. $(-6, 1)$, $(-2, 1)$

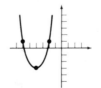

17. $V(-3, -4)$. $A: x = -3$
Pts. $(-5, 0)$, $(-1, 0)$

19. $V(2, 2)$. $A: y = 2$
Pts. $(6, 4)$, $(6, 0)$

21. $V(-3, 2)$. $A: y = 2$
Pts. $(-5, 3)$, $(-5, 1)$

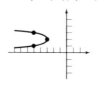

23. $V(-5, 1)$. $A: y = 1$
Pts. $(-4, 2)$, $(-4, 0)$

25. $V(12, -2)$. $A: y = -2$
Pts. $(0, 0)$, $(0, -4)$

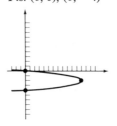

27. $V(0, 5)$. $A: y = 5$
Pts. $(1, 6)$, $(1, 4)$

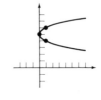

29. $V(-4, 0)$. $A: y = 0$
Pts. $(-6, 1)$, $(-6, -1)$

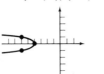

31. $V(-9, -3)$. $A: y = -3$
Pts. $(0, 0)$, $(0, -6)$

33. $V(0, -3)$. $A: y = -3$
Pts. $(-4, -1)$, $(-4, -5)$

35.

37.

39.

41.

43.

45.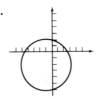

47. 50 ft by 25 ft

49. $t = 4$ sec; max. height $= 256$ ft

EXERCISES 5-5,
page 139

1. $\{x \mid x < -3 \text{ or } x > 4\}$ **3.** $\{x \mid -1 \le x \le 3\}$ **5.** $\{x \mid x \le 1 \text{ or } x \ge 3\}$
7. $\{x \mid x < -\sqrt{3} \text{ or } x > \sqrt{3}\}$ **9.** $\{x \mid x < 0 \text{ or } x > 9\}$
11. $\{x \mid 0 \le x \le 4\}$ **13.** $\{x \mid x < -\sqrt{2} \text{ or } x > \sqrt{2}\}$
15. $\{x \mid x \le 0 \text{ or } x \ge 2\}$ **17.** $\{x \mid -4 < x < -2\}$ **19.** $\{z \mid -5 \le z \le 3\}$
21. $\{t \mid t < -5 \text{ or } t > \frac{2}{3}\}$ **23.** $\{x \mid \frac{2}{3} < x < \frac{6}{5}\}$ **25.** $\{t \mid t < 0 \text{ or } t > \frac{4}{3}\}$
27. $\{r \mid -\frac{7}{3} < r < \frac{5}{2}\}$ **29.** $\{x \mid -\frac{7}{4} < x < 3\}$
31. $\{t \mid (-1 - \sqrt{2})/2 \le t \le (-1 + \sqrt{2})/2\}$
33. $\{x \mid 1 < x < 2 \text{ or } x > 3\}$ **35.** $\{x \mid \frac{1}{2} < x < \frac{4}{5}\}$ **37.** $\{x \mid -2 < x < \frac{3}{2}\}$
39. $\{t \mid -5 \le t < -3\}$ **41.** $\{z \mid -3 < z < 2 \text{ or } z > 7\}$
43. $\{w \mid -2 < w < -\frac{1}{2} \text{ or } w \ge 0\}$ **45.** $\{t \mid -\frac{3}{2} < t \le -1 \text{ or } t > 0\}$
47. $\{t \mid t < 0 \text{ or } 0 < t < \frac{1}{3} \text{ or } t > 1\}$ **49.** $\{z \mid z < \frac{4}{3} \text{ or } z > 8\}$
51. $\{z \mid -1 < z < 0\}$ **53.** $\{s \mid s \neq -\frac{1}{2}\}$

EXERCISES 6-2,
page 148

1. 5   **3.** 8   **5.** 5   **7.** 29   **9.** $3\sqrt{5}$ **11.** $2\sqrt{85}$ **13.** $|a - b|\sqrt{2}$
15. 4   **17.** $10 - f(x)$ **19.** Circle with center $(0, 0)$ and radius 4
21. Circle with center $(-1, 0)$ and radius 1
23. Circle with center $(0, -4)$ and radius $\sqrt{3}/5$
25. Circle with center $(-3, -5)$ and radius 5
27. Circle with center $(2a, -2a)$ and radius $2|a|$
29. Circle with center $(-1, -3)$ and radius $\sqrt{10}$
31. Circle with center $(\frac{3}{2}, 0)$ and radius $\frac{3}{2}$ **33.** Not a circle
35. Circle with center $(2, -8)$ and radius 0 (point-circle)
37. Circle with center $(-1, -2)$ and radius 3   **39.** $x^2 + (y - 1)^2 = 4$
41. $(x - 2)^2 + (y + 2)^2 = \frac{1}{4}$ **43.** $(x + 3)^2 + (y + 4)^2 = 49$
45. $(x + 3)^2 + (y - 5)^2 = 1$ **47.** $(x - a)^2 + (y - 1)^2 = 9$

49.

51.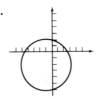

53.

55.

57.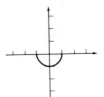

59.

61.

63.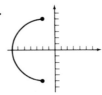

65. The lengths of the sides of the triangle formed using the given points as vertices are $3\sqrt{2}$, $4\sqrt{2}$, and $5\sqrt{2}$. Since these lengths satisfy the Pythagorean Theorem, the triangle is a right triangle.

67. All four sides of the quadrilateral formed using the given points have the same length ($\sqrt{109}$), and the two diagonals have the same length ($\sqrt{218}$). Thus the given points are vertices of a square.

69. Two opposite sides of the quadrilateral formed using the given points have length 5, and the remaining two sides have length 10. The two diagonals have the same length ($5\sqrt{5}$). Thus the given points are vertices of a rectangle.

EXERCISES 6-3,
page 154

1.

3.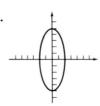

5.

7.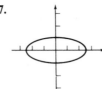

9.

11.

13.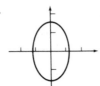

15.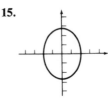

17.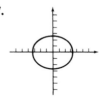

19.

21.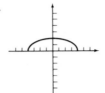

23.

25.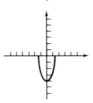

27.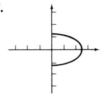

29.

EXERCISES 6-4,
page 161

1. $V: (\pm 3, 0)$
   $A: y = \pm 4x/3$

3. $V: (0, \pm 2)$
   $A: y = \pm 2x/5$

5. $V: (0, \pm 3)$
   $A: y = \pm x$

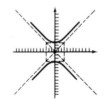

7. $V: (\pm 3, 0)$
   $A: y = \pm 2x/3$

9.

11.

13. $V: (\pm 2, 0)$
   $A: y = \pm x/2$

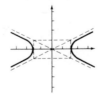

15.

17. $V: (0, \pm \frac{5}{4})$
   $A: y = \pm 15x/8$

19.

21. $V: (\pm 2, 0)$
$A: y = \pm\sqrt{2}\,x$

23. $V: (0, \frac{3}{5})$
$A: y = \pm 2x/5$

25. $V: (2, 0)$
$A: y = \pm 2x$

27. $V: (-\frac{5}{2}, 0)$
$A: y = \pm x/2$

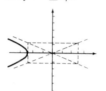

EXERCISES 6-5,
page 169

1. $\{(-2, 3)\}$ **3.** $\{(3, -1)\}$ **5.** $\{(3, 4)\}$ **7.** $\{(-3, 1)\}$ **9.** $\varnothing$
11. $\{(\frac{1}{2}, 2), (7, \frac{1}{7})\}$ **13.** $\{(\frac{3}{2}, -\frac{4}{3}), (-\frac{3}{2}, \frac{4}{3})\}$
15. $\{(1, 2), (-1, -2), (\sqrt{2}, \sqrt{2}), (-\sqrt{2}, -\sqrt{2})\}$
17. $\{(2, 2), (-2, -2)\}$ **19.** $\{(1, 1), (-1, -1), (5, -3), (-5, 3)\}$

21. $\{(-2, 2), (6, 18)\}$

23. $\varnothing$

25. $\{(0, -2), (\frac{8}{5}, \frac{6}{5})\}$

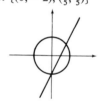

27. $\{(8, 4), (\frac{8}{25}, -\frac{44}{25})\}$

29. $\{(-1, -1), (-\frac{1}{7}, \frac{5}{7})\}$

31. $\varnothing$

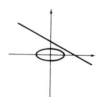

33. $\{(0, 3), (0, -3)\}$

35. $\{(-1, 0), (4, 5)\}$

37. $\{(4, 3), (4, -3), (-4, 3), (-4, -3)\}$

39. $\pm\frac{4}{3}$

EXERCISES 6-6,
page 175

1. A: $(-2, 2)$
  B: $(6, 18)$

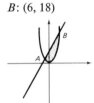

3. No solution

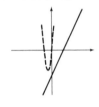

5. A: $(0, -2)$
  B: $(\frac{8}{5}, \frac{6}{5})$

7. A: $(8, 4)$
  B: $(\frac{8}{25}, -\frac{44}{25})$

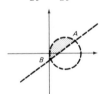

9. A: $(-1, -1)$
  B: $(-\frac{1}{7}, \frac{5}{7})$

11. No solution

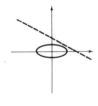

13. A: $(0, 3)$
  B: $(0, -3)$

15. A: $(-1, 0)$
  B: $(4, 5)$

17. A: $(4, 3)$, B: $(-4, 3)$
  C: $(-4, -3)$, D: $(4, -3)$

19. A: $\left(\frac{9}{4}, \frac{3}{4}\right)$

21. A: $\left(\frac{10}{3}, \frac{1}{3}\right)$

23. A: $(-1, 0)$, B: $(1, 4)$,
C: $(5, -6)$

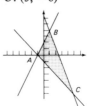

25. A: $(2, -1)$
B: $(5, 2)$

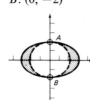

27. A: $(0, 2)$
B: $(0, -2)$

29. A: $(4, 1)$, B: $(-4, 1)$
C: $(-4, -1)$, D: $(4, -1)$

31. A: $\left(\frac{3\sqrt{65}}{13}, \frac{4\sqrt{26}}{13}\right)$, B: $\left(\frac{-3\sqrt{65}}{13}, \frac{4\sqrt{26}}{13}\right)$
C: $\left(-\frac{3\sqrt{65}}{13}, \frac{-4\sqrt{26}}{13}\right)$, D: $\left(\frac{3\sqrt{65}}{13}, -\frac{4\sqrt{26}}{13}\right)$

33. A: $(4\sqrt{2}, 4)$, B: $(-4\sqrt{2}, 4)$
C: $(-4\sqrt{2}, -4)$, D: $(4\sqrt{2}, -4)$

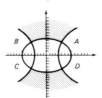

EXERCISES 6-7,
page 181

1. $f^{-1} = \{(x, y) \,|\, x = y^2 + 2y\}$; f is a function, f^{-1} is not a function.
3. $g^{-1} = \{(x, y) \,|\, x^2 + y^2 = 4\} = g$; g and g^{-1} are not functions.
5. $h^{-1} = \{(x, y) \,|\, x = -\sqrt{4 - y^2}\}$; h is a function, h^{-1} is not a function.
7. $f^{-1} = \{(x, y) \,|\, x = \sqrt{y^2 + 2y}\}$; f is a function, f^{-1} is not a function.

9. (a) $y = x/2 + 2$

(b)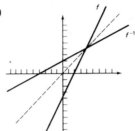

(c) f^{-1} is a function.

13. (a) $x = |2y + 6|$

(b)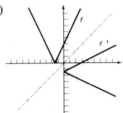

(c) f^{-1} is not a function.

17. (a) $x = \sqrt{16 - y^2}$

(b)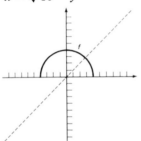

(c) f^{-1} is not a function

11. (a) $y = -3x/4 + 3$

(b)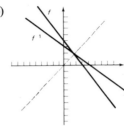

(c) f^{-1} is a function.

15. (a) $x = \sqrt{y - 1}$

(b)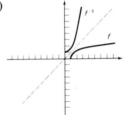

(c) f^{-1} is a function.

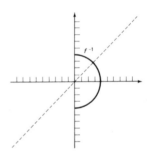

19. f and g are inverse functions.   **21.** f and g are inverse functions.

23. f and g are not inverse functions.   **25.** f and g are inverse functions.

27. f and g are not inverse functions.

29. (a) Domain of f = Range of $f^{-1} = \Re$
Range of f = Domain of $f^{-1} = \Re$

(b) $y = x/2 + 3$

31. (a) Domain of f = Range of $f^{-1} = \Re$
Range of f = Domain of $f^{-1} = \Re$

(b) $y = 16 - 4x/3$

33. (a) Domain of $f = \{x \mid x \neq -1\}$
Range of $f = \{y \mid y \neq 0\}$
Domain of $f^{-1} = \{x \mid x \neq 0\}$
Range of $f^{-1} = \{y \mid y \neq -1\}$

(b) $y = (1 - x)/x$

35. (a) Domain of $f = \{x \mid x \neq 0\}$
Range of $f = \{y \mid y \neq 0\}$
Domain of $f^{-1} = \{x \mid x \neq 0\}$
Range of $f^{-1} = \{y \mid y \neq 0\}$
(b) $y = 1/x$

37. (a) Domain of $f = \{x \mid x \geq 2\}$
Range of $f = \{y \mid y \geq 0\}$
Domain of $f^{-1} = \{x \mid x \geq 0\}$
Range of $f^{-1} = \{y \mid y \geq 2\}$
(b) $y = x^2 + 2$, $x \geq 0$

39. (a) Domain of f = Range of $f^{-1} = \Re$;
Range of f = Domain of $f^{-1} = \Re$
(b) $y = \sqrt[3]{x}$

41. The equation of the line between (a, b) and (b, a) is $(y - b) = \dfrac{a - b}{b - a}(x - a)$, or $y = -x + a + b$. The slope of this line is -1, and the slope of the line $y = x$ is 1. Thus the lines $y = -x + a + b$ and $y = x$ are perpendicular lines. The length of the line segment between (a, b) and the points (x, x) on the line $y = x$ is $d_1 = \sqrt{(x - a)^2 + (x - b)^2}$. The length of the line segment between (b, a) and the point (x, x) on the line $y = x$ is $d_2 = \sqrt{(x - b)^2 + (x - a)^2}$. Since $d_1 = d_2$, the line $y = x$ is the perpendicular bisector of the line segment joining (a, b) and (b, a).

EXERCISES 7-1,
page 189

1. 2×3 **3.** 3×1, column matrix **5.** 2×4
7. 1×1, column matrix, row matrix, square matrix, diagonal matrix
9. 1×2, row matrix **11.** 3×3, square matrix, diagonal matrix
13. 3×3, square matrix **15.** 2×1, column matrix **17.** $\begin{bmatrix} 3 & 4 & 5 & 6 \\ 5 & 6 & 7 & 8 \end{bmatrix}$

19. $\begin{bmatrix} -1 & -2 \\ 1 & 2 \\ -1 & -2 \\ 1 & 2 \end{bmatrix}$ **21.** $\begin{bmatrix} 0 & 1 & 1 & 1 \\ 0 & 0 & 1 & 1 \end{bmatrix}$ **23.** Not equal **25.** Equal

27. Not equal **29.** Not equal **31.** Equal only if $x = 2$ and $y = 3$
33. Equal only if $x = 2$ **35.** Not equal **37.** $\begin{bmatrix} 5 & 0 & 1 \\ 3 & 10 & -5 \end{bmatrix}$

39. Not possible **41.** $\begin{bmatrix} 2 & -8 \\ 4 & 7 \end{bmatrix}$ **43.** Not possible **45.** Not possible

47. $\begin{bmatrix} 0 & 0 \\ 0 & 0 \\ 0 & 0 \\ 0 & 3 \end{bmatrix}$ **49.** $\begin{bmatrix} 2x - 4 & x - 2 \\ -x & 4 \\ 2x & 3 - x \end{bmatrix}$ **51.** $\begin{bmatrix} 3a & 0 & 0 \\ 0 & 3b & 0 \\ 0 & 0 & 3c \end{bmatrix}$

53. Since A, B, and C are $m \times n$ matrices, the sums $A + (B + C)$ and $(A + B) + C$ are defined by Definition 7-4. Let a_{ij}, b_{ij}, and c_{ij} denote the elements in row i and column j of the matrices A, B, and C, respectively. The element in row i and column j of $B + C$ is $b_{ij} + c_{ij}$, so the element in row i and column j of $A + (B + C)$ is $a_{ij} + (b_{ij} + c_{ij})$. Similarly, the element in row i and column j of $A + B$ is $a_{ij} + b_{ij}$, and the element in row i and column j of $(A + B) + C$ is $(a_{ij} + b_{ij}) + c_{ij}$. But $a_{ij} + (b_{ij} + c_{ij}) = (a_{ij} + b_{ij}) + c_{ij}$, since addition of real numbers is associative. Therefore, $A + (B + C) = (A + B) + C$.

1. Conformable, 2×7 **3.** Not conformable   **5.** Conformable, 1×1
7. Conformable, 2×7 **9.** Conformable, 8×1
11. Conformable, 3×3 **13.** Not conformable

15. Conformable, 1×12 **17.** $\begin{bmatrix} 7 & -1 \\ 7 & -5 \end{bmatrix}$ **19.** $\begin{bmatrix} -1 & 2 \\ 5 & 20 \\ -11 & -13 \end{bmatrix}$

21. Not possible   **23.** Not possible   **25.** $[22]$ **27.** $\begin{bmatrix} -8 & -22 & 4 \\ 4 & 11 & -2 \\ 4 & 11 & -2 \end{bmatrix}$

29. Not possible   **31.** $\begin{bmatrix} 0 & 0 \\ 0 & 0 \\ 0 & 0 \end{bmatrix}$ **33.** Not possible   **35.** $\begin{bmatrix} 1 & 0 \\ 0 & 1 \end{bmatrix}$

37. $\begin{bmatrix} 1 & 0 & 0 \\ 0 & 1 & 0 \\ 0 & 0 & 1 \end{bmatrix}$ **39.** $BA = \begin{bmatrix} 9 & -17 \\ -1 & 3 \\ 0 & 10 \end{bmatrix}$, $CB = \begin{bmatrix} -4 & -2 \\ 5 & -2 \\ 1 & 4 \end{bmatrix}$

41. $AC = \begin{bmatrix} -3 & 2 \\ 5 & 0 \\ 1 & -3 \\ 3 & -2 \end{bmatrix}$, $BC = \begin{bmatrix} -1 & 8 \\ 9 & 7 \end{bmatrix}$, $BA = \begin{bmatrix} 0 & -2 & 2 & 8 \\ 1 & 4 & 2 & 29 \end{bmatrix}$,

$CB = \begin{bmatrix} 1 & 3 & 22 & 12 \\ 4 & -4 & 8 & 0 \\ 1 & 1 & 12 & 6 \\ 0 & -1 & -5 & -3 \end{bmatrix}$

43. $A = \begin{bmatrix} 1 & 2 \\ 0 & 1 \\ 1 & 1 \end{bmatrix}$, $B = \begin{bmatrix} 2 & 1 & 1 \\ -1 & 2 & 0 \end{bmatrix}$ is a possible answer.

45. $A = \begin{bmatrix} 1 & 2 & -1 \\ 0 & 1 & 2 \\ 1 & 0 & 0 \end{bmatrix}$, $B = \begin{bmatrix} 1 & 0 & 0 \\ 0 & -1 & 2 \\ 1 & 1 & 1 \end{bmatrix}$ is a possible answer.

47. $A = \begin{bmatrix} 1 & 0 \\ 1 & 2 \end{bmatrix}$, $B = \begin{bmatrix} 0 & 0 \\ 2 & 2 \end{bmatrix}$ is a possible answer.

49. $AB + AC = A(B + C) = \begin{bmatrix} -16 & -7 \\ -14 & -8 \\ 4 & 3 \end{bmatrix}$

51. $(A - B)(A + B) = \begin{bmatrix} 42 & -15 \\ 2 & 5 \end{bmatrix}$, $A^2 - B^2 = \begin{bmatrix} 39 & -14 \\ -5 & 8 \end{bmatrix}$

53. \$3.44 at store A, \$3.24 at store B, \$3.55 at store C.

55. Let $A = \begin{bmatrix} a_{11} & a_{12} \\ a_{21} & a_{22} \end{bmatrix}$, $B = \begin{bmatrix} b_{11} & b_{12} \\ b_{21} & b_{22} \end{bmatrix}$, $C = \begin{bmatrix} c_{11} & c_{12} \\ c_{21} & c_{22} \end{bmatrix}$. Then

$A(BC) = \begin{bmatrix} a_{11} & a_{12} \\ a_{21} & a_{22} \end{bmatrix} \begin{bmatrix} b_{11}c_{11} + b_{12}c_{21} & b_{11}c_{12} + b_{12}c_{22} \\ b_{21}c_{11} + b_{22}c_{21} & b_{21}c_{12} + b_{22}c_{22} \end{bmatrix}$

$= \begin{bmatrix} a_{11}b_{11}c_{11} + a_{11}b_{12}c_{21} & a_{11}b_{11}c_{12} + a_{11}b_{12}c_{22} \\ \quad + a_{12}b_{21}c_{11} + a_{12}b_{22}c_{21} & \quad + a_{12}b_{21}c_{12} + a_{12}b_{22}c_{22} \\ a_{21}b_{11}c_{11} + a_{21}b_{12}c_{21} & a_{21}b_{11}c_{12} + a_{21}b_{12}c_{22} \\ \quad + a_{22}b_{21}c_{11} + a_{22}b_{22}c_{21} & \quad + a_{22}b_{21}c_{12} + a_{22}b_{22}c_{22} \end{bmatrix}$

and

$$(AB)C = \begin{bmatrix} a_{11}b_{11} + a_{12}b_{21} & a_{11}b_{12} + a_{12}b_{22} \\ a_{21}b_{11} + a_{22}b_{21} & a_{21}b_{12} + a_{22}b_{22} \end{bmatrix} \begin{bmatrix} c_{11} & c_{12} \\ c_{21} & c_{22} \end{bmatrix}$$

$$= \begin{bmatrix} a_{11}b_{11}c_{11} + a_{12}b_{21}c_{11} & a_{11}b_{11}c_{12} + a_{12}b_{21}c_{12} \\ \quad + a_{11}b_{12}c_{21} + a_{12}b_{22}c_{21} & \quad + a_{11}b_{12}c_{22} + a_{12}b_{22}c_{22} \\ a_{21}b_{11}c_{11} + a_{22}b_{21}c_{11} & a_{21}b_{11}c_{12} + a_{22}b_{21}c_{12} \\ \quad + a_{21}b_{12}c_{21} + a_{22}b_{22}c_{21} & \quad + a_{21}b_{12}c_{22} + a_{22}b_{22}c_{22} \end{bmatrix}$$

Thus $A(BC) = (AB)C$.

57. Let $B = \begin{bmatrix} b_{11} & b_{12} \\ b_{21} & b_{22} \end{bmatrix}$, $C = \begin{bmatrix} c_{11} & c_{12} \\ c_{21} & c_{22} \end{bmatrix}$, and $D = \begin{bmatrix} d_{11} & d_{12} \\ d_{21} & d_{22} \end{bmatrix}$. Then

$$(B + C)D = \begin{bmatrix} b_{11} + c_{11} & b_{12} + c_{12} \\ b_{21} + c_{21} & b_{22} + c_{22} \end{bmatrix} \begin{bmatrix} d_{11} & d_{12} \\ d_{21} & d_{22} \end{bmatrix}$$

$$= \begin{bmatrix} b_{11}d_{11} + c_{11}d_{11} & b_{11}d_{12} + c_{11}d_{12} \\ \quad + b_{12}d_{21} + c_{12}d_{21} & \quad + b_{12}d_{22} + c_{12}d_{22} \\ b_{21}d_{11} + c_{21}d_{11} & b_{21}d_{12} + c_{21}d_{12} \\ \quad + b_{22}d_{21} + c_{22}d_{21} & \quad + b_{22}d_{22} + c_{22}d_{22} \end{bmatrix}$$

$$= \begin{bmatrix} b_{11}d_{11} & b_{11}d_{12} \\ \quad + b_{12}d_{21} & \quad + b_{12}d_{22} \\ b_{21}d_{11} & b_{21}d_{12} \\ \quad + b_{22}d_{21} & \quad + b_{22}d_{22} \end{bmatrix} + \begin{bmatrix} c_{11}d_{11} & c_{11}d_{12} \\ \quad + c_{12}d_{21} & \quad + c_{12}d_{22} \\ c_{21}d_{11} & c_{21}d_{12} \\ \quad + c_{22}d_{21} & \quad + c_{22}d_{22} \end{bmatrix}$$

$$= BD + CD.$$

59. Let $A = [a_{ij}]_{(m, n)}$. The elements in row i and column j of aA and bA are aa_{ij} and ba_{ij}, respectively. Therefore, the element in row i, column j of $aA + bA$ is $aa_{ij} + ba_{ij} = (a + b)a_{ij}$. But this is the element in row i, column j of $(a + b)A$. Therefore, $aA + bA = (a + b)A$.

**EXERCISES 7-3,
page 205**

1. $x = 0, y = 2, z = 5$ **3.** $x_1 = 1, x_2 = 3, x_3 = 5$

5. $x = 0, y = 0, z = 0$ **7.** $\begin{bmatrix} 3 & -1 & | & 0 \\ -1 & 1 & | & 1 \end{bmatrix}$ **9.** $\begin{bmatrix} 3 & -2 & 5 & | & 0 \\ 4 & 7 & -1 & | & 0 \\ 1 & 0 & 1 & | & 0 \end{bmatrix}$

11. $\begin{bmatrix} 1 & -1 & 0 & 0 & | & 0 \\ 0 & 0 & 1 & 1 & | & 0 \\ 3 & 0 & 0 & 2 & | & 0 \\ 0 & 5 & -1 & 0 & | & 0 \end{bmatrix}$ **13.** $\begin{array}{l} x + 2y = 5 \\ 3x + 4y = 6 \end{array}$ **15.** $\begin{array}{l} x = a \\ z = b \\ y = c \end{array}$

17. $\begin{array}{l} x \quad + z \quad\quad = 0 \\ 2y + z + 3w = 7 \\ 3x + y + z + w = 1 \\ -3x + y - z + 2w = 5 \end{array}$ **19.** $x = 5, y = -4$

21. $a = -3, b = -2$ **23.** No solution   **25.** $x = -8, y = -5$
27. $x = 0, y = 0$ **29.** $x = 3, y = 1, z = -1$ **31.** $x = 4, y = 4, z = 1$
33. $x = 1, y = 3, z = -2$ **35.** $r = 4, s = -6, t = 3$ **37.** No solution
39. $x = 1, y = -1, z = 2, t = -2$
41. There are many solutions of the form $a = 5 - 3r$, $b = 10 - 7r$, $c = -4r$, $d = r$, where r is any real number.
43. The last row corresponds to the equation $0 = 3$, which is impossible.
45. (a) No solution for $c = -2$,   (b) exactly one solution for $c \neq 2$ and $c \neq -2$,   (c) many solutions for $c = 2$.

47. (a) Solutions exist for all values of c, (b) exactly one solution for $c \neq 0$, (c) many solutions for $c = 0$.

1. $\begin{bmatrix} 1 & -\frac{1}{2} \\ 0 & \frac{1}{2} \end{bmatrix}$ **3.** $\begin{bmatrix} 2 & -3 \\ -1 & 1 \end{bmatrix}$ **5.** $\begin{bmatrix} -1 & \frac{3}{2} \\ -2 & \frac{5}{2} \end{bmatrix}$ **7.** Does not exist

9. Does not exist **11.** $\frac{1}{20}\begin{bmatrix} 1 & -5 \\ 2 & 10 \end{bmatrix}$ **13.** $\begin{bmatrix} \frac{1}{2} & 0 \\ 0 & -\frac{1}{4} \end{bmatrix}$ **15.** $\begin{bmatrix} \frac{2}{3} & -\frac{1}{2} \\ \frac{1}{3} & 0 \end{bmatrix}$

17. $-\frac{1}{8}\begin{bmatrix} 2 & -3 \\ -2 & -1 \end{bmatrix}$ **19.** $\frac{1}{45}\begin{bmatrix} 3 & 15 \\ -2 & 5 \end{bmatrix}$ **21.** Does not exist

23. $-\frac{1}{29}\begin{bmatrix} 7 & -3 \\ -5 & -2 \end{bmatrix}$ **25.** $\begin{bmatrix} -7 & 1 & 1 \\ 1 & 0 & 0 \\ 3 & 0 & -1 \end{bmatrix}$ **27.** $\begin{bmatrix} 11 & -6 & 2 \\ 3 & -2 & 1 \\ 1 & -1 & 1 \end{bmatrix}$

29. $\begin{bmatrix} \frac{5}{4} & \frac{1}{4} & -\frac{1}{2} \\ \frac{3}{4} & \frac{3}{4} & -\frac{1}{2} \\ -\frac{1}{4} & -\frac{1}{4} & \frac{1}{2} \end{bmatrix}$ **31.** $\begin{bmatrix} -\frac{1}{3} & -\frac{4}{3} & 1 \\ \frac{1}{3} & \frac{1}{3} & 0 \\ 1 & 1 & -1 \end{bmatrix}$ **33.** $\begin{bmatrix} 0 & -\frac{1}{2} & \frac{1}{2} \\ \frac{1}{2} & \frac{1}{2} & 0 \\ -\frac{1}{2} & 0 & \frac{1}{2} \end{bmatrix}$

35. $\frac{1}{15}\begin{bmatrix} 5 & -10 & 30 \\ -1 & 5 & 0 \\ 1 & -5 & 15 \end{bmatrix}$ **37.** $\begin{bmatrix} 3 & 2 & -4 \\ 1 & 1 & -1 \\ -\frac{5}{3} & -\frac{4}{3} & \frac{7}{3} \end{bmatrix}$ **39.** Does not exist

41. Does not exist **43.** $\begin{bmatrix} \frac{1}{2} & 0 & 0 & \frac{1}{2} \\ 0 & \frac{1}{4} & \frac{1}{4} & 0 \\ 0 & \frac{1}{6} & -\frac{1}{6} & 0 \\ -\frac{1}{8} & 0 & 0 & \frac{1}{8} \end{bmatrix}$ **45.** $\begin{bmatrix} -23 & 4 & 5 & 2 \\ -97 & 15 & 22 & 8 \\ -50 & 8 & 11 & 4 \\ -13 & 2 & 3 & 1 \end{bmatrix}$

47. $\begin{bmatrix} 1/a & 0 & 0 \\ 0 & 1/b & 0 \\ 0 & 0 & 1/c \end{bmatrix}$

49. By problem 48, $(AB)^{-1}$ exists and $(AB)^{-1} = B^{-1}A^{-1}$.
$$
\begin{aligned}
(ABC)^{-1} &= [(AB)C]^{-1} && \text{(Associative property, } \cdot\text{)} \\
&= C^{-1}(AB)^{-1} && \text{(Problem 48)} \\
&= C^{-1}(B^{-1}A^{-1}) && \text{(Problem 48)} \\
&= C^{-1}B^{-1}A^{-1} && \text{(Associative property, } \cdot\text{)}
\end{aligned}
$$

1. $x = 2, y = -5$ **3.** $a = 1, b = -2$ **5.** $x = \frac{1}{3}, y = \frac{1}{6}$
7. $x_1 = 5, x_2 = 1$ **9.** $x = 10, y = 4$ **11.** $x = 1, y = 3, z = 5$
13. $r = 10, s = -12, t = -2$ **15.** $x = 0, y = -\frac{1}{2}, z = \frac{1}{2}$
17. $x = 3, y = -4, z = -3, t = 1$ **19.** $x = 5, y = -3$
21. $x_1 = 3, x_2 = 3$ **23.** $x = 6, y = 2$ **25.** $x = -\frac{3}{4}, y = \frac{7}{4}$
27. $x_1 = -4, x_2 = 0, x_3 = -4$ **29.** $x = 2, y = 1, z = -3$
31. $x = 5, y = 3, z = 1$ **33.** $x = 3, y = 1, z = -1$
35. $x = 1, y = 3, z = -2, w = 0$ **37.** $X = BA^{-1}$
39. $x = \pm 3\sqrt{2}/2, y = \pm\sqrt{2}$ **41.** $x = 0, y = \pm 3$ **43.** $x = 0, y = 0$

1. -5 **3.** -1 **5.** -21 **7.** 0 **9.** $-\frac{1}{9}$ **11.** $5x$ **13.** 6 **15.** 26
17. 12 **19.** 7 **21.** -3 **23.** -1 **25.** 4 **27.** -3 **29.** -1 **31.** 0
33. -2 **35.** 1 **37.** -1 **39.** 2 **41.** -2 **43.** -3 **45.** $2, -3$
47. $-1, -2$ **49.** $5, -3, 2$ **51.** $0, 4, -3$

1. $x = -4$ **3.** $a = -1$ **5.** $x = 1$ **7.** $x = -6$ **9.** $x = 2$
11. $x = -2$ **13.** 15   **15.** 10   **17.** -20 **19.** -1 **21.** 1   **23.** 0
25. -1 **27.** 4   **29.** 23   **31.** 4

1. $x = -3, y = 1$ **3.** $x = 1, y = -2$ **5.** $a = -2, b = 3$
7. Dependent system   **9.** Inconsistent   **11.** $x = 2, y = 1, z = -1$
13. $x = 5, y = -10, z = 4$ **15.** $x = -21, y = -12, z = 39$
17. $a = 1, b = 2, c = 3$ **19.** $x = 2, y = -1, z = 4$
21. Dependent system   **23.** $x_1 = \frac{28}{13}, x_2 = -\frac{6}{13}, x_3 = \frac{53}{13}$
25. $x = \frac{1}{6}, y = \frac{1}{4}, z = \frac{1}{3}$ **27.** Inconsistent
29. $x = -\frac{1}{2}, y = \frac{15}{14}, z = -\frac{25}{14}$ **31.** $x = -1, y = -1, z = 1$
33. $x = -2, y = 5, z = 1$ **35.** $x = 2, y = -1, z = 0, w = 1$
37. $x = w = 1, y = z = 0$ **39.** $m = 2, b = -4$
41. $a = 2, b = 3, c = -4$

1. $x = 2, y = -6$ **3.** $x = 0, y = -7$ **5.** $x = \pm 2, y = 0$ or $y = -$
7. $x = \frac{7}{2}, y = 2$ **9.** $4i$ **11.** $-7i$ **13.** $-\sqrt{5}\,i$ **15.** $9 - i$
17. $15 + 67i$ **19.** $-4 - 2i$ **21.** $(2x - 7) + (6y - 3)i$
23. $(x^2 + y^2) - 8i$ **25.** $9 + 7i$ **27.** $-54 - 10i$ **29.** $27 - 36i$
31. $40i$ **33.** $(3x - 2y^2) + (-xy - 6y)i$ **35.** $2 - \frac{7}{3}i$ **37.** $\frac{9}{5} + \frac{3}{5}i$
39. $\frac{27}{37} - \frac{23}{37}i$ **41.** $\frac{7}{2}$ **43.** $-i$ **45.** $\frac{3}{25} - \frac{4}{25}i$ **47.** $\frac{5}{169} + \frac{12}{169}i$
49. $\frac{15}{289} - \frac{8}{289}i$ **51.** 1   **53.** i **55.** $-i$ **57.** i **59.** -1 **61.** $-i$
63. (a) $\overline{z_1 + z_2} = \overline{(a_1 + a_2) + (b_1 + b_2)i}$
$\qquad\qquad = (a_1 + a_2) - (b_1 + b_2)i$
$\qquad\qquad = a_1 - b_1 i + a_2 - b_2 i$
$\qquad\qquad = \bar{z}_1 + \bar{z}_2$
$\quad$ (b) $\overline{z_1 \cdot z_2} = \overline{(a_1 a_2 - b_1 b_2) + (a_1 b_2 + a_2 b_1)i}$
$\qquad\qquad = (a_1 a_2 - b_1 b_2) - (a_1 b_2 + a_2 b_1)i$
$\qquad\qquad = (a_1 - b_1 i)(a_2 - b_2 i)$
$\qquad\qquad = \bar{z}_1 \cdot \bar{z}_2$
65. $\overline{(z^n)} = \underbrace{\overline{(z \cdot z \cdot z \cdots z)}}_{n \text{ factors } z}$ (Definition of z^n)

$\qquad = \underbrace{\bar{z} \cdot \bar{z} \cdot \bar{z} \cdots \bar{z}}_{n \text{ factors } \bar{z}}$ [Problem 64(b)]

$\qquad = (\bar{z})^n$ (Definition of $(\bar{z})^n$)

1. $-\dfrac{1}{4} \pm \dfrac{\sqrt{7}}{4}i$ **3.** $-2 \pm i$ **5.** $\dfrac{3}{8} \pm \dfrac{\sqrt{7}}{8}i$ **7.** $4i, -i$ **9.** $2i, -\frac{1}{2}i$
11. $\frac{1}{2} + \frac{1}{2}i, -1 - i$ **13.** 7   **15.** $4\sqrt{2}$ **17.** $-3i$ **19.** 9
21. $6 + 6\sqrt{2}$ **23.** $-16 + 3i$ **25.** $-16 - 22i$ **27.** No   **29.** Yes
31. No   **33.** Yes   **35.** Yes   **37.** No   **39.** Yes   **41.** $x = 3, x = \pm i$
43. $k = \frac{14}{3}$

1. $P(x) = Q(x) = x^2 + 2x - 15$ **3.** $P(x) = x - 2i, Q(x) = x^2 + 4$
5. $P(x) = x^2 - (3 + 2i)x + 6i, Q(x) = x^3 - 3x^2 + 4x - 12$
7. $P(x) = x^3 - (7 + i)x^2 + (17 + 2i)x - 15 + 3i,$
$\quad Q(x) = x^5 - 11x^4 + 51x^3 - 119x^2 + 140x - 78$

9. 2 positive, 2 negative; 2 positive, 2 complex; 2 negative, 2 complex; 4 complex

11. 2 negative, 4 complex; 6 complex   **13.** 4 complex   **15.** 6 complex

17. $3i$ **19.** $1 - 2i$ **21.** $2i, -4i$ **23.** $3i, -\dfrac{1}{2} \pm \dfrac{\sqrt{3}}{2}i$

25. $1 - i$ of multiplicity 2, 2

27. If all the coefficients of $P(x)$ are positive, then there are no variations in sign. By Descartes' rule of signs, there are no positive real zeros.

29. Let $P(x)$ be a polynomial of odd degree which has real coefficients. By Theorem 9-13, any complex zeros of $P(x)$ occur in conjugate pairs ($a + bi$ and $a - bi$). This means that the number of complex zeros of $P(x)$ is an even number. Since the total number of zeros is odd, at least one zero must be real.

EXERCISES 9-4,
page 269

1. (a) 3   (b) -3 **3.** (a) 2   (b) -1 **5.** (a) 2   (b) -3

7. (a) 2   (b) -3 **9.** $3, 2i, -2i$ **11.** $-3, -1, \frac{1}{2}$

13. $2, (-1 \pm \sqrt{23}i)/6$ **15.** $-2, 1, 1 \pm i$ **17.** $-1, \frac{3}{2}, \pm\sqrt{5}$

19. $-3, \frac{1}{2}, \pm\sqrt{2}$ **21.** No rational zeros   **23.** $-\frac{3}{2}$ **25.** $2, -\frac{3}{2}$

27. The set of possible rational zeros is $\{\pm 1, \pm 3\}$. But none of the elements of this set are zeros of $P(x) = x^2 - 3$. Thus any zero must be irrational. Since $P(\sqrt{3}) = 0$, then $\sqrt{3}$ is an irrational zero of $P(x)$.

EXERCISES 9-5,
page 273

1. 2.2   **3.** 0.5   **5.** 2.7   **7.** -0.8 **9.** 1.2   **11.** $-1.5, 0.3, 2.2$

13. $0.2, 1, 1.8$ **15.** $-2.3, -1.5, 4.8$ **17.** $1, 2.2$

EXERCISES 9-6,
page 278

1.

3.

5.

7.

9.

11.
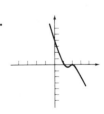

13.
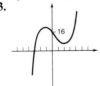

15.
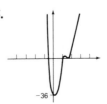

17.
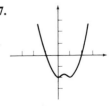

19.

21.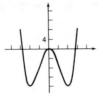

23.

25.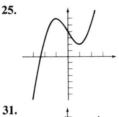

27.

29.

31.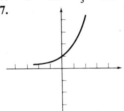

EXERCISES 10-1,
page 287

1. 1 **3.** $\frac{9}{4}$ **5.** $\frac{2}{3}$ **7.** $\sqrt{6}/3$ **9.** 3 **11.** -4 **13.** 0 **15.** -3
17. -2 **19.** $-\frac{4}{3}$ **21.** -4 **23.** $-\frac{3}{2}$ **25.** $-\frac{5}{8}$ **31.**

27.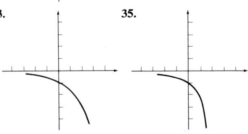

29.

31.

33.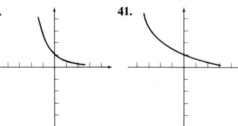

35.

37.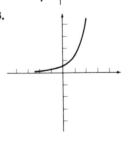

39.

41.

43.

45. 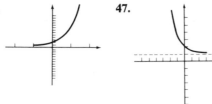 **47.** **49.**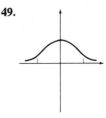

51. 256,000; every 10 years

EXERCISES 10-2,
page 294

1. $\log_4 16 = 2$ **3.** $\log_3 81 = 4$ **5.** $\log_3 \frac{1}{9} = -2$ **7.** $\log_{10} \frac{1}{100} = -2$
9. $\log_5 1 = 0$ **11.** $\log_4 \frac{1}{2} = -\frac{1}{2}$ **13.** $2^6 = 64$ **15.** $3^{-3} = \frac{1}{27}$
17. $(\frac{3}{4})^2 = \frac{9}{16}$ **19.** $(10)^{-2} = 0.01$ **21.** $y = 0$ **23.** $y = 3$ **25.** $y = 2$
27. $x = 9$ **29.** $x = \frac{1}{81}$ **31.** $y = -2$ **33.** $y = -2$ **35.** $a = 2$
37. $a = \frac{1}{5}$ **39.** $y = -3$ **41.** $y = 2$ **43.** 1.1761   **45.** 1.4313
47. 0.3495   **49.** −0.2219   **51.** −0.4438
53. **55.** **57.**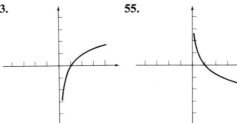

59.

61. $\log_a x + \log_a y$ **63.** $2\log_a x + 3\log_a y - 4\log_a z$
65. $(\log_a x)/5 - 3(\log_a z)/5$ **67.** $\log_a(x\sqrt{z}/y^2)$ **69.** $\log_a(5x^2y^5/2)$
71. $\log_a(36x^{8/3}y^2z^5)$
73. Let $u = a^x$, $v = a^y$, so that $\log_a u = x$ and $\log_a v = y$. Then $uv = a^x \cdot a^y$
$= a^{x+y}$, by Theorem 10-1(a). Therefore, $\log_a uv = x + y = \log_a u + \log_a v$.
75. Let $x = b^y$, so that $\log_b x = y$. Then $\log_a x = \log_a(b^y)$, and $\log_a x = y\log_a b$. Therefore $y = \log_a x/\log_a b$. But $y = \log_b x$, so we have
$$\log_b x = \log_a x/\log_a b = \frac{1}{\log_a b} \cdot \log_a x.$$

EXERCISES 10-3,
page 301

1. $1.6191 - 10$ **3.** 1.4871   **5.** 0.6542   **7.** 1   **9.** 6.0294
11. $7.0294 - 10$ **13.** 7.18   **15.** 283,000,000   **17.** 0.00476
19. 0.001   **21.** 1.0047   **23.** 0.6202   **25.** 5.6210   **27.** $7.8875 - 10$
29. 1.292   **31.** 87,760   **33.** 0.004954   **35.** 0.2444

1. 65.3 **3.** 0.190 **5.** 2130 **7.** 1.96 **9.** 2.04 **11.** 0.00280
13. 0.04424 **15.** 0.8474 **17.** 0.7993 **19.** 557.1 **21.** 4,285,000,000
23. 0.4601 **25.** 20.18 **27.** 8270

1. 2.58 **3.** −1.43 **5.** 7 **7.** −1.13 **9.** ±1.02 **11.** 1.778 **13.** 1.74
15. $\frac{1}{3}$ **17.** 9 **19.** $\frac{10}{3}$ **21.** $\frac{8}{7}$ **23.** 0.329 **25.** 0.028 or −8.03
27. 19.8 **29.** $1270 **31.** 12.54 yr

1. $\frac{1}{2}, \frac{1}{3}, \frac{1}{4}, \frac{1}{5}, \frac{1}{6}$ **3.** $0, \frac{1}{2}, \frac{2}{3}, \frac{3}{4}, \frac{4}{5}$ **5.** $-2, 4, -8, 16, -32$
7. 2, 2, 2, 2, 2 **9.** $-1, \frac{1}{4}, -\frac{1}{9}, \frac{1}{16}, -\frac{1}{25}$ **11.** $1, \frac{2}{3}, \frac{3}{5}, \frac{4}{7}, \frac{5}{9}$
13. 1, 4, 7, 10, 13
15. $x - 1, (x - 1)^2, (x - 1)^3, (x - 1)^4, (x - 1)^5$
17. $-x^2, x^4, -x^6, x^8, -x^{10}$ **19.** 1, 2, 4, 8, 16, 32
21. $-5, 5, -5, 5, -5, 5$ **23.** 2, 3, 8, 19, 46, 111
25. $2 + 4 + 8 + 16 + 32 = 62$
27. $\frac{1}{4} + \frac{2}{5} + \frac{3}{6} + \frac{4}{7} + \frac{5}{8} + \frac{6}{9} + \frac{7}{10} = \frac{3119}{840}$
29. $0 + 1 + 3 + 6 + 10 + 15 + 21 = 56$
31. $-1 + 1 - 1 + 1 - 1 + 1 - 1 = -1$
33. $2 + 2 + 2 + 2 + 2 + 2 = 12$
35. $4 - 9 + 16 - 25 + 36 - 49 = -27$
37. $-1 + 0 + 1 + 8 + 27 + 64 + 125 = 224$
39. $(1 - \frac{1}{2}) + (\frac{1}{2} - \frac{1}{3}) + (\frac{1}{3} - \frac{1}{4}) + (\frac{1}{4} - \frac{1}{5}) + (\frac{1}{5} - \frac{1}{6}) + (\frac{1}{6} - \frac{1}{7}) = \frac{6}{7}$
41. $\frac{1}{4} - \frac{1}{8} + \frac{1}{16} - \frac{1}{32} = \frac{5}{32}$

1. 1, 2, 3, 4, 5, 6 **3.** 13, 11, 9, 7, 5, 3 **5.** 3, −1, −5, −9, −13, −17
7. 2, 6, 10, 14, 18, 22 **9.** $d = 5, a_n = 5n - 22$
11. $d = -3, a_n = -3(n + 1)$ **13.** Not arithmetic
15. $d = -14, a_n = 21 - 14n$ **17.** $d = y, a_n = x - y + ny$
19. Not arithmetic unless $x = 0$ or $x = 1$. If $x = 0$, $d = 0$ and $a_n = 0$. If
$x = 1$, $d = 0$ and $a_n = 1$.
21. $d = x, a_n = nx$ **23.** $a_{11} = -25, S_{13} = -221$
25. $a_{11} = 7, S_{13} = -65$ **27.** $a_{11} = -13, S_{13} = -13$
29. $a_{11} = m - 10x, S_{13} = 13(m - 6x)$
31. $a_{11} = x - 10k, S_{13} = 13(x - 2k)$ **33.** 53 **35.** −60 **37.** 23
39. $-\frac{3933}{4}$ **41.** $a_1 = -7, d = \frac{1}{2}$ **43.** $a_1 = 17, d = -5$ **45.** 91
47. $\frac{110}{3}$ **49.** 2550 **51.** −117
53. Let $a_1, a_2, a_3, \ldots$ be a sequence such that $a_n = a \cdot n + b$. Then $a_{n+1} = a(n + 1) + b$, and $a_{n+1} - a_n = [a(n + 1) + b] - [an + b] = a$, for all positive integers n. This difference, a, is a constant, so the sequence is an arithmetic progression with first term $a_1 = a + b$ and common difference $d = a$.
55. 1065 **57.** −20, −25, −30, −35 **59.** 4510 **61.** $n = 36$

1. $-1, 2, -4, 8, -16$ **3.** $4, 1, \frac{1}{4}, \frac{1}{16}$ **5.** $\frac{3}{4}, 3, 12, 48$
7. $\frac{1}{4}, -\frac{1}{2}, 1, -2, 4$ **9.** $-3, -6, -12$ or $-3, 6, -12$
11. $a_5 = 4, a_n = 2^n/8, S_5 = \frac{31}{4}$

13. $a_5 = 27, a_n = -(-3)^n/9, S_5 = \frac{61}{3}$
15. $a_5 = -8, a_n = (-2)^n/4, S_5 = -\frac{11}{2}$ **17.** $r = \frac{1}{2}, a_n = 14/2^n$
19. $r = \sqrt{2}/2, a_n = (\sqrt{2})^{5-n}$ **21.** Not geometric   **23.** Not geometric
25. $r = -\frac{1}{2}, a_n = (-4)(-\frac{1}{2})^{n-1}$ **27.** $r = -\frac{1}{7}, a_n = (-1)^{n-1}/7^{n-4}$
29. 31   **31.** $\frac{4323}{3125}$ **33.** $\frac{275}{81}$ **35.** $\frac{2101}{4}$ **37.** $\frac{63}{4}$
39. $r = \pm\sqrt{3}, a_1 = -\frac{1}{3}$
41. $r = 1.01, a_1 = 4/1.01$, or $r = -1.01, a_1 = -4/1.01$
43. \$163.84, \$327.67   **45.** \$768,000

EXERCISES 11-4,
page 331

1. -2 **3.** $\frac{27}{8}$ **5.** Sum does not exist   **7.** Sum does not exist
9. $2(2 + \sqrt{2})$ **11.** $\frac{50}{9}$ **13.** 5   **15.** 30   **17.** 3
19. Sum does not exist. Geometric series with $|r| = |-1.02| > 1$.
21. Sum does not exist. Geometric series with $|r| = |-\frac{100}{99}| > 1$.
23. $\frac{4096}{7}$ **25.** Sum does not exist. Geometric series with $|r| = |6| > 1$.
27. 10   **29.** Sum does not exist. Geometric series with $|r| = |-3| > 1$.
31. $\frac{13}{3}$ **33.** $-\frac{4}{3}$ **35.** 1   **37.** 1   **39.** $\frac{1}{99}$ **41.** 131/10,000   **43.** $\frac{28}{9}$
45. $-\frac{4579}{1998}$ **47.** 10,033/990   **49.** $4\frac{2}{3}$ meters

EXERCISES 12-1,
page 341

1. 60   **3.** 12   **5.** 480   **7.** 1024   **9.** 16   **11.** 120
13. 1,000,000,000   **15.** 7999   **17.** 108

EXERCISES 12-2,
page 346

1. 0; 1; 0, 1; 1, 0   **3.** 1, 2; 2, 1; 3, 1; 4, 1; 5, 1;   **5.** 6   **7.** 1260
  1, 3; 2, 3; 3, 2; 4, 2; 5, 2;
  1, 4; 2, 4; 3, 4; 4, 3; 5, 3;
  1, 5; 2, 5; 3, 5; 4, 5; 5, 4;
9. 20   **11.** 24   **13.** 210   **15.** 1   **17.** $n(n - 1)$ **19.** $n!/2$
21. $n(n - 1)$ **23.** $1/(n + 2)(n + 1)$ **25.** (a) 720   (b) 5040
27. 479,001,600   **29.** 144   **31.** 1440   **33.** 120
35. (a) 2520   (b) 5040   (c) 3   (d) 180   **37.** 630; 630   **39.** 3150
41. 144   **43.** 24   **45.** 144, 576

EXERCISES 12-3,
page 351

1. $\{0\}, \{1\}, \{0, 1\}$
3. $\{1, 2\}, \{1, 3\}, \{1, 4\}, \{1, 5\}, \{2, 3\}, \{2, 4\}, \{2, 5\}, \{3, 4\}, \{3, 5\}, \{4, 5\}$
5. 6   **7.** 1   **9.** 1   **11.** 15   **13.** n **15.** 1   **17.** 495   **19.** 5,005; 3,003
21. 280   **23.** 3,300   **25.** 44,100   **27.** 10,090,080
29. (a) 70   (b) 10   (c) 70   (d) 0   **31.** 36
33. (a) 8,211,173,256   (b) 169   (c) 2,944,656
35. By Theorem 12-5,
$$C(n, n - 4) = \frac{n!}{[n - (n - r)]!\,(n - r)!} = \frac{n!}{r!\,(n - r)!} = C(n, r).$$

EXERCISES 12-4,
page 356

1. $x^7 + 7x^6y + 21x^5y^2 + 35x^4y^3 + 35x^3y^4 + 21x^2y^5 + 7xy^6 + y^7$
3. $16x^4 + 32x^3y + 24x^2y^2 + 8xy^3 + y^4$
5. $x^{10} + 10x^8y + 40x^6y^2 + 80x^4y^3 + 80x^2y^4 + 32y^5$
7. $x^8 - 4x^6y^2 + 6x^4y^4 - 4x^2y^6 + y^8$

9. $16x^4 - 32x^3/y + 24x^2/y^2 - 8x/y^3 + 1/y^4$

11. $81x^4 + 36x^3y + 6x^2y^2 + 4xy^3/9 + y^4/81$

13. $x^{18} - 3x^{17} + 15x^{16}/4 - 5x^{15}/2 + 15x^{14}/16 - 3x^{13}/16 + x^{12}/64$

15. $a^5/32 - 15a^4b^2/16 + 45a^3b^4/4 - 135a^2b^6/2 + 405ab^8/2 - 243b^{10}$

17. 45   **19.** 80   **21.** $1760x^3y^9$ **23.** $-29,120x^{12}y^6$

25. $3003p^{12}q^{24}$ **27.** $8192y^{13}$

29. $1 + x/4 - 3x^2/32 + 7x^3/128 - 77x^4/2048 + \cdots, -1 < x < 1$

31. $1 - 2y/3 - y^2/9 - 4y^3/27 - 7y^4/243 + \cdots, -1 < y < 1$

33. $1 - x + 3x^2/2 - 5x^3/2 + 35x^4/8 - \cdots, -\frac{1}{2} < x < \frac{1}{2}$

35. $1 - 2x^2 + 3x^4 - 4x^6 + 5x^8 - \cdots, -1 < x < 1$

37. $1 + a + 3a^2/4 + a^3/2 + 5a^4/16 + \cdots, -2 < a < 2$

39. $x^4 + 4x^3y + 6x^2y^2 + 4xy^3 + y^4 + 4x^3 + 12x^2y + 12xy^2 + 4y^3$
$+ 6x^2 + 12xy + 6y^2 + 4x + 4y + 1$

41. $x^3 - y^3 + z^3 - w^3 - 3x^2y + 3xy^2 + 3x^2z + 3xz^2 + 3y^2z - 3yz^2$
$- 3x^2w + 3xw^2 - 3y^2w - 3yw^2 - 3z^2w + 3zw^2 - 6xyz + 6xyw$
$- 6xzw + 6yzw$

43. 1.061520   **45.** 0.922744   **47.** 27.270901   **49.** 1.0099505

EXERCISES 12-5,
page 362

1. $\dfrac{11}{x-3} - \dfrac{8}{x-2}$ **3.** $\dfrac{4}{x+2} - \dfrac{1}{x}$ **5.** $\dfrac{1}{x-1} + \dfrac{1}{x-2} - \dfrac{2}{x}$

7. $\dfrac{\frac{1}{2}}{x} + \dfrac{\frac{1}{6}}{x-2} - \dfrac{\frac{2}{3}}{x+1}$ **9.** $\dfrac{7}{3x+2} - \dfrac{2}{x+1} + \dfrac{1}{x+2}$

11. $\dfrac{1}{x-2} - \dfrac{1}{(x-2)^2}$ **13.** $\dfrac{1}{x-1} + \dfrac{2}{(x-1)^2} + \dfrac{2}{(x-1)^3}$

15. $\dfrac{\frac{1}{2}}{x+1} - \dfrac{1}{(x+1)^2} + \dfrac{\frac{1}{2}}{x-1}$ **17.** $\dfrac{-\frac{1}{4}}{x^2} + \dfrac{\frac{1}{16}}{x-2} - \dfrac{\frac{1}{16}}{x+2}$

19. $\dfrac{1}{x-1} + \dfrac{1}{(x-1)^2} - \dfrac{1}{x} - \dfrac{2}{x^2}$ **21.** $\dfrac{2}{x-2} + \dfrac{9}{(x-2)^2} + \dfrac{2}{3x-1}$

23. $\dfrac{x+2}{x^2+x+2} - \dfrac{1}{x-1}$ **25.** $\dfrac{4}{x-3} + \dfrac{x-1}{x^2+x-1}$

27. $\dfrac{2}{x-1} - \dfrac{1}{x+1} + \dfrac{4x}{x^2+1}$ **29.** $\dfrac{x}{x^2+1} - \dfrac{x+1}{(x^2+1)^2}$

31. $\dfrac{2x+1}{2x^2-x+1} - \dfrac{1}{(2x^2-x+1)^2}$ **33.** $x + \dfrac{x}{x^2-2x+3}$

35. $2x - 1 - \dfrac{3}{x} + \dfrac{1}{(x-1)^2}$

EXERCISES 12-6,
page 370

1. $V: x = -4;$
$H: y = 0$

3. $V: x = 3;$
$H: y = 2$

5. $V: x = -\frac{5}{3};$
$H: y = -\frac{4}{3}$

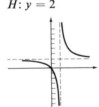

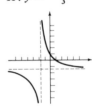

7. $V: x = 0, x = 3;$
   $H: y = 0$

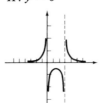

9. $V: x = -1;$
   $H: y = 0$

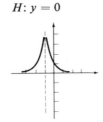

11. $H: y = 0$

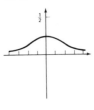

13. $V: x = -2;$
   $H: y = 0$

15. $V: x = 0, x = -1;$
   $H: y = 0$

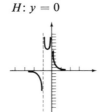

17. $V: x = -2, x = 3;$
   $H: y = 0$

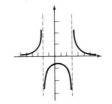

19. $V: x = 0, x = 1;$
   $H: y = 1$

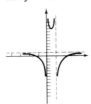

21. $V: x = -1, x = \frac{3}{2};$
   $H: y = \frac{1}{2}$

23. $V: x = -1;$
   $O: y = 2x$

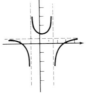

25. $V: x = 1, x = 2;$
   $H: y = 0$

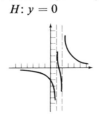

27. $V: x = 0, x = -2;$
   $H: y = 0$

29. $V: x = 2;$
   $O: y = x + 2$

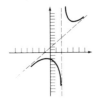

31. $V: x = 0;$
   $H: y = 0$

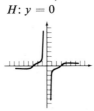

33. $V: x = 1$

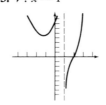

35.

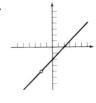

37.

39.

Index

A

Abscissa, 110
Absolute value, 24, 71
  equations, 71, 81
  inequalities, 81
Addition, 8
  of complex numbers, 243
  of fractions, 15
  of matrices, 187
  of polynomials, 36
Addition postulates, 8
Addition property:
  of equality, 8, 70
  of inequalities, 79
Additive cancellation law, 13

Additive identity, 8
  for complex numbers, 243
  for matrices, 188
Additive inverse, 8
  for complex numbers, 243
  for matrices, 188
Algebraic expression, 30
Algebraic method of solution of
    an inequality, 134
Antilogarithm, 299
Approximate equality, 18
Arithmetic progression, 316
Associative property, 8–9
  of addition of matrices, 187
  of multiplication of matrices,
    194
Asymptotes:

Asymptotes *(cont.)*
　horizontal, 364, 366
　of a hyperbola, 157–58
　oblique, 368
　vertical, 364, 366
Augmented matrix, 200
Axis, 110
　of symmetry, 125–27

Base, 31, 281
Binary operation, 8
Binomial, 36
Binomial coefficients, 352, 354–56
Binomial theorem, 354
Boundary, 171
Bounds for zeros, 267

Cancellation law:
　for addition, 13
　for multiplication, 14
Cartesian coordinate system, 109
Characteristic of a logarithm, 297
Characteristic values, 226
Circle, 142, 144–46
Closure properties, 8, 11–12, 19
Coefficient matrix, 214
Cofactor, 222
Cofactor expansion, 223
Column matrix, 186
Column operations, 227
Combinations, 348
Combined variation, 86
Common difference, 316
Common logarithm, 295
Common ratio, 322
Commutative property, 8–9
　of matrix addition, 188
Complement, 4
　of one set in another, 4

Completeness property, 18
Completing the square, 92
Complex fraction, 50
Complex numbers, 242
　addition of, 243
　additive identity for, 243
　additive inverse for, 243
　conjugate of, 245
　difference of, 243
　division of, 245
　equality of, 242
　multiplication of, 244
　multiplicative inverse for, 245
　standard form for, 246
　subtraction of, 243
Components, 106
Composition function, 109
Compound amount, 308
Compound inequality, 79
Cone, 142
Conformable matrices, 192
Conic sections, 142
Conjugate of a binomial, 63
Conjugate of a complex number,
　245
Conjugate pairs of zeros, 261
Constant matrix, 214
Constant of variation, 86
Coordinates, 110
Counting numbers, 5
Counting principle, 339
Cramer's Rule, 235, 237
Cube root, 54

Degree:
　of a term, 35, 37
　of a polynomial, 35, 37
De Morgan's Theorem, 4
Dependent system, 235, 237
Descartes' rule of signs, 259
Determinant, 220–21, 223
　cofactors of, 222
　expansion of, 223

Determinant *(cont.)*
  order of, 220
  properties of, 228–29
Diagonal matrix, 186
Difference, 9
  common, 316
  of complex numbers, 243
  of matrices, 189
Dimension of a matrix, 185
Direct variation, 86
Discriminant, 97
Disjoint sets, 3
Distance formula, 144
Distributive property, 9
Dividend, 42
Division, 10
  by detached coefficients, 45
  of polynomials, 42-43
  synthetic, 45
Divisor, 42
Domain, 106
  of an exponential function, 284
  of a logarithmic function, 292

E

Eigenvalues, 226
Element:
  of a matrix, 185
  of a set, 1
Elimination method, 164
Ellipse, 142, 149
  foci of, 149
  standard equations for, 151
Empty set, 3
Equality, 7–8
  of complex numbers, 242
  of fractions, 15
  of matrices, 186
  properties of, 8
  of sets, 2
Equations:
  absolute value, 71, 81
  equivalent, 69
  exponential, 306

  linear, 69
  logarithmic, 307
  quadratic, 90
Equivalent equations, 69
Equivalent inequalities, 79
Exponent, 31
  definition of, 31, 33
  laws of, 31
Exponential equation, 306
Exponential form, 58
Exponential function, 282
Expressions:
  algebraic, 30
  rational integral, 47
Extraneous solutions, 101

F

Factor, 39
Factorial, 343
Factoring by grouping, 41
Factorization formulas, 40
Factor theorem, 252
Field, 8
Field properties, 8–9
Finite sequence, 312
Focus:
  of an ellipse, 149
  of a hyperbola, 155
Fractions:
  complex, 50
  operations with, 15
Function, 108
  composition, 109
  exponential, 282
  logarithmic, 289
Fundamental operations, 10
Fundamental theorem of algebra,
  257

G

Gaussian elimination:
  to find an inverse, 209

Gaussian elimination *(cont.)*
  to solve a system, 203
General term:
  of an arithmetic progression, 319
  of a geometric progression, 325
  of a sequence, 312
Geometric progression, 322
Graph:
  of an equation, 112
  of an inequality, 171
  of a relation, 111
Graphical method of solution of a quadratic inequality, 132
Greater than, 18

H

Half-plane, 171
Hierarchy of operations, 10
Horizontal asymptote, 364, 366
Horizontal line, 117
Hyperbola, 142, 155
  asymptotes of, 157–58
  standard equations for, 158

I

Identity elements, 8–9
Identity matrix, 195
Imaginary part of a complex number, 242
Inconsistent system, 235, 237
Index:
  of a radical, 55
  of summation, 313
Induction hypothesis, 334
Induction principle, 333
Inequalities:
  absolute value, 81
  compound, 79
  equivalent, 79

linear, 79, 171
quadratic, 131
systems of, 170
Infinite sequence, 312
Integers:
  even, 5
  positive, 5
Intercepts of a graph, 116
Interpolation, 299
Intersection, 3
Inverse:
  of a complex number, 243, 245
  of a matrix, 208
  of a relation, 113, 176
Inverse elements, 8–9
Inverse variation, 86

J

Joint variation, 86

L

Laws of exponents, 31, 282
Least common denominator, 49
Least common multiple, 48
Less than, 19
Linear equation, 69, 120
  point–slope form of, 121
  slope–intercept form of, 121
  standard form of, 120
Linear function, 115
Linear inequality, 79, 171
Linear interpolation, 299
Linear relation, 115
Lines:
  equations of, 120–21
  parallel, 122
  perpendicular, 122
  straight, 116
Location theorem, 270
Logarithm, 288
  characteristic, 297

Logarithm *(cont.)*
  common, 295
  mantissa, 298
Logarithmic equation, 307
Logarithmic function, 289
Lower bound, 267
Lowest terms, 47

Mantissa, 298
Mathematical induction, 333
Matrix, 185
  addition of, 187
  augmented, 200
  coefficient, 214
  column, 186
  constant, 214
  determinant of, 220-21, 223
  diagonal, 186
  dimension of, 185
  equality of, 186
  identity, 195
  multiplication, 194
  multiplicative inverse of, 208
  row, 186
  square, 186
  unknown, 214
  zero, 188
Method:
  of completing the square, 92
  of elimination, 164
  of substitution, 162
Minor, 222
Monomial, 36
Multiplication:
  of complex numbers, 244
  of fractions, 15
  of matrices, 191
  of polynomials, 36
Multiplicative cancellation law, 14
Multiplicative identity, 9
Multiplicative inverse, 9

of a complex number, 245
of a matrix, 208
Multiplicity of zeros, 258

nth root, 54
nth term:
  of an arithmetic progression, 319
  of a geometric progression, 325
  of a sequence, 312
Number line, 18
Numbers:
  complex, 242
  counting, 5
  irrational, 6
  natural, 5
  rational, 5
  real, 6

Oblique asymptote, 368
Operations:
  with fractions, 15
  on radicals, 60–61
Order:
  of a determinant, 220
  of a matrix, 186
  of a radical, 55
Ordered pair, 106
Ordering of real numbers, 17
Ordinate, 110
Origin, 17, 110

Parabola, 125, 142
Parallel lines, 122

Partial fractions, 356
Permutations, 343
Perpendicular lines, 122
Point-circle, 146
Points of intersection, 162
Point-slope form, 121
Polynomial, 35
    addition of, 36
    degree of, 35, 37
    division of, 42–44
    irreducible, 41
    monic, 35
    multiplication of, 36
    prime, 41
    rational zeros of, 263
    zeros of, 253
Positive integers, 5
Powers of i, 247
Principal, 308
Principal n^{th} root, 54
Principal square root, 54
Principle of Mathematical Induc-
        tion, 333
Product:
    of complex numbers, 244
    of matrices, 191
    of polynomials, 36
Progression:
    arithmetic, 316
    geometric, 322
Proper subset, 2
Properties:
    of absolute value, 25
    of equality, 8
    of inequalities, 19–20
    of radicals, 55
Proportion, 85
Proportionality constant, 86
Pure imaginary number, 242
Pythagorean Theorem, 143

Q

Quadrants, 110

Quadratic equation, 90
Quadratic formula, 96, 250
Quadratic function, 124
Quadratic inequality, 131
Quotient, 10
    of complex numbers, 245
    of polynomials, 42

R

Radical, 55
    index of, 55
    order of, 55
Radicand, 55
Range, 107
    of an exponential function, 284
    of a logarithmic function, 292
Ratio, 84, 322
Rational exponents, 58
Rational function, 363
Rational integral expression, 47
Rationalizing the denominator,
        63
Rational number, 5
Rational zeros, 263
Real part of a complex number,
        242
Rectangular coordinate system,
        109
Reflexive property, 8
Relations, 106
    linear, 115
    quadratic, 129
Remainder theorem, 251
Root, n^{th}, 54
Root, principal square, 54
Roots of an equation, 90
Row matrix, 186
Row operations:
    on a determinant, 227
    on a matrix, 202
r^{th} term of a binomial expansion,
        355

Scientific notation, 296
Sequence:
  finite, 312
  infinite, 312
  terms of, 312
Series, 313
Set-builder notation, 2
Sets, 1
  disjoint, 3
  empty, 3
  equal, 2
  intersection of, 3
  union of, 2
  universal, 3
Sigma notation, 313
Simplest radical form, 64
Slope of a line, 117, 119
Slope-intercept form, 121
Solutions:
  extraneous, 101
  of inequalities, 79, 132, 170
  of linear equations, 69
  of linear inequalities, 79
  of nonlinear inequalities, 131
  of quadratic equations, 90, 96, 250
  of quadratic inequalities, 132
  of systems of equations, 162
  of systems of inequalities, 170
  of systems of linear equations, 199, 215, 235, 237
Special products, 37
Square matrices, 186
Square root, 54
Standard form:
  for a complex number, 246
  for the equation of an ellipse, 151
  for the equation of a hyperbola, 158
  for the equation of a line, 120
Straight line, 116
Subset, 2

proper, 2
Substitution method, 162
Substitution property, 8
Subtraction, 9
  of complex numbers, 243
  of matrices, 189
Sum:
  of an arithmetic progression, 319
  of complex numbers, 243
  of an infinite geometric progression, 329
  of matrices, 187
  of n terms of a geometric progression, 325
Summation notation, 313
Symmetric property, 8
Synthetic division, 45
Systems of equations, 162
Systems of inequalities, 170
Systems of linear equations:
  dependent, 235, 237
  inconsistent, 235, 237

Terms of a polynomial, 35
Terms of a sequence, 312
Transitive property:
  of equality, 8
  of order, 19
Tree diagram, 339
Triangular inequality, 26
Trichotomy property, 19
Trinomial, 36

Union of sets, 2
Universal set, 3
Unknown matrix, 214
Upper bound, 267

Variable, 30
Variation:
  combined, 86
  constant of, 86
  direct, 86
  inverse, 86
  joint, 86
  of sign, 259
Venn diagram, 4
Vertex of a parabola, 125
Vertical asymptote, 364, 366
Vertical line, 117
Vertical line test, 112
Vertices:
  of an ellipse, 152
  of a hyperbola, 156
  of parabolas, 125–27

x-axis, 110
x-coordinate, 110
x-intercept, 116

y-axis, 110
y-coordinate, 110
y-intercept, 116

Zero exponent, 33
Zero matrix, 188
Zeros of a polynomial, 253
  rational, 263

Four-Place Common Logarithms (Base 10)

N	0	1	2	3	4	5	6	7	8	9
1.0	.0000	.0043	.0086	.0128	.0170	.0212	.0253	.0294	.0334	.0374
1.1	.0414	.0453	.0492	.0531	.0569	.0607	.0645	.0682	.0719	.0755
1.2	.0792	.0828	.0864	.0899	.0934	.0969	.1004	.1038	.1072	.1106
1.3	.1139	.1173	.1206	.1239	.1271	.1303	.1335	.1367	.1399	.1430
1.4	.1461	.1492	.1523	.1553	.1584	.1614	.1644	.1673	.1703	.1732
1.5	.1761	.1790	.1818	.1847	.1875	.1903	.1931	.1959	.1987	.2014
1.6	.2041	.2068	.2095	.2122	.2148	.2175	.2201	.2227	.2253	.2279
1.7	.2304	.2330	.2355	.2380	.2405	.2430	.2455	.2480	.2504	.2529
1.8	.2553	.2577	.2601	.2625	.2648	.2672	.2695	.2718	.2742	.2765
1.9	.2788	.2810	.2833	.2856	.2878	.2900	.2923	.2945	.2967	.2989
2.0	.3010	.3032	.3054	.3075	.3096	.3118	.3139	.3160	.3181	.3201
2.1	.3222	.3243	.3263	.3284	.3304	.3324	.3345	.3365	.3385	.3404
2.2	.3424	.3444	.3464	.3483	.3502	.3522	.3541	.3560	.3579	.3598
2.3	.3617	.3636	.3655	.3674	.3692	.3711	.3729	.3747	.3766	.3784
2.4	.3802	.3820	.3838	.3856	.3874	.3892	.3909	.3927	.3945	.3962
2.5	.3979	.3997	.4014	.4031	.4048	.4065	.4082	.4099	.4116	.4133
2.6	.4150	.4166	.4183	.4200	.4216	.4232	.4249	.4265	.4281	.4298
2.7	.4314	.4330	.4346	.4362	.4378	.4393	.4409	.4425	.4440	.4456
2.8	.4472	.4487	.4502	.4518	.4533	.4548	.4564	.4579	.4594	.4609
2.9	.4624	.4639	.4654	.4669	.4683	.4698	.4713	.4728	.4742	.4757
3.0	.4771	.4786	.4800	.4814	.4829	.4843	.4857	.4871	.4886	.4900
3.1	.4914	.4928	.4942	.4955	.4969	.4983	.4997	.5011	.5024	.5038
3.2	.5051	.5065	.5079	.5092	.5105	.5119	.5132	.5145	.5159	.5172
3.3	.5185	.5198	.5211	.5224	.5237	.5250	.5263	.5276	.5289	.5302
3.4	.5315	.5328	.5340	.5353	.5366	.5378	.5391	.5403	.5416	.5428
3.5	.5441	.5453	.5465	.5478	.5490	.5502	.5514	.5527	.5539	.5551
3.6	.5563	.5575	.5587	.5599	.5611	.5623	.5635	.5647	.5658	.5670
3.7	.5682	.5694	.5705	.5717	.5729	.5740	.5752	.5763	.5775	.5786
3.8	.5798	.5809	.5821	.5832	.5843	.5855	.5866	.5877	.5888	.5899
3.9	.5911	.5922	.5933	.5944	.5955	.5966	.5977	.5988	.5999	.6010
4.0	.6021	.6031	.6042	.6053	.6064	.6075	.6085	.6096	.6107	.6117
4.1	.6128	.6138	.6149	.6160	.6170	.6180	.6191	.6201	.6212	.6222
4.2	.6232	.6243	.6253	.6263	.6274	.6284	.6294	.6304	.6314	.6325
4.3	.6335	.6345	.6355	.6365	.6375	.6385	.6395	.6405	.6415	.6425
4.4	.6435	.6444	.6454	.6464	.6474	.6484	.6493	.6503	.6513	.6522
4.5	.6532	.6542	.6551	.6561	.6571	.6580	.6590	.6599	.6609	.6618
4.6	.6628	.6637	.6646	.6656	.6665	.6675	.6684	.6693	.6702	.6712
4.7	.6721	.6730	.6739	.6749	.6758	.6767	.6776	.6785	.6794	.6803
4.8	.6812	.6821	.6830	.6839	.6848	.6857	.6866	.6875	.6884	.6893
4.9	.6902	.6911	.6920	.6928	.6937	.6946	.6955	.6964	.6972	.6981
5.0	.6990	.6998	.7007	.7016	.7024	.7033	.7042	.7050	.7059	.7067
5.1	.7076	.7084	.7093	.7101	.7110	.7118	.7126	.7135	.7143	.7152
5.2	.7160	.7168	.7177	.7185	.7193	.7202	.7210	.7218	.7226	.7235
5.3	.7243	.7251	.7259	.7267	.7275	.7284	.7292	.7300	.7308	.7316
5.4	.7324	.7332	.7340	.7348	.7356	.7364	.7372	.7380	.7388	.7396
N	0	1	2	3	4	5	6	7	8	9